Proceedings of
THE FOUNDING CONVENTION OF
THE MARS SOCIETY

Part I

Proceedings of
THE FOUNDING CONVENTION OF THE MARS SOCIETY

Part I

Edited by
Robert M. Zubrin
Maggie Zubrin

*Proceedings of the Founding Convention of
The Mars Society held August 13-16, 1998,
Boulder, Colorado.*

Published for The Mars Society by
Univelt, Incorporated, P.O. Box 28130,
San Diego, California 92198

Copyright 1999

by

Univelt, Incorporated, Publishers
P.O. Box 28130
San Diego, California 92198

First Printing 1999

ISBN 0-912183-12-8 (Part I - Soft Cover)
ISBN 0-912183-13-6 (Part II - Soft Cover)
ISBN 0-912183-14-4 (Part III - Soft Cover)

*Published for The Mars Society
by Univelt, Incorporated, P.O. Box 28130, San Diego, California 92198*

Printed and Bound in the U.S.A.

CONTENTS

	Page
Foreword, Robert and Maggie Zubrin	xv
The Founding Convention of the Mars Society, Richard Wagner	xvii
The Founding Declaration of the Mars Society	xxi

Part I

CHAPTER 1: PLENARY TALKS — 1

Opening Address to The Founding Convention of The Mars Society (MAR 98-001)
    Robert Zubrin 3

Mars: The Case for Life (MAR 98-003)
    Christopher P. McKay 17

CHAPTER 2:
HISTORICAL AND PHILOSOPHICAL SIGNIFICANCE OF MARS — 25

Civilizations at the Crossroads: Spain and China in the 15th Century (MAR 98-004)
    Wayne H. Bowen 27

On the Legitimization, Importance and Duty of Colonizing Mars (MAR 98-005)
    Josef Oehmen 33

The Race to Settle Mars - Earth Having a Baby - What's it Worth to the Parent Civilization? (MAR 98-006)
    Peter Perrine 37

Mars Exploration: The Survival of Our Civilization (MAR 98-007)
    E. G. Petrakakis 47

Reassessing the Human Condition: Philosophical Aspects of Mars Exploration (MAR 98-008)
    Richard L. Poss 51

CHAPTER 3: HISTORICAL LESSONS — 61

The Outward Course of Empire: The Hard, Cold Lessons from Euro-American Involvement in the Terrestrial Polar Regions (MAR 98-009)
    Marilyn Dudley-Rowley 63

	Page
A Shining City on a Higher Hill: Lessons From the Last Colonization of a 'New World' (MAR 98-010)	
James D. Heiser	73

CHAPTER 4: MOBILIZING THE PUBLIC — 83

Is There a Short-Term Economic and Social Justification for Human Exploration and Settlement of Mars? (MAR 98-011)
   Robert E. Becker 85

Mars: Fostering Public Support (MAR 98-012)
   Sam Burbank 101

Planet Mars Home Page©: Some Results After Three Years Operation (MAR 98-013)
   Thomas A. Gunn 109

How (Why and Under Which Conditions) Could International Cooperation Reinforce the Case for Mars (MAR 98-014)
   Richard Heidmann 117

Why NASA Might Never Launch a Manned Mission to Mars - The Devil's Advocate (MAR 98-015)
   Fred Kelly 127

Mars Needs Guitars (MAR 98-016)
   M. R. Jardin 133

Yes, But Will the People Support Us? - Engaging Our Customers in the Mars Exploration Adventure (MAR 98-017)
   Humboldt C. Mandell, Jr. 143

Human Mission from Planet Earth: Technology Assessment and Social Forecasting of Moon/Mars Synergies (MAR 98-018)
   Eligar Sadeh and Evan Vlachos 151

Making it Happen (MAR 98-019)
   Jonathan Stabb 169

A Socially Supportable Mars Colonization Program: Earth Mars Ambassadors (MAR 98-020)
   Philip A. Turek 175

The Gen-X Rallying Cry? To Mars? Gen-X Needs a Cause, and That Cause Should Be Space (MAR 98-021)
   George T. Whitesides 187

CHAPTER 5: VOICES OF YOUTH — 191

Our Future on Mars from My Perspective as a Twelve Year Old (MAR 98-022)
   Kathleen B. Bohné 193

	Page
The Hakluyt Prize Letter (MAR 98-023)	
Adrian Hon	195

CHAPTER 6: EDUCATION AND THE ARTS — 197

Secondary Mars Education: Faster, Better, Cheaper (MAR 98-024)
    Thomas W. Becker 199

Growing the Future (MAR 98-025)
    Charmin P. Gerardy 223

The Frontier of Mars as an Agriscience Classroom: Terrafarming Mars (MAR 98-026)
    Larry Payne 231

Teaching from Mars (MAR 98-027)
    Gabriel F. Rshaid 237

Preparing for the Journey: An Introduction to Mars Education (MAR 98-028)
    Donald M. Scott 243

From Bradbury to Blamont: The Science of Mars in the Arts (MAR 98-029)
    Michael Carroll 253

CHAPTER 7: PRIVATE FUNDING FOR MARS MISSIONS — 257

A Sponsoring Concept for Manned Missions to Mars (MAR 98-030)
    Michael Bosch 259

Conducting Mars Exploration on a Private Basis by Reviving the "East India Company" Financing Model (MAR 98-031)
    John Q. Coston 269

Private Sector Mars Wake-Up Call for Non-Government Participation (MAR 98-032)
    Thomas A. Gunn 275

Thinking About Martian Economics (MAR 98-033)
    Edward L. Hudgins 285

The Business of Commercializing Space (MAR 98-034)
    David M. Livingston 289

Funding The First Human Expedition to Mars (MAR 98-035)
    George Osorio 299

Promoting Privately Funded Settlement of Mars (MAR 98-036)
    Alan B. Wasser 311

CHAPTER 8: ROBOTIC EXPLORATION — 315

Navigation and Mobility Systems Architecture for Planetary Rovers (MAR 98-037)
    Pablo Flores 317

	Page
On the Development of Airborne Science Platforms for Martian Exploration (MAR 98-038)	
David W. Hall and Robert W. Parks	323
Reducing Risk and Complexity of Rover and Robotic Operations on Mars (MAR 98-039)	
Russell R. Mellon and Thomas R. Meyer	337
Soil Sampling on Mars (MAR 98-040)	
John L. Paterson	343
Marsplane - Flying on Mars with Existing Aircraft (MAR 98-041)	
Fabrizio Pirondini	353
Autonomous Rovers for Human Exploration of Mars (MAR 98-042)	
John Bresina, Gregory A. Dorais, Keith Golden, David E. Smith and Richard Washington	369

Part II

	Page
CHAPTER 9: SOFTWARE AND AUTONOMY	379

Using COTS Software for Mars Missions (MAR 98-043)
  Ned Chapin 381

Adjustable Autonomy for Human-Centered Autonomous Systems on Mars (MAR 98-044)
  Gregory A. Dorais, R. Peter Bonasso, David Kortenkamp,
  Barney Pell and Debra Schreckenghost 397

Model-Based Autonomy for Robust Mars Operations (MAR 98-045)
  James A. Kurien, P. Pandurang Nayak and Brian C. Williams 421

CHAPTER 10: THE QUESTION OF LIFE 435

Life on Mars: Evidence Within Martian Meteorites (MAR 98-046)
  Everett K. Gibson, Jr., David S. McKay and Kathie Thomas-Keprta . . . 437

ESA Exobiology Activities (MAR 98-047)
  Gerhard Kminek 449

Preserving Possible Martian Life (MAR 98-048)
  Mark Lupisella 457

The Ethical Ramifications of Discovering Life on Mars (MAR 98-049)
  Katherine Osborne 481

Interplanetary Biological Transfer of Bacteria Entrapped in Small Meteorites: Analysis of Bacterial Resistance to Impact in Ballistic Experiments (MAR 98-050)
  C.-A. H. Roten, A. Galluser, G. D. Borruat, S. D. Udry, G. Niederhäuser,
  A. Croxatto, O. Blanc, S. De Carlo, C. K. Mubenga-Kabambi
  and D. Karamata 485

CHAPTER 11: TECHNOLOGIES FOR HUMAN EXPLORATION 501

MARSSAT: Assured Communication with Mars (MAR 98-051)
  Thomas Gangale 503

Mobility of Large Manned Rovers on Mars (MAR 98-052)
  George William Herbert 515

Boosters for Manned Missions to Mars, Past and Present (MAR 98-053)
  Scott Lowther 529

	Page
An RLV / Shuttle Compatible Habitation System (MAR 98-054)	
Kurt Anthony Micheels	555
Aresam: Student Concept of Future Mars Space Station (MAR 98-055)	
Jonathon Smith and Jim Bishop	567
Current Progress in Water Reclamation Technology (MAR 98-056)	
Bradley S. Tice	573
Design of a Nuclear-Powered Rover for Lunar or Martian Exploration (MAR 98-057)	
Holly R. Trellue, Rachelle Trautner, Michael G. Houts, David I. Poston, Kenji Giovig, J. A. Baca and R. J. Lipinski	577

CHAPTER 12: POWER ON MARS — 585

Surviving on Mars Without Nuclear Energy (MAR 98-058)
  George James, Gregory Chamitoff and Donald Barker 587

Near-Term, Low-Cost Space Fission Systems (MAR 98-059)
  Michael G. Houts, David I. Poston, Marc V. Berte and
  William J. Emrich, Jr. 611

CHAPTER 13: ACCESSING MARTIAN RESOURCES — 621

Artesian Basins on Mars: Implications for Settlement, Life-Search and Terraforming (MAR 98-060)
  Martyn J. Fogg 623

Drilling Operations to Support Human Mars Missions (MAR 98-061)
  Brian M. Frankie, Frank E. Tarzian, Scott Lowther and Trevor Wende . . . 637

Extraction of Atmospheric Water on Mars in Support of the Mars Reference Mission (MAR 98-062)
  M. R. Grover, M. O. Hilstad, L. M. Elias, K. G. Carpenter,
  M. A. Schneider, C. S. Hoffman, S. Adan-Plaza and A. P. Bruckner . . . 659

A Comparison of *In Situ* Resource Utilization Options for the First Human Mars Missions (MAR 98-063)
  Kristian Pauly 681

Producing a Brick from a Martian Soil Simulate (MAR 98-064)
  David Seymour 695

The Case for a Mars Base ISRU Refinery (MAR 98-097)
  Kelly R. McMillen and Thomas R. Meyer 699

CHAPTER 14: HUMAN FACTORS — 709

Coping With Effects of Enforced Intimacy on Long Duration Space Flight (MAR 98-065)
  Lara Battles 711

	Page

On Our Best Behavior: Optimization of Group Functioning on the Early Mars Missions (MAR 98-066)
    Vadim I. Gushin and Marilyn Dudley-Rowley 717

Mars Mission Operations (MAR 98-067)
    Kenneth E. Peek 723

At What Risk is it Acceptable to Commit to a Manned Mars Mission? (MAR 98-068)
    Dennis G. Pelaccio and Joseph R. Fragola 729

The Case for Nurses as Key Contributors to Mars Exploration Teams (MAR 98-069)
    Mary Ellen Symanski 743

Man and Extended Space Flight: Mental and Physiological Factors (MAR 98-070)
    Bradley S. Tice 751

Who Should Go to Mars (MAR 98-071)
    Paul VanSteensburg 755

Part III

	Page
CHAPTER 15: MEDICAL ISSUES	761

No Means They Can Go (MAR 98-072)
  Thomas J. Burke and Michael C. Trachtenberg 763

The Effects of Variable Gravity on the Life Cycle of Tenebrio Molitor
(MAR 98-073)
  Amy M. Davis 767

Nutritional Supplements as Radioprotectors: A Review and Proposal
(MAR 98-074)
  Anthony C. Muscatello 773

Running to Mars: Exercise Countermeasures for Mars Astronauts
(MAR 98-075)
  Erik Seedhouse 787

CHAPTER 16: MISSION STRATEGIES 791

New Directions: Reevaluating the Lunar Refueling Option (MAR 98-076)
  J. D. Beegle and H. L. Beegle 793

A Stepped Approach to the Moon and Mars (MAR 98-077)
  James A. Bickford 799

A Novel Space Transportation Concept Designed to Reduce Per-Mission
Costs for Repeated Travel to/from a Celestial Body (MAR 98-078)
  Stephen Heppe 809

OneWay and Back: An Introduction to Comparative Missionology
(MAR 98-079)
  George William Herbert 817

Phobos on the Cheap (MAR 98-080)
  George William Herbert 823

Free Return Trajectories for Mars Missions (MAR 98-081)
  Chris Hirata 831

Polar Landing Site for a First Mars Expedition (MAR 98-082)
  Geoffrey A. Landis 835

	Page
CHAPTER 17: LAW AND SOCIETY ON MARS	843

Martian Law (MAR 98-083)
   Edward L. Hudgins 845

Legislation and Space Law Concepts Proposed for the Eventual
Industrialization of Mars by Man (MAR 98-084)
   James J. Hurtak 853

Martian Equality (MAR 98-085)
   Richard A. Jones 865

Mars Governance (MAR 98-086)
   Declan J. O'Donnell 873

The Politics of a Mars Colony (MAR 98-087)
   Kevin Archbold, Randall Hessler and Blaine Thompson 881

The Rights of Mars (MAR 98-088)
   Robert Zubrin 889

CHAPTER 18: TERRAFORMING MARS	893

Physiological Ecology of Terrestrial Microbes on a Terraformed Mars
(MAR 98-089)
   James M. Graham and Linda E. Graham 895

Successional Stages in Terraforming Mars (MAR 98-090)
   James M. Graham and Linda E. Graham 901

Terraformation of Mars (MAR 98-091)
   Charles R. Hancox 905

An Ecological Approach to Terraforming, Mapping the Dream
(MAR 98-092)
   Richard W. Miller 937

Ethics of Terraforming: A Practical System (MAR 98-093)
   George A. Smith 985

Terraforming Mars - Waterfield Reservoir Management (MAR 98-094)
   Patrick Whittome 1003

CHAPTER 19: CALENDARS AND TIMEKEEPING	1019

The Darian Calendar (MAR 98-095)
   Thomas Gangale 1021

The Millennium Mars Calendar (MAR 98-096)
   James M. Graham and Kandis Elliot 1031

	Page
APPENDIX	**1035**
Conference Sponsors	1036
Conference Schedule	1037
Conference Abstracts	1044
INDEX	**1123**
Numerical Index	1125
Author Index	1130

FOREWORD

Between the 13th and 16th of August, 1998, something special happened in Boulder Colorado. From all over the world, people came to be part of the beginning of a great enterprise – the founding of a society dedicated to opening Mars to humanity. The event was unique, everyone attending could feel it – there was something electric in the air. It was more than a turning point; it was akin to being present at the creation.

Seven hundred people came, and over 180 talks were given. Every aspect of the profoundly human endeavor of the development of a new world was discussed and debated -from engineering to economics to ethics – and from nearly every point of view. There was no consensus, nor could there be, for new life requires no consensus, but the riotous play of diversity seeking light through myriads of unexpected paths.

No proceedings could do this conference justice. The exciting debates in the halls, the breakout sessions, and the late-night bars, cannot be found in these pages. Most profoundly absent are the proceedings of the general meeting of Mars Society members, which occurred after the Saturday night banquet, during which the Founding Declaration was ratified and the Society formed. To get the full experience of the conference, you had to be there, or failing that, to speak with those who were.

Registration was furious on the first day. Maggie Zubrin (far right) keeps things moving.

Nevertheless, what paper, ink, and the best efforts of nearly 100 contributors can do to bring this conference to you and preserve it for the future, has been done. If nearly half the presentations are unrecorded, more than half are published here, as are some of the most important plenary talks. The article by Richard Wagner gives the blow-by-blow history of the convention, and accompanied by a fair number of photos, gives some feel for the actual flavor of the event.

As a personal note, we would like to say that the organizing of this conference was one of the most engaging and moving events of our lives. Going into the conference, we were a handful of people with no organization and no staff. On the evening of August 12, when the first large numbers of attendees showed up to pre-register, total chaos appeared unavoidable. But it took just a moment for many of those who were there to register for the conference to move to the other side of the registration tables as volunteers, and the pattern of volunteers stepping forward to fill in the innumerable gaps remained constant for the next four hectic days. Those who helped are too numerous to name here, but we thank you all. We also wish to thank the generous donations of Fisher Space Pen, the Bushnell Corporation, Robert Bigelow, the Longview Foundation, and the National Geographic Society, without which the event would have been impossible.

Since the convention, the Mars Society has grown exponentially. More volunteers with skills of every description have stepped forward, and the Society is making its presence known in over 30 countries. The seed planted in Boulder has sprouted, and a new force for life is loose in the world.

It gives us great pleasure to be able to present to the interested reader these proceedings of the Founding Convention of the Mars Society.

Robert and Maggie Zubrin
June 30, 1999

THE FOUNDING CONVENTION OF THE MARS SOCIETY

Richard Wagner

Against a backdrop of the Front Range's ruddy face, close to 700 Mars enthusiasts filled — and sometimes overflowed — the University of Colorado's Memorial Center in Boulder for the Founding Convention of the Mars Society. Over the course of four days, they heard from researchers and scientists involved in Mars exploration and shared their own ideas concerning humanity's future on the Red Planet. But, more than that, they helped create a popular movement that will promote the cause of human Mars exploration.

The buzz rocketing through the Glen Miller ballroom on the morning of August 13 was undeniable. While all at the conference were well aware of their personal, private passion for Mars, few were perhaps ready for the display of shared passion they found on walking through the ballroom doors. More than 600 people filled the seats of the auditorium. With camera crews scurrying about in back and on stage, audience members gazed around in something close to wonder at the multitude at hand. In due time, Robert Zubrin took to the stage and opened the convention with the simple words, "Hello, I'm Robert Zubrin, president of the Mars Society, and am I glad to see you," which the audience greeted with an eruption of raucous applause and cheers.

The Convention begins.

Anyone reviewing the convention schedule was well aware that this was a nearly overwhelming affair — four days of meetings with mornings devoted to Plenary sessions, afternoons to five different subject tracks, and evenings to further plenaries and society events. With more than 200 talks by roughly as many speakers on subjects ranging from terraforming to the art of Martian landscapes, it was sometimes hard to keep up, but no one seemed to mind. It added to the energy, the insights, and the sheer exuberant fun of the convention. The mix of researchers and engineers mingling with the likes of artists, writers, professionals, and students produced surprises throughout the convention. While it possibly came as no surprise that Zubrin's convention opening talk, "Humans to Mars within a Decade," brought the audience to its feet for an extended standing ovation, those lucky enough to hear Kathleen Bohné's talk "Our Future on Mars from My Perspective as a 12 Year Old" were, as one audience member remarked, "blown away" by the honesty and vision of her presentation.

Morning plenary sessions allowed all to come together to hear from world-class Mars researchers, astronauts, and others. Thursday morning's opening plenary offered up a good example of what was to come in the following days. In addition to Zubrin, the morning featured Everett Gibson of the ALH 84001 "Mars Meteorite Team" presenting a vigorous defense of recent attacks on the team's biologic interpretation of the stone's more intriguing features; Pascal Lee of NASA Ames Research Center with an overview of recent research in the Haughton Impact Crater on Devon Island in the Canadian Arctic; Rob Manning of the Jet Propulsion Lab with a review of NASA's current robotic Mars exploration program; and Jacques Blamont of the French space agency CNES with an overview of a U.S./French program to explore Mars by balloon. Afternoons were given over to the various tracks, which offered a smorgasbord of subjects for both those interested in the technical aspects of Mars exploration as well as the social aspects, while evenings were reserved for Society events or further plenary sessions.

In addition to offering a forum for Mars researchers and aficionados, the Founding Convention provided a sense of place and purpose for establishing an organization and movement dedicated to the human exploration of Mars. On Saturday afternoon, members of the Mars Society Steering Committee, along with interested convention attendees, wrestled with the questions of how to build a movement from the energies unleashed by the convention, and what projects to undertake to forward the cause of human Mars exploration.

The afternoon meeting was fast-paced and focused, with all agreeing that the Mars Society had to be a results oriented organization focusing on both politics and programs. By consensus, members of the steering committee mapped a project task list that focuses on the near, mid, and long-term. The first approved project is the design and construction of a full-scale Mars base simulation in the Canadian Arctic. Pascal Lee's aforementioned talk on the Haughton Impact Crater and the Haughton-Mars 1998 Project proved the impetus for this concept. Field methods and hab designs will be tested at this base, while providing a long-term base of operations for geological explorations in the crater. The Mars Society aims to have the base up and running by the summer of 2000. Second on the project list, a payload to Mars during the 2003 opportunity. While there is a wide range of options for this project, most envisioned it as a hitchhiker payload on a scheduled Mars mission or Ariane launch. The Society will release a request for proposals with an eye to selecting a mission that is affordable, quick, and sure to catch public attention. Finally, Steering Committee members agreed that a long term goal should be its own mission to Mars.

On an organizational level, there was quick agreement on the necessity of raising $100,000 within the next few months to enable the hiring of an executive director for the Society. In addition, the near future will see a series of local chapters and task forces established to provide both grassroots and managerial direction for the Society. Finally, all agreed that funding for the Arctic project will be a priority item over the course of the next year, as will organizing for a Mars Society convention in 1999.

Everything came together during Saturday evening's banquet. Highlights of the evening included some folk and filk singing entertainment (bravely led, at one point, by Bob Zubrin) and words and inspiration from Adrian Hon (winner of the Society's Hakluyt Prize) and Kathleen Bohné (an addition to the schedule, owing to the rousing reception her track talk received). Society President Zubrin later read the Founding Declaration of the Mars Society and, when he asked for acceptance of the document by acclamation, the energy and enthusiasm rippling through the convention burst forth in roar of approval. With that, the evening was transformed into an open discussion on the future of the Mars Society with a stream of individuals approaching two microphones to offer suggestions and ideas, ranging from project ideas to management suggestions to long-term goals. While offering ideas is relatively painless, volunteering to execute them is something else entirely. Yet, time and again, folks at the microphone would offer a suggestion and then volunteer to see it through. It became obvious that Society members were ready to take action as the final fifteen minutes of the evening involved on-the-spot chapter organization as members from various regions across the country came together to meet one another and set early plans for activities back home after the Convention.

By Sunday's final session, it became apparent that the Founding Convention had been more than just a meeting of Mars enthusiasts. True, the convention offered memorable appearances by such luminaries as astronaut John Young and Mars Pathfinder project scientists Matt Golombek; it provided reams of information for anyone interested in the planet Mars and its exploration; it gave Mars enthusiasts a forum in which to meet one another and swap ideas and tales. But just about any decent conference on any subject will offer up a similar collection of services. The Mars Society Founding Convention did more, though, by allowing individuals to become part of an organization; by allowing them to channel their energies and their own talents into a movement that will promote human Mars exploration. It would be hard to find more appropriate words to end the Convention than the ones spoken by Robert Zubrin to close the final session of the day.

"We're gonna win. On to Mars!"

Former NASA Associate Administrator for Exploration, Mike Griffin, signs the Founding Declaration of The Mars Society.

FOUNDING DECLARATION OF THE MARS SOCIETY

The time has come for humanity to journey to Mars.

We're ready. Though Mars is distant, we are far better prepared today to send humans to Mars than we were to travel to the Moon at the commencement of the space age. Given the will, we could have our first teams on Mars within a decade.

The reasons for going to Mars are powerful.

We must go for the knowledge of Mars. Our robotic probes have revealed that Mars was once a warm and wet planet, suitable for hosting life's origin. But did it? A search for fossils on the Martian surface or microbes in groundwater below could provide the answer. If found, they would show that the origin of life is not unique to the Earth, and, by implication, reveal a universe that is filled with life and probably intelligence as well. From the point of view learning our true place in the universe, this would be the most important scientific enlightenment since Copernicus.

We must go for the knowledge of Earth. As we begin the twenty-first century, we have evidence that we are changing the Earth's atmosphere and environment in significant ways. It has become a critical matter for us better to understand all aspects of our environment. In this project, comparative planetology is a very powerful tool, a fact already shown by the role Venusian atmospheric studies played in our discovery of the potential threat of global warming by greenhouse gases. Mars, the planet most like Earth, will have even more to teach us about our home world. The knowledge we gain could be key to our survival.

We must go for the challenge. Civilizations, like people, thrive on challenge and decay without it. The time is past for human societies to use war as a driving stress for technological progress. As the world moves towards unity, we must join together, not in mutual passivity, but in common enterprise, facing outward to embrace a greater and nobler challenge than that which we previously posed to each other. Pioneering Mars will provide such a challenge. Furthermore, a cooperative international exploration of Mars would serve as an example of how the same joint-action could work on Earth in other ventures.

We must go for the youth. The spirit of youth demands adventure. A humans-to-Mars program would challenge young people everywhere to develop their minds to participate in the pioneering of a new world. If a Mars program were to inspire just a single extra percent of today's youth to scientific educations, the net result would be tens of millions more scientists, engineers, inventors, medical researchers and doctors. These people will make innovations that create new industries, find new medical cures, increase income, and benefit the world in innumerable ways to provide a return that will utterly dwarf the expenditures of the Mars program.

We must go for the opportunity. The settling of the Martian New World is an opportunity for a noble experiment in which humanity has another chance to shed old baggage and begin the world anew; carrying forward as much of the best of our heritage as possible and leaving the worst behind. Such chances do not come often, and are not to be disdained lightly.

We must go for our humanity. Human beings are more than merely another kind of animal, - we are life's messenger. Alone of the creatures of the Earth, we have the ability to continue the work of creation by bringing life to Mars, and Mars to life. In doing so, we shall make a profound statement as to the precious worth of the human race and every member of it.

We must go for the future. Mars is not just a scientific curiosity; it is a world with a surface area equal to all the continents of Earth combined, possessing all the elements that are needed to support not only life, but technological society. It is a New World, filled with history waiting to be made by a new and youthful branch of human civilization that is waiting to be born. We must go to Mars to make that potential a reality. We must go, not for us, but for a people who are yet to be. We must do it for the Martians.

Believing therefore that the exploration and settlement of Mars is one of the greatest human endeavors possible in our time, we have gathered to found this Mars Society, understanding that even the best ideas for human action are never inevitable, but must be planned, advocated, and achieved by hard work. We call upon all other individuals and organizations of like-minded people to join with us in furthering this great enterprise. No nobler cause has ever been. We shall not rest until it succeeds.

www.marssociety.org

Chapter 1
PLENARY TALKS

Shortly after his opening address, Mars Society President Robert Zubrin (center) shares thoughts with Harold Miller (left) and James Oberg.

Chris McKay discusses the search for life with conference participants. At the far right is Pascal Lee, leader of the Society's Arctic Base task force.

OPENING ADDRESS TO THE FOUNDING CONVENTION OF THE MARS SOCIETY

Robert Zubrin[*]

Hello, I'm Robert Zubrin of the Mars Society, and I am very glad to see you....[applause]. I'd like to welcome you to the beginning of the Founding Convention of the Mars Society. I believe that this is an historic occasion. Over the next four days what we're going to attempt to do is found an international association dedicated to the objective of getting humans on Mars. We have people here from every NASA center and from every national lab of the United States—from most of the major top universities and we have people from Canada, we have people from Western Europe, [the microphone produces bad feedback.] Can people hear me without the mike? [crowd: Yes. He tosses away mike] We have people here from every major country in Western Europe, and most of the rest. We have people here from Japan, from China, from Indonesia, from Australia and New Zealand, from Argentina and Brazil and somebody here from Mozambique! And it's a powerful message that people all over the world are interested in this objective and see its importance. We have the eyes of the world upon us as well. You may be interested to know that most of the leading elements of the world press are in attendance. We have the Washington Post. We have the Boston Globe. We have the New York Times. We've got the London Times. We have the Associated Press. We have Agence France Presse. We have the CBC, National Public Radio and the BBC. We have Discover Channel, we have ABC News, we have Discover Magazine, Reason Magazine, Popular Mechanics, the Rocky Mountain News, the Denver Post, the Boulder Camera, and ... Space News too! [applause] And I think it's great they're here, both to let the world know what we're doing, but also because I think that historians are going to want to look over their shoulders at what we initiate here today.

I believe we are at an important historical juncture. We live in a unique time. We've just really concluded a period of warfare that's dominated the entire middle and second half of the 20th century. Starting with World War II and going through the Cold War, literally trillions of dollars have been spent to build up masses of military industrial technological capabilities, largely for defense purposes. Those capabilities have now been liberated from that purpose and are now available to undertake a major new set of objectives for humanity. I believe that this juncture won't last forever. For one of two reasons. Either the peace will hold, in which case, if there is no other use for these capabilities they will tend to wither away and not be available; or peace will not hold, and they will have to be rededicated for military purposes. So we have a unique opening right now. We have a prosperous world, a world which has never been as rich as it is, better prepared to take on a venture of this sort. We have the tools required to do it—we have essentially massive amounts of engineering talent is available to do it. And we still have with us the memory of the last time we undertook a great project of this type, mainly of course, the Apollo project. The veterans of that are still with us, you know, making clear that it

---

[*] President, Pioneer Astronautics, P.O. Box 273, Indian Hills, Colorado 80454.

is possible things can be done. And I believe that our greatest chance for actually getting the humans to Mars program initiated will occur in the year 2001, when the new administration of the United States—the first year of the first American administration of the next millennium Europe will have just completed major milestones towards unification,—putting in place those institutions that should tend to guarantee that a another general European war will not be repeated. So that we will be moving out of the most dangerous century of human history into a different kind of epoch. It will be similar to the situation that occurred in Europe after the turn of the last millennium, where people realized that they had survived the barbarian invasions, the Vikings, the Moors, and the threats of supernatural destruction of the world at the turn of year 1000, and realized that there would be a future and they started building the cathedrals. Well you know, I think with the turn of the next millennium there's going to be that feeling that as well that it will be time to build cathedrals again. And no more profound statement of faith in the future could be better expressed by taking those steps and initiating humanity as a multi-planet species. [applause].

So now we have three years to intercept that process to educate politicians, congress and parliamentary bodies both here and in Europe and elsewhere. To educate the people forming around these various camps of the different presidential candidates as they develop. To reach public opinion in general—to make it clear that this is the next major step for humanity; that it is a challenge we must be willing to embrace, in order to have the decisions converge at that time.

Now it is sometimes said that it would be great to send humans to Mars—but we can't do it, that it is beyond our technology, that it is beyond our capabilities. That it would cost an incredible amount of money, even if it were possible. This is not true. There is an impression that was created in that direction in 1989 or so, when President Bush, if you recall, in July 1989, got up on the steps of the National Air and Space Museum, flanked by Armstrong, Aldrin and Collins—the Apollo 11 crew—and he announced; "This is the 20th anniversary of the Apollo Moon landing—that was great, that is what America is all about, and therefore, I as the President, am committing this nation to go back to the Moon and on Mars, and this time, to stay!" It was great stuff. He thus announced the initiation of the program that became known as the "Space Exploration Initiative". Now you may recall at that time what happened. NASA, the space agency, went off and conducted a study on how this might be accomplished. And they came back three months later with a report which therefore became known, in fact officially, as the 90 Day Report, which said; "George—we can do it, we can get you to Mars. However we will require a budget of 450 billion dollars and 30 years in order to prepare. But if you can give us that, we are ready to rock." [laughter]. And of course the immediate result of this proposal was that the Space Exploration Initiative went completely down in congress, because 400 billion dollars is truly an immense amount of money, and no one was willing to spend it to send humans to Mars.

Now this is the sort of thing that NASA came up with at that time, although this particular design was not in the 90 Day Report, it came out immediately afterwards. [shows chart depicting giant spacecraft] This thing here is an interplanetary spaceship. I call it The Death Star [laughter]. It weighs a thousand tons, which is about the payload of the United States has launched cumulatively to orbit since 1975. It's a hundred meters long—as big as a football field. It's got one, two, three, four, five, thick tanks of liquid hydrogen, each weighing 150 tons, so each of those have to be launched into orbit separately by a launch vehicle with the capability of a Saturn V Moon rocket. That would take about a year, during which time a substantial part of the liquid hydrogen would boil away. But we'll put that problem aside, because we have other problems including the need for two nuclear rocket engines, that do not exist, each with a 75,000 pounds of thrust, so that this truss here, which would have to be built in space, would have to take a load on it equal to a truss standing vertically on the surface of the Earth with 75 cars stacked on top it, with all the cryogenic plumbing and electrical lines in it and everything that you would need. Then you've got the somewhat complicated payload over here—we won't even talk about it, except to note that it's inside of an aeroshell 100 feet diameter—much too big to fit inside the payload fairing of any launch vehicle that anyone is willing to propose. So obviously it would have to be constructed on orbit, with all the integrity required of an aeroshell that is going to hit the Martian atmosphere at Mach 30. In fact the entire vehicle would clearly have to be constructed on orbit because if you built a booster big enough to launch this in one piece you'd blow away Orlando when you took off [laughter]. So you know they came up with a plan to do it, next chart. [shows a chart depicting a complex orbital assembly sequence] So they came up with a scheme to do it in 10 easy steps as illustrated here. You can build this on-orbit, provided you have available to you on-orbit- orbiting hangers, construction docks, cryogenic fuel depots, power generation stations, checkout points, crew construction shacks,—an entire range of paraphernalia that collectively I call the "parallel universe" [laughter]. So that's the 90 Day Report plan; spend 400 billion dollars, build a parallel universe, use it to construct a Deathstar and sail it off to Mars.

Now I was working Martin Marietta at that time designing interplanetary missions for them, and a number of us engineers in that line went to management there and said you know, look, this is not going to work. We can give you a whole bunch of technical reasons why this won't work but you wouldn't understand them—actually we left that part out of the briefing. [laughter]. But YOU can certainly understand that nobody is going to give them 400 billion dollars and 30 years in order to do this. And you know, the management of the major aerospace companies, as I think most of you know is composed of exclusively brilliant visionaries with lightning-quick minds. [laughter] So, what they said was, you know, we'll lets wait and see. Well they waited and they saw and the 90 Day Report went down, and then they said, well what do you want to do? And what we said is what we wanted to do was to put together our own team, with a clean sheet of paper and a charter to come up with a totally alternative approach for sending humans to Mars. But here's the deal: we don't want to have a bunch of managers or marketeers or people like this coming in and telling us we have to sign a mission or this way or that way—because we want to please this guy down at Johnson Space Center, or Marshall Spaceflight Center, or Jet Propulsion Lab, or somewhere or other, you know, who really wants to see this technology or that technology included in the mission. Because you see that kind of thinking is what drove the 90 Day Report so berserk. They had literally thousands of people involved with this design and they all wanted to see their favorite technology that they were working on, included as mission critical. So they designed the most complicated mission that they possibly could. And we felt—we knew—that it would be impossible—it would be difficult enough to design an attractive Mars program without being driven by such imperatives. And the management to their credit said, "yes, you're on". So we put together a team, called the Scenario Development Team, composed of 12 people from the whole big Martin company, I was one of them, to come up with an alternative plan. And because there was a lot of creative spirits on this team we could not agree with each other. So in fact, we came up with three different plans, and we floated all three to NASA during the spring of 1990. And it rapidly became clear that one of the plans, the Mars Direct plan- that I was largely responsible for along with another engineer named David Baker, had the most potential to overthrow the equation, and was by far the most radical break from this kind of thinking. Because it involved no on-orbit assembly at all, and no advanced propulsion, and it fact if you looked at this plan in the spring of 1990, you could say this is a shot for humans to Mars by 1999. And the plan immediately became extremely controversial. A lot of people in NASA rallied to it - they became very supportive of it - extremely so - immediately. Others became vehemently oppositional for any number of reasons; the dominant one being simply that it was a massive paradigm shift from the way people had thought about Mars missions before. However, over the next couple of years, more and more people became convinced of it, so that I was given a chance to speak to Mike Griffin, who you're going to hear from tomorrow, who was then the Associate Administrator of NASA for Exploration, who immediately became a supporter of the plan and he briefed Goldin, who then became supportive in looser terms. So with Griffin's support I went back to Johnson Space Center - briefed it again, and they finally got a good understanding. And at this point - I'll go into it later, a derivation of this plan is now the basis for NASA's thinking for how do you send humans to Mars.

But how can you do it? Okay. How can we do humans to Mars, with our own technology, without the need the parallel universe or a Deathstar? Next chart. [shows a chart depicting a shuttle-derived heavy lift launch vehicle.] By the way, let me just say this, Mars Direct plan is not the end of thinking about Mars—it's the beginning. It's an "Existence Proof". I'm wide open; if someone comes up with a better plan, and their people at this conference proposing alternative plans, if we can improve it in any way, I'm all for it. But this is an existence proof, that if we can do it this way we can at least do it this good.

Well what do you need? In my opinion, you do need a heavy lift vehicle. So we designed one, called Ares, after the Greek name for Mars. Now the Ares is basically a junkyard special. That is, it's designed to be built out of components that can be found in junkyards today, or should be. [laughter] Which is to say Shuttle technology. [laughter] The reason why we designed it as a junkyard special, to be thrown together with stuff we have now, is not that it's necessarily the best way to design a booster, but that it is the fastest way to design a booster, that you can have in a minimal amount of time. And the thinking behind the Mars Direct plan is very much that you have to get to Mars fast. In the thinking that we had in designing Mars Direct, is that the fundamental strategic situation facing something like a humans to Mars program is basically similar to that of a tactical situation facing the Children of Israel attempting to cross the Red Sea in the Book of Exodus. What do I mean by that? You want to get to the Promised Land. But you can't do it because there's an impossible obstacle in front on you. Then suddenly, a miracle happens—Moses parts the waters—Bush gives a speech. [laughter] So the waters part, and you've got these two cliffs of water standing there with a path of dry land in between. So now you can cross, but you can't do this on a 30 year timeline. Because the Egyptians are behind you. And God's patience isn't infinite. And the U.S. Congress is worse. [laughter]. If you take 30 years to do this, I guarantee the walls are going to come together on you, and you will drown, because a consensus cannot put together for that amount of time. No. If you want to get to Mars, you can't do it in 30 years. You can't do it in 20 years. If you want to get to Mars, you got to get there in 10 years or less, or the waters are going to come together on you. Thus the need for a junkyard special.

So. What have we got? We've got a booster whose primary element is a Shuttle external tank, that they make down there in Louisiana, except not with this conical tops,—we don't need for this application. Then you've got four Space Shuttle main engines. Those things are available. They're lying around in crates near the Rocketdyne factory in Canoga Park, California. And you can actually get them for free. If you go at night. [laughter] …because they have laid off the night watchmen, the earthquake took down the fence, it's not really a problem. Then you've got two solids like they used on the shuttle. People in Utah would be happy to sell them to you. Then on top you have a hydrogen oxygen upper stage with 250,000 pounds of thrust. That's the same amount of thrust that they had in the J-2 engine that powered the upper stages of the Saturn 5 that they used in Apollo. That stage does not exist as a piece of integrated technology right now, but everything in it does, and we could develop it fairly readily.

Now, people ordinarily rate boosters by what they can lift to Low Earth orbit or leo, L–E–O, and if you want to know what Ares can lift,—it's a hundred and twenty metric tons of LEO which is pretty good. It's a little less than a Saturn 5 that could do a hundred and forty and a little more than the Russian Energia that could do a hundred. But it's much better than the Shuttle or Titan 4 or Ariane 5, which can do about twenty each. Now, however, like the Saturn 5, but unlike for example the Space Shuttle, this booster has additional capabilities. Because it is a staged booster, it can use its upper stage to throw payloads far beyond Earth orbit. In particular it can throw 47 tonnes on a direct trajectory to Mars—or 59 to the Moon. And if you really want to do the mission—if you really want to do it in our time, that's the way you want to do it. The way we've launched every real robotic interplanetary mission to date is simply using the upper stages of the same booster that lifted the payload to LEO to throw the payload the payload to the planet. And we can do manned missions the same way. Or put it this way; if you can do manned missions that way, you've gone 90% of the way right there towards taking the mission out of the realm of the parallel universe and putting it in our universe.

But how could you? The Deathstar spacecraft that I showed you weighed a thousand tons. This can only lift a hundred and twenty. Well what could you do? Well, you could divide the mission up into chunks—a thousand by a hundred and twenty eight is eight. You could fire the different parts of the mission out of to Mars in convoy, rendezvous along the way, or in Mars orbit, or on the surface or somewhere—and get the mission together that way. Well it might be possible to do a Mars mission that way, but there's an awful lot of risk involved. Because if you've got 8 launches, and they are all mission critical, and one goes in the drink, you lose the whole mission. So I don't really support that. But I could support dividing it into two launches, counting on two launches both working—I think that's reasonable. But that would still leave us 500 tons of a launch, which is much too much. Well what else could you do? Well you could invoke advanced propulsion. Nuclear propulsion. Ion drives. Fusion propulsion. Anyone for anti-matter? [laughter] There's teleportation. Warp drive. Pixie dust. [laughter] There are a lot of options that offer considerably greater capability than our current chemical propulsion. And some of it is absolutely possible. Nuclear propulsion, as you'll hear from several talks in this conference, is absolutely possible—we had test stand engines using it in the 1960's. The ion drive engine is certainly possible. Someday we'll even see fusion propulsion and it will help us a great deal with colonizing Mars. But it's not clear at all that even the nearest term of these, nuclear propulsion, can be available within a ten year time frame. If you want to design this mission to be prepared for launch in ten years you got to do it with propulsion that's either available now or very close to being off the shelf. Otherwise you are guaranteeing the waters are going to come together on you. So what's your alternative? The alternative is to adopt a same kind of strategy for supporting Mars that explorers and pioneers have consistently used for exploring Earth for centuries. Which is; live off the land. In other words, reduce the mass you got to send to Mars by using the mass that is already there.

How would this work in practice. Next chart. This is the mission sequence chart for the Mars Direct plan. [The mission sequence chart is displayed.] You see on the left, year one, year three, year five. You get a launch opportunity to Mars every other year. Realistically right now, year one could be 2005, year three could be 2007, year five could be 2009.

So year one, 2005, you launch one of these boosters off the Cape, and use its upper stage to deliver to Mars an unmanned payload weighing about 40 metric tonnes. This flies out to Mars on a minimum energy trajectory, it takes 8 months to get there, then it uses this umbrella-shaped aeroshell to aerocapture into Mars orbit. Then it is landed on Mars with a parachute and retrorockets, just like we did with the Viking mission in 1976.

Now what is this we have landed on Mars? Next chart. [A chart showing the Earth Return Vehicle is displayed.] The primary object is an Earth Return Vehicle, or ERV. Now the Earth Return Vehicle, basically, is a little rocketship that contains Spartan quarters for a crew of 4 to take a 6-month voyage from Mars back to Earth in the terminal phase of the mission. But nobody is in it now. Then below that cabin, you've got two methane/oxygen chemical propulsion stages, which however, are unfueled. They've got to be unfueled or this thing would weight four times as much as it does, and be much too heavy for anything in the class of an Ares to throw to Mars. However in some of the lower stage tanks that are later going to contain methane, we have about 6 tonnes of liquid hydrogen probably in gelled form. And then slung below the vehicle, not shown in this picture, we've got a little light truck, like a little pick-up truck, that runs on a methane/oxygen internal combustion engine, and in the back of that truck we've got a little nuclear reactor with a power of 100 kilowatts. So the reactor is not like Fort St. Vrain or Vermont Yankee, it's just a little put-put nuke sitting in the back of a truck. And it's there for the ride.

So it lands, and then we drive the truck a few hundred yards away from the landing site, unwinding a cable off the back of it as we go. We get a couple of hundred yards away and then we lower the reactor into a crater or a ditch, and then we turn it on. Now we have power at the ship. What do we do then? Next chart.

We go hunting, for what's available on Mars. [A chart showing chemical reactions for propellant synthesis is displayed.] And what's available, everywhere, is the atmosphere. Mars has got an atmosphere, it's 95% CO2. CO2 is the ideal feedstock for making rocket fuel. We took hydrogen with us to Mars; we can react that with CO2 to make methane and water. In the presence of a catalyst, that reaction will just go. Methane is natural gas, it's great rocket fuel. We store that in our tank. the water, we condense, and electrolyze into hydrogen and oxygen. The oxygen we store, that's our oxidizer to burn the methane with. The hydrogen we recycle to make more methane and oxygen. Using just those two reactions, you take your 6 tonnes of hydrogen and turn it into 72 tonnes of methane/oxygen propellant on the surface of Mars. But you can do better than that. Because the ideal mixture ratio for burning these two involves additional oxygen, which we can make by running a third reactor in which we take CO2 and break it down into carbon monoxide and oxygen. The oxygen we store, the carbon monoxide we vent, as waste. You can do that on Mars—there is no EPA there. [laughter.] And when you are all done, what you've managed to do is turn your 6 tonnes of hydrogen into 108 tonnes of methane/oxygen on the surface of Mars. That's a leverage of 18 to 1; it's like a pioneer being able to acquire the useful mass of a bison for the transported mass of several bullets and cartridges. And that's what makes the whole mission sing; 95% of your return propellant is coming from Mars. In fact, we can make so much propellant that we make extra propellant, which we use not only to support operation of the Earth Return Vehicle, but ground cars that use combustion engines. We want those, because combustion engines have a much higher power-to-mass ratio than you can get with battery powered vehicles or those operating on fuel cells. That's why they are so popular here on Earth. So the ability to make fuels on Mars not only allows you to do the mission cheaply, it allows you to do it effectively, because it gives you mobility on the surface which is what you really need to explore the planet. Next chart.

By the way, a lot of times it is easy to write chemical equations on a chart and something else entirely to make it work in practice. That's not the case here. I can vouch for that, because when I was at Martin, we built a machine that does exactly the chemical synthesis that I just showed you. Here it is. [A photo of a prototype Mars chemical synthesis unit is displayed.] It is a full-scale unit. We did it in 3 months at a cost of $47,000, which is the cheapest thing that has ever been built at Martin Marietta. [laughter and applause]. It was 94% efficient, and not a single person on this project, frankly, was a real chemical engineer. We were all aerospace engineers dinking around with chemistry to prove to NASA that wheels roll. And since then, at Pioneer Astronautics, my own company, we have built more advanced machines that can not only make methane, they can make methanol, they can make ethylene, propylene, lots of stuff. We can make fuels on Mars. Let's have the mission sequence chart again.

So, it took 8 months for this to fly to Mars, it took 10 months to make the propellant, that's 18 months. There are 26 months between launch windows to Mars. So long before the next launch window opens up, in year three—2007, we know we have a fully fueled Earth Return Vehicle sitting, waiting for us on the Martian surface. That being the case, at that time we launch two more vehicles off the Cape. One is another Earth Return Vehicle/fuel factory object, the other is a habitat with a crew of 4 astronauts in it.

Now because our return ride is waiting for us on Mars, we don't have to fly to Mars in the Deathstar. We don't even have to fly to Mars in the Millennium Falcon. We can fly to Mars in, next chart, ... a tuna can [laughter. A chart showing a tuna-can shaped habitat spacecraft is shown] And this is a very good thing, because we know how to build tuna cans, and they have been proven in commerce to be as extremely efficient shape for volumetric packaging purposes. Now this one is a little bigger than the standard Chicken of the Sea model, it's around 27 feet in diameter and 16 feet tall. It's got two decks, each with 8 feet of head room. The upper deck is where people live. the lower deck is a kind of garage/cargo hold area. Next chart.

This is the upper deck of the hab. [A diagram showing the layout of the hab upper deck is displayed.] What you can see is that you have a stateroom for each of the 4 astronauts. You've got public areas for science, exercise, and a galley. And in the center, you have a solar flare storm shelter.

Now a few words are needed about this, because certain people that are trying to say that a Mars mission is impossible are using radiation as a dress or skirt to hide behind. And it's just baloney. There are two kinds of radiation that can get you in interplanetary space; solar flares and cosmic rays. Solar flares are events that occur in a very unpredictable way, but you might have a nice big one about once a year. Or you might have two in one year and none in the next. But a hefty solar flare can happen suddenly, and in the course of a few hours deliver several thousand Rems of radiation to an unprotected astronaut, which is enough to kill. However, the kind of radiation that solar flares are made out of is protons, with energies of about a million volts. And those can be stopped by 5 inches of water, or things that from a nuclear point of view are fundamentally the same thing as water, such as food, or things that water and food become as the mission proceeds. And we have enough of that stuff on board the ship that we can pack it in around a central cylinder shaped pantry, creating a zone in which you are entirely shielded from solar flares. So if a solar flare happens, the alarm bell rings, and people go in there. They are packed like commuters on the New York subway for a few hours, except they probably don't have panhandlers coming through. If they do, you have another problem. This happens once, maybe twice, in the whole mission, and you're safe.

Now the other kind of radiation are cosmic rays. Those are very different from solar flares. They don't come from the Sun—in fact people are really not clear on where they do come from. And they have energies, not of millions of volts, but billions of volts, thousands of times more energy particle for particle than solar flare particles. So you can't stop them with 5 inches of water. You would need meters of water to stop them, and you can't afford the mass to do it. So you're going to take the dose. However the magnitude of that dose is around 50 Rem for every year you are in space in the Earth-Mars region of the solar system. What's 50 Rem? It represents a 1% risk of getting a fatal cancer at some point later in your life, assuming no advance in medical technique. Now that sounds pretty bad. But you already have a 20% chance of dying of cancer, right now, assuming you don't smoke. If you are an average smoker, your chance of dying of cancer are 40%. So that, in fact, if we recruited the crew out of smokers, [laughter] and sent them to Mars without their tobacco, we would be reducing their chance of getting cancer. So one reason to send people to Mars would be for their health. [laughter]

Now the other thing you'll notice about this hab, is that in certain respects, it has a very conventional layout. There is a table, and chairs, and shelves, and a sink. This thing is designed to be used in a gravity environment. It can function in zero-g, and it does for a few hours in different phases of the mission. But it mostly operates in gravity. We are going to land it on Mars, where it will be our house on Mars, where of course there is gravity at 38% Earth levels. And on the way to Mars, next chart, we can make gravity by tethering off the burnt-out upper stage of the booster that threw us to Mars. [A chart showing the hab connected by an extended tether to the Ares upper stage is shown.] We spin this assembly up at 1 rpm, which creates gravity in here, and by so doing, you avoid the ill health effects of long-duration exposure to zero gravity. So you don't have to do 30 years of Nazi doctor experiments on astronauts watching their bones thin out before we send people to Mars. It's not necessary.

Now it happens to be the case, however, that we do have a new data point since this mission was designed. Because, as I observed earlier, this mission was designed in 1990. Next chart.

The new data point, of course, is Shannon Lucid. [A photo of Shannon Lucid shaking the hand of President Clinton is displayed.] She flew on the Russian space station Mir. In 1996 she flew for 6 months. And unlike the Russian cosmonauts who went up for that duration and have to be hauled off on stretchers, and reported that none of their anti-zero gravity countermeasures worked, Shannon Lucid, when she landed after 6 moths in space, walked off the Shuttle. She walked off the Shuttle. And here she is, one day after she landed, walking around hot muggy Johnson Space Center, shaking hands with Bill Clinton, and she is still not sick! [laughter and applause] Now how did she pull that off? Was is sheer moxie? Well, in part, she has a lot of it, there's no doubt about it. But in fact it was because she actually implemented the strenuous exercise program that had been designed by the flight surgeons at Johnson Space Center, which involve two hours of strong exercises every day. In consequence, when she landed, she was in better athletic condition than when she took off, because she had never exercised like that before in her life. It turned out that the Russian cosmonauts never implemented their exercise programs. They are undisciplined, and their mission control would wink at it. So if you do have astronauts that are as self-disciplined as Shannon Lucid, you can send people to Mars zero-g. She was up for 6 months, which is as long as it takes to fly from Earth to Mars. But I still prefer artificial gravity, in order to put a lower burden on the crew. Go back to the mission sequence chart.

So, these people take 6 months to fly out to Mars. when they get close to Mars, they fire a pyro bolt that cuts the cable, they aerocapture into Mars orbit, and then they land at site number one, where a fully fueled Earth Return Vehicle is waiting for them.

Now we have been on the ground at site number one for two years. We have thoroughly explored the area with little robotic rovers, and taken pictures of everything nearby to use to train the crew to do the landing. We've got a radar beacon on the ground to draw them in. We've got an ace flying this thing; we should be able to land, pow, right on the spot. Some of you may recall during Apollo, we actually landed an Apollo lunar lander within 200 yards of a Surveyor spacecraft that had been put on the Moon several years prior. And we have much better guidance systems today. But let's say that doesn't happen; let's say we land 10, 20, 30 miles away. Well, we're still OK. Because we have with us in the lower deck of the hab a pressurized ground rover, about the size of a 4 by 4, run by a methane/oxygen engine, which has a one-way driving range of 600 miles. So it would really take piss-poor piloting to land outside of the range of action of that vehicle. But let's say that happens. Let's say they land on the wrong side of the planet- which would represent a very severe problem with the pilot selection process at Johnson Space Center. [laughter] If that were to occur, we can still save the mission because we have a second Earth Return Vehicle following us out to Mars, and we can always bring that one down to land wherever we did land, and that landing will be done accurately, because it will be automated. [laughter] In that case, we would be depending upon the chemical synthesis gear to work real-time instead of past-time, which is the baseline of the mission. But this is, after all, a third level mission back up, and we do have a crew on the scene to adjust the chemical synthesis gear should it malfunction—although you probably wouldn't want the pilot to be involved in that.

Finally, we have a fourth-level mission backup. You've got the whole crew landed on Mars, where they have natural radiation protection, they have natural gravity, and they have enough supplies with them to last for three years. So that if all the above fails, they can just tough it out on the surface of Mars until another launch window opens up in 2009 and more supplies and another Earth Return Vehicle can be fired out to them at that time. So what we have here is a four-layer defense-in-depth on the mission, and each layer involves successfully carrying off the mission.

But let's say that it works the way it is really supposed to work. They land here at site number one, they look the Earth Return Vehicle over, they kick the tires, it is in great shape. What do we do then with the second Earth Return Vehicle? We land it somewhere else, perhaps a couple of hundred miles away, site number two, where it starts making propellant which it will use to support the next piloted mission which will fly there in 2009, along with another Earth Return Vehicle, which is their backup, but which otherwise be used to open up site number three.

So the idea is that every two years you launch two boosters off the Cape; one to open up a new site, the other to exploit the previously opened site. Two boosters every two years is an average of one per year to support a continuous program of human exploration of Mars. [applause] And that is something could easily afford to do, and could even better afford to do if it was an international program. Next chart.

Now this, by the way is an actual photograph of the Mars base. [laughter. An artist's conception of the Mars Direct surface base is shown] What you see here is the Earth Return Vehicle, the cabin, the two stages, the chemical unit. Here is the reactor in the crater in the background. Here is the habitat module; upper deck is where they live, lower deck is the garage.

Here are some solar panels they bring out as backup power in case they have to turn the reactor off. Here is the light truck, you may not be able to see it well. It was used to deploy the reactor. Now it is a backup car for the pressurized ground rover, which is their main field exploration vehicle. And here is an inflatable greenhouse that they are using to learn how to grow crops in Martian soils. It is not a mission critical element, but it's for the future.

Now they are going to be on Mars for a year and a half. They are going to be on Mars to try to find the answers to two fundamental sets of questions. The first set of questions is: Was there, or perhaps, is there, life on Mars. All of you here, unless you were on Mars yourself during the summer of 1996, will have heard about the Mars meteorite—which in fact Everett Gibson is going to talk about in detail immediately following me, so I won't go into it—which shows significant evidence for life on Mars in the past.

However, even before that meteorite turned up, next chart, Mars was a suspect for life. [A Viking photo mosaic showing dry river channels on Mars is displayed.] It was a big suspect, because we have images of water erosion features all over the surface of Mars. And it is now extremely clear, from Viking pictures, and Surveyor pictures making it more clear, and Pathfinder pictures making it abundantly clear from ground level, that there was liquid water and lots of it on Mars, for a long period of time in the distant past. So if the theory is correct, that life is a natural evolution of chemical processes wherever you have a liquid-water temperate environment for significant periods of time, then life should have originated on Mars, even if it subsequently went extinct when conditions on the planet deteriorated. And if we can go to Mars, and find fossils of life on Mars, let alone extant life, what we will have proven is not merely that life once existed on Mars. That's a fascinating fact in itself, but it would just be a footnote in science. What we will have proven is that processes that lead to life are highly probable; that they are general. And since we now know that planets are highly populous, because we have discovered a number of solar systems outside of ours—and once again, we haven't just discovered a dozen solar systems; we've discovered that the processes that lead to the creation of solar systems are non-exceptional in nature. That means that solar systems are very common. So that if we can show that life is common where planets are common, that means that life is everywhere, and probably intelligent life too. Because the whole history of life on Earth is a history of evolution to greater degrees of complexity, capable of ever greater activity and intelligence. So if life is everywhere, then intelligence is everywhere. It means we are not alone. It means we are part of something a lot bigger than most people suspect right now. It's worth finding out. [applause.]

And I have to emphasize, that while robotic exploration is a very good first step towards Mars and finding the answer to those types of questions, and we in the Mars Society must absolutely support the existing robotic program and its expansion, robotic exploration by itself won't give us the final answers to this. Because, look, we are right here in the Rockies, I live near here. I live in a place near Morrison, which is famous for its dinosaur fossils. It's dinosaur heaven around here. I don't know how many of you have discovered some dinosaur fossils since you have been here. I haven't. Trained paleontologists come out here and they take years exploring the environment, doing rockhounding, before they make any significant finds. You could parachute a thousand Sojourners into the Rockies, and I guarantee you would never find a dinosaur fossil,—until the next ice age came and your rovers would be run over by the advancing glaciers which they could not outrun. No, if you want to find the answers to this, we absolutely must have human explorers on the surface to do it.

Now we are going to be on the surface for a year and a half, exploring all over the place, and at the end of that time we get in the Earth Return Vehicle, put the key in the ignition, start her up, and shoot home direct for Earth. We leave behind on Mars the habitat, the reactor, the greenhouse, the solar panels, the ground car, the light truck, all this stuff. Is that a tragedy leaving all this stuff behind on Mars? No. We don't need to bring the stuff back to Earth—we have a lot of stuff right here, right now, as it is. [laughter] The idea of Mars missions is to bring as much stuff to Mars as we can, and come back with as little as we can, because everything ewer leave behind on Mars is available for use by future missions. Next chart.

So here's the situation that might emerge after 8 missions have occurred. [a map of Mars, with 8 circles depicting landing areas, and an icon of Texas for scale, is displayed] These are a series of landings, which are at the centers of this set of circles. The circles have a radius of 300 miles, which is the sortie radius of the ground vehicle. As you can see, each mission is able to explore an area about the size of Texas. So we are exploring some pretty substantial areas in this sort of way. But after a certain number of these missions have occurred, and we put these landings within driving range of each other, we have a string of warming huts, defining an extended area of terra cognita available for exploration on Mars. After a series of these missions have occurred, and I don't know whether that number is 8, or 10, or 6 or whatever, maybe 8 is a good guess, we'll have gotten the answer to this question. We will know whether there was ever life on Mars, and we'll have a fairly good idea of how complexly it managed to manifest itself. And at that point, the fundamental question dealing with Mars is going to shift. It's going to shift, from question number one: was there, is there, life on Mars, to the second question, which is ultimately far, far more important. And that second question is not WAS there life on Mars, it is WILL there be life on Mars? [applause] Next chart. [A chart depicting a Mars base under construction is displayed.]

Because the most fundamental fact about Mars everybody needs to know is that Mars is not merely an object of scientific inquiry. It is that, it is an incredibly important object of scientific inquiry, but it is more than that. It's a world. It is a planet that has a surface area equal to all the continents of Earth put together, and which has on it all the resources needed to support life, and not just life, but someday a new branch of human civilization. And if we can go to Mars, and establish a permanent presence on Mars, and learn how to use those resources; learn how to take the water and carbon dioxide that exists on Mars and grow plants in Martian soils,—which we can do because the sunlight on Mars comes in 24 hour cycles which equivalent intensity to that which they have in Norway. You can grow plants in Martian sunlight, and we should be able to grow plants in Martian soil, which means we can make food on Mars, in a way that is simply not possible on the Moon because of the extreme rarity of water, and carbon, and nitrogen. Nitrogen is also present on Mars—it's the minority constituent of the atmosphere. And not only that, on Mars, we have the hydrological, and volcanic, and perhaps the biogenic processes that have formed mineral ore on Earth, that have concentrated geochemically rare elements and made them usable. and you have on Mars all the elements of industry; sulfur, and calcium, and phosphorus, and iron, titanium, and aluminum, and silicon, and all the rest of it. Everything you need to make whatever you have does in fact exist on Mars. So if you can go to Mars, and learn the craft of transforming those materials into useful substances, we can make Mars habitable. [applause]

Now, what do I mean by 'make Mars habitable?' Do I mean terraform Mars, actually transform it into a new Earth, as will be discussed in many sessions here at this conference? Well, in the long run, yes. I do. Ii think that humans will terraform Mars, because it is the nature of life to take barren environments and transform them into those that can support life. That's the entire history of life on Earth, and it would unnatural for human life not to do the same. But that is not what I am talking about us doing—us, the people in this room doing. I believe that is further in the future than our own personal horizons. However, what we can do to make Mars habitable is to transform the planet intellectually. Because the thing that determines whether an environment is habitable of not is the ingenuity you bring to it. That is, two people can be stranded the woods, and one can starve to death in 3 weeks, while the other can live there indefinitely in relative comfort. The difference, of course, is that one perceives and knows how to use the resources that are available in that environment, while to the other they are invisible.

So if we go to Mars, and establish a permanent base, and learn that craft, what we can do is transform Mars into a place that human beings can settle, because they can become self sufficient there. Next chart. [A chart is displayed showing a lunar base built with Mars Direct hardware.]

I won't go into this at length, because it is a talk in itself, but the same elements, the same hardware elements that we use to develop a Mars base can in fact be used to develop a lunar base. I don't think a lunar base is necessary to send people to Mars. A lunar base is useful though to support astronomy, objectives totally independent of Mars. And it just happens to be a good thing that the hardware we develop to go to Mars will also allow us to do whatever we want to do on the Moon. Next chart.

So, this is the complete set of hardware that we need to open up two new worlds. [A chart showing the limited array of Mars and Luna Direct hardware is displayed.] A good booster with a good throw stage; no giant interplanetary space ships, no floating interplanetary spaceports. Two fundamental hardware elements; a habitat module that can be used on either the Moon or Mars, and an Earth Return Vehicle, to use on either the Moon or Mars. The Mars one uses two stages to come back, the lunar vehicle only needs to use the upper one, although we can make the propellant to come back more easily on Mars. Finally, there is an aeroshell module that we use on the Moon...I mean on Mars. You wouldn't want to use an aeroshell on the Moon, unless perhaps it was made in a politically significant district. [laughter]

Now why should we do this? There are a lot of reasons, some of them are listed in the Founding Declaration there. We all have our own reasons. But in the deepest sense, if you look at human history in the broad sweep of time, what do we see? Humanity, homo sapiens, the current human species of which every person alive today, be they Norwegians or Australian Aborigines are members, originated in Africa about 200,000 years ago. And we persisted in Africa, in East Africa, for the next 150,000 years; in the tropics in which we originated and to which we are naturally adapted. Humans, homo sapiens, are tropical animals, that's why we have no fur, and we have long, thin limbs for getting rid of heat. And during that 150,000 years that we stayed in our natural habitat where we originated, the human tool kit did not change in any substantial way. It was an incredible period of technological stagnation.

Then, for some reason, around 50,000 years ago, small groups of humans started migrating out of Africa right into the teeth of the ice age in Europe and Asia. And suddenly, they found themselves in a hostile environment, relatively speaking. They found themselves in an environment where you couldn't get food all year round. They found themselves in an environment where you had to have shelter or clothing or both to survive, as well as efficient use of fire. The shelters had to be either built, or won from 3,000-lb cave bears, so they had to have weapons that could kill at a distance and better means of group cooperation. And they had to have fishing, which is an essential winter survival skill, as well as big game hunting. And they had to have sewing in order to make those clothes. And all of a sudden, what you find 50,000 years ago in Europe and Asia is an incredible revolution in the tool kit that continues radically from there on. All of a sudden you find finely polished tools and weapons. You find fishing kits. You find sewing kits made out of bone, very complex things. And you find cave art, with nothing primitive about it. It's beautiful advanced art right from the get-go, which implies the ability to carry out symbolic communication and probably the development of language in the sense that we know it today.

So what happened was,—it's kind of like the Bible has it right but has its times backwards. Human beings did not leave paradise because they ate of the tree of knowledge; they ate of the tree of knowledge because they left paradise. And all of a sudden our fundamental relationship with nature changed. We became a creative species. We became not merely the smartest of all animals, but became a species who fundamental relationship with nature is that of inventor; homo technologicus. And on the basis of that we have succeeded in colonizing the entire Earth. We have transformed ourselves from a local African species into a global species, that colonized the temperate zones and the Arctic, that has been able to cross seas with our boats, our ships, and our airplanes. And we transformed ourselves from a local species in Kenya into what the Russian space visionary Kardashev calls a Type I civilization, a civilization that has full mastery of the resources of its planet.

Well Kardashev has other types. Type II is a civilization that has full mastery of the resources of its solar system, and Type III of its galaxy. The entire history of humanity for the past 50,000 years since the trek out of Africa is the history of that transformation, and our achievement, which is right now just about completed, of having attained the status of a true Type I species. And not merely in terms of conquering the environment physically, but linking the world together with electronic communications, jet aircraft, and now, in fact, even unifying the world politically. This has brought about a situation that a scholar, Professor Fukuyama, a number of years back, put out as book saying that we had reached the end of history, because we have done it all, we're there. And it was very interesting, in the book *The End of Science*, the writer James Horgan interviewed Prof. Fukuyama and asked what he thought about those people who disagreed and did not believe we have reached the end of history. To which Fukuyama, responded, and I quote; "they must be a bunch of space travel buffs." [laughter and applause]

And the fundamental question, really is, that we face for ourselves right now, at this point when we have achieved Type I and we have mastered our own environment, and we have unified ourselves, is; do we settle for that? And then, stagnate and decay on that basis. Or do we reach our further, and become a Type II civilization? Do we reach our further, and take on challenges offered by still more challenging environments, on Mars, and after that the asteroids, where once again, we are going to take ourselves out of the environment to which we have adapted and force ourselves to be creative by moving into a new area in which we are both forced to be creative and free to be creative because we are separated from the dominant ways of doing things back home.

That is the fundamental issue of Mars. It is the question of whether we have reached the end of history or have simply reached the end of the first stage of human history and are ready to begin another and far more promising and wonderful stage.

I have to tell you; when I was a boy, I used to read a lot of classical history, about the Greeks and the Romans. And I can still remember this speech that Pericles, the Athenian leader gave at the funeral of the Athenian war dead two years into their desperate all-out war with militaristic Sparta. And he brought everybody together for the funeral, and he said; These, your sons, your relatives, are dead, and I understand that you are sad, but let's take a look at what they died for. They died for Athens, and what is that but an incredible city where the people rule themselves, a city of art, a city of great philosophers who have advanced the incredible radical proposition that the human mind is capable of understanding the laws of the universe – which is the basis of all science ever since. And he said: "Future ages will wonder at us, even as the present age wonders at us now."

Now, it's been 2000 years, and despite the fact that Athens lost the war, the Spartans defeated them, Pericles was right; people still do wonder at Athens. And what I'm saying is this; that if we here succeed in doing our mission, if we pull together the organization that forces the initiation of a humans to Mars program, and therefore forces humanity to take that next step, out of this Earth, into the universe, which is the fist important step in becoming a multi-planet, interplanetary, and ultimately interstellar species; then what I believe is this: That 2000 years from now, there will be people not only on Earth and Mars and the asteroids, there will be people on hundreds of civilized planets orbiting stars in this region of the galaxy. And they will have technologies that would appear as magical and incredible to us as ours would to an inhabitant of Periclean Athens. But nevertheless, if we pull this off, they will look back at this time, and they will wonder at us. Thank you. [The audience rose in cheers and an extended standing ovation.]

MARS: THE CASE FOR LIFE

Christopher P. McKay[*]

It would be most interesting to live in a solar system that was home to a multitude of planets with life: worlds full of life in all its forms. Unfortunately, we don't. Our solar system has just one planet with life: Earth. But there is another planet that still has the potential to teach us about life: Mars.

In my opinion the most compelling scientific question about Mars is life: Did it, or does it, have life? Beyond that, can life from Earth survive there — what is the future of life on Mars? I'm going to review what we know about life and Mars and how we've looked for it in the past and will, or ought to, in the future.

The search for life on Mars began in earnest with the Viking missions in 1976. Viking went to search the sands of Mars for LGMs — little green microbes. Actually, of three Viking biology experiments, only one searched for green (photosynthetic) microbes. The other two experiments tried to detect bacteria capable of consuming organic material. Interestingly, all the biology experiments yielded some positive results. Indeed, the Labeled Release (LR) experiment gave precisely the result that would have been expected had there been life on Mars. In this experiment, nutrients added to the Martian soil were decomposed to carbon dioxide, but when the soil was heated to sterilizing temperatures, the decomposition did not occur — all consistent with life. However, one of the other biology experiments, the Gas Exchange (GEx) experiment, also showed activity, but not in a way suggestive of biology. This experiment was able to detect a variety of gases released from the soil when a nutrient solution was added. Researchers found that merely moistening the soil with water released oxygen, and that eating the soil did not deactivate the oxygen release. The abrupt release of the oxygen; the fact that the sample was in the dark; and the inability to quench the reaction by sterilization all imply a non-biological cause. The case for a chemical explanation was sealed by the organic analysis instrument on the Viking landers. Neither lander detected any complex organic material in the Martian soil at levels of parts per billion.

The standard explanation for the reactivity seen by the Viking biology experiments and the absence of organics in the Martian soil centers on the presence of one or more chemical oxidants in the soil. These oxidants are presumably produced by ultraviolet light acting on the soil or creating hydrogen peroxide from water in the atmosphere. These oxidants are what has — over billions of years — made Mars red. I can summarize what Viking told us about the surface of Mars: "It's dead Jim."

Why is Mars do dead? The answer is not that Mars lacks the elements necessary to support life. Its atmosphere contains carbon dioxide, nitrogen, and water. These are the basic compounds of a biosphere. In addition, Martian soil holds other elements needed for life. The prob-

---

[*] Scientist, Space Science Division, NASA Ames Research Center, Mail Stop 245-3, Moffett Field, California 94035-1000.

lem with Mars from a biological perspective is its low atmospheric pressure and the resultant lack of liquid water. The pressure on Mars averages about 120 times less than the pressure at sea level on Earth. At this low pressure water can barely exist as a liquid. As pressure decreases water boils at lower temperatures. In Denver — a mile high — water boils at 95°C while at sea level it boils at 100°C. The pressure on Mars corresponds to an Earthly altitude of nearly 20 miles, so water boils at a temperature only a few degrees above its freezing point. Thus, any liquid water on Mars would evaporate rapidly since it would be near its boiling point. This evaporation would cool the water quickly, and since it would be so near its freezing point as well, the water would freeze. For this reason liquid water essentially does not exist on Mars, at any place, at any season. Without liquid water it's not surprising that Mars has no life on its surface.

Mars has not always been a dry world. There is direct evidence that early in its history Mars had liquid water flowing on its surface long enough to carve impressive canyons and channels. Orbital images from Mariner 9, Viking and now from the Mars Global Surveyor show these fluvial features. The image that, to me, is the most compelling evidence of long term stable flow on Mars is shown in Figure 1. Here we see a typical-sized Martian canyon that is about the size of the Grand Canyon on Earth. On the floor of the canyon, we can see what appears to be a small riverbed. Clearly, for the comparatively small river seen here to have carved the much larger canyon would have taken a long period of time — millions of years at least. The observation that at one time Mars had liquid water for sustained periods is the fundamental motivation for the search for past life on Mars.

The large volcanoes on Mars — all now extinct — provide further evidence that Mars was more Earth-like early in its history. The increased volcanism and the presence of stable liquid water both indicate that the atmosphere must have been thicker early in Martian history.

We can determine when Mars experienced its early Earth-like period from the association of the fluvial channels with Mars' ancient cratered terrain. We know from the analysis of the lunar samples returned by the Apollo program that the Moon's period of intense cratering ended 3.8 billion years ago. Assuming that this was true on Mars as well, we then can infer that the main epoch of stable liquid water flow on Mars was around 3.8 billion years ago. Computer simulations of how the Martian climate would have deteriorated after the end of the heavy bombardment suggest that liquid water remained on Mars until about 3.5 billion years ago. The total duration of liquid water on Mars may have been several hundred million years. Long compared to the life of an organism but short compared to the life of a planet.

The question of life becomes most interesting when we compare the early history of Earth and Mars. This comparison is shown in Figure 2.

The fossil record on Earth from 3.5 to 4 billion years ago is incomplete, since most of the rocks that old have been destroyed or heavily altered. But still we have direct evidence of life on Earth at 3.5 billion years ago. This evidence is in the form of stromatolites (fossilized layers of microbes) and microfossils that are similar to modern microbes. At 3.9 billion years ago, we do not find clear fossils, but there is chemical evidence for life in the form of a characteristic enrichment of the lighter isotope of carbon in organic sediments. Throughout Earth's history this characteristic enrichment has been due to biological processes. These dates are shown in Figure 2 and the comparison to Mars suggests that if Mars had liquid water at this time, it could also have had life.

The scenario I just described — of an initially warm and wet Mars that quickly became the cold dry world we see today — is based on data from spacecraft that went to Mars. However, the same basic outline of Martian history can be deduced from meteorites on Earth that have come from Mars.

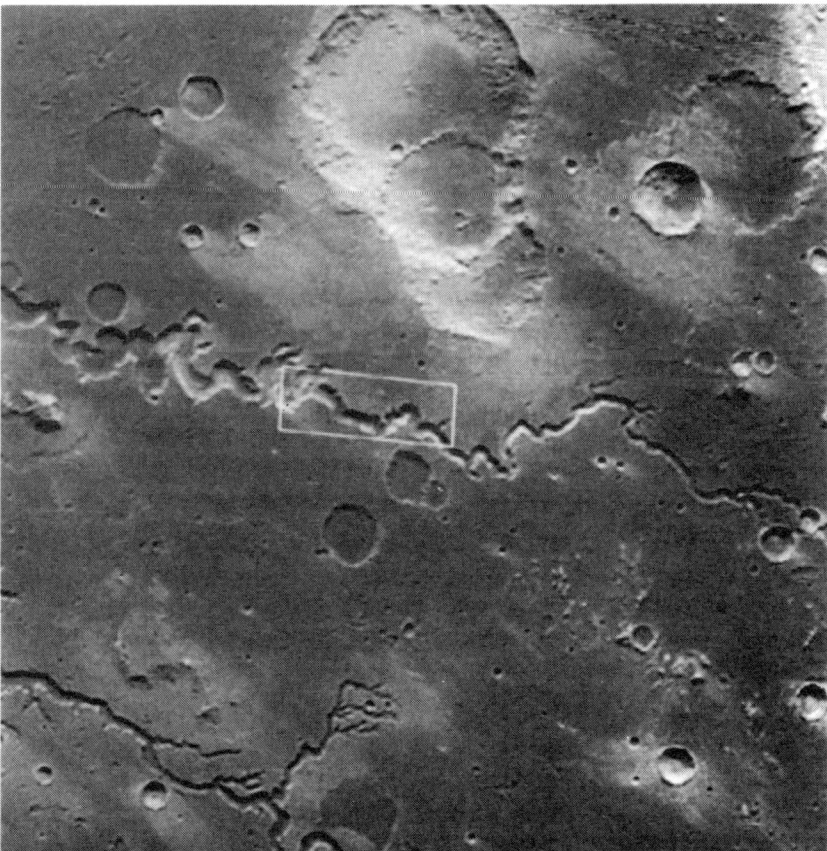

Figure 1   This Mars Global Surveyor image of Nanedi Canyon on Mars shows what appears to be a river channel on the bottom of the canyon. The canyon is about 2.5 km across. (NASA/MSS Photo).

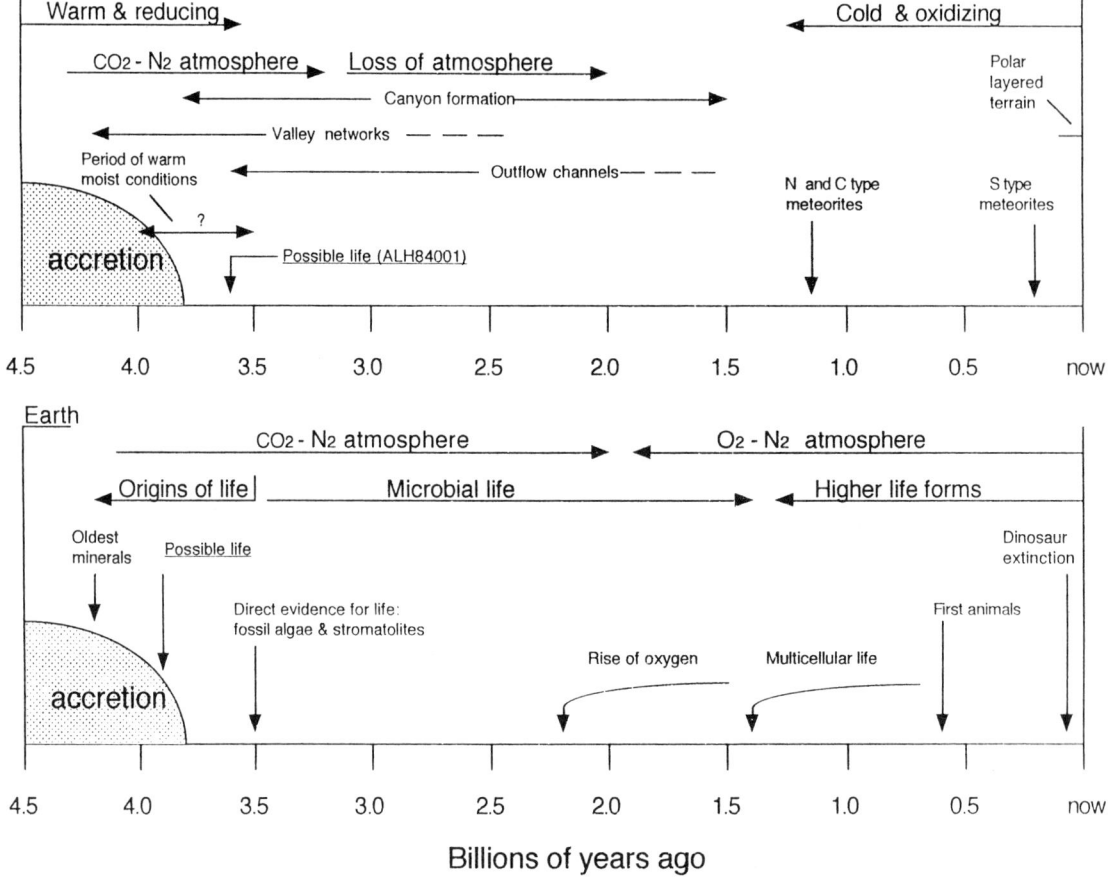

Figure 2 Mars and Earth comparison timeline from McKay (*Origins Life Evol. Biosph.*, **27**, 263-289, 1997). Note that the main epoch of liquid water on Mars is from 3.9 to 3.5 billion years ago and that the oldest evidence for life on Earth is of this same age.

Among the many thousands of meteorites known, we have about a dozen rocks that came from Mars. The evidence for their Martian origin is compelling. There is oxygen isotope data indicating that all these rocks came from the same parent body. These data do not by themselves show conclusively that Mars was that parent body. Direct evidence of a Martian origin comes from gas inclusions in the youngest of the Martian meteorites. The relative concentrations of different gases in the bubbles in this meteorite compare exactly (over a range of concentrations that span nine orders of magnitude) with the Martian atmosphere as measured by the Viking landers.

The Martian meteorites also indicate that Mars started out more Earth-like and then became the cold and dry planet we know today. The oldest Martian meteorite — formed over 4 billion years ago — reflects formation in warm, wet environmental conditions, while the younger Martian meteorites — formed on Mars less than 1.3 billion years ago — reflect formation in today's cold, dry environment.

Our quest then is to go back in time to Mars 3.8 billion years ago during its Earth-like phase and search for evidence of life. Where on Mars should we search?

We have developed an approach for searching for Martian fossils based on studies in the coldest, driest, most Mars-like place on Earth: the dry valleys of Antarctica.

The mean temperature in the dry valleys is -20°C and the precipitation — all as snow — is ten times less than in Death Valley. The valleys are so cold and dry that they appear lifeless, but life is there, hidden a millimeter or two beneath the surface of sandstone rocks and beneath the thick ice covers of lakes on the valley floors. These perennially ice-covered lakes provide an analog for past life on Mars and indicate what might be the best way to search for evidence of this past life.

Although the mean temperatures in the dry valleys are 20°C below freezing, the summertime temperatures can rise to slightly above freezing. When this happens, glaciers ringing the valleys melt and their waters flow down the valley floor and into the lakes. As this meltwater freezes, under the ice in the lake, it releases the latent heat of fusion. This heat release is the fundamental energy source that maintains the liquid water in the lakes despite average ambient temperatures below freezing. The thickness of the ice cover (about 4 -6 meters) is set by the balance between heat added to the lake by the freezing of meltwater and heat loss from the lake by conduction. In steady state, the former is equal to the rate of ablation from the top of the ice cover and the latter is proportional to the thickness of the ice cover. Beneath the ice, the lakes vary from 30 to over 100 meters deep.

Sunlight penetrating the thick ice covers allows for photosynthesis. Simple life forms — algae, diatoms, and bacteria — are found in the lakes, but no higher life forms are present.

These ice-covered lakes are an example of how liquid water can be present and how life can survive when temperatures are well below freezing. A similar type of habitat could have been the main reservoir of life on Mars while average temperatures were below freezing.

I would suggest that the place to search for fossils on Mars is an ancient lake bed. Not only would such a lake bed have been a potential site for life as Mars became colder (as shown by the Antarctic dry valley lakes), but the sediments on the bottom of the lake would have provided a means to preserve fossil evidence of that life.

Note that a river or canyon might have had liquid water, but such site would not usually be a location in which sedimentary material would have trapped fossil evidence of the life that had been present. Dried lake beds are the main target for a fossil hunt on Mars.

Gusev Crater (shown in Figure 3) has a river flowing into it and the bottom of the crater appears to be filled with sediments. While there may be a mantle of windblown dust on the crater floor as well, it is likely that most of the sediments in Gusev were deposited while it was full of liquid water. Within these sediments there may be fossil imprints of the life that lived in this lake.

Finding fossils on Mars that provide unambiguous evidence of past life will be a most interesting event, and is certainly a worthy goal for the robotic rover and sample return program currently underway. However, fossils alone will not answer the main question we have about life on Mars: Is it truly a second genesis, a separate and independent origin of life? It could well be the case that life from Earth and Mars share a common origin. The planets may have exchanged biological material throughout their history via meteorites. The Martian meteorites show that this is possible, even likely. To determine if life on Mars was a separate genesis will require more than fossils. We will need to analyze the biochemistry of actual Martian organisms — whether dead or alive. The likely place to find Martian organisms is frozen in the ancient permafrost near the southern polar regions. There lies undisturbed ground from 3.8 billion years ago, frozen and possibly containing life from that warmer, wetter state. It is likely, however, that any organisms in the permafrost are dead due to accumulated low level radiation. Even buried well beneath the surface and hence shielded from cosmic radiation, these dormant Martians would receive radiation from the natural radio-nucleotides uranium, thorium, and potassium that are present in all material in the solar system. Even if the uranium and thorium were somehow depleted from the Martian sediments, the potassium found in the life forms themselves would provide a radiation dose over 3 billion years more than sufficient to kill the dormant microorganisms. Though dead, the frozen Martian microbes would be biochemically preserved and would allow for direct comparison to the biochemistry of life on Earth. We could determine quite easily if they were genetically related to us.

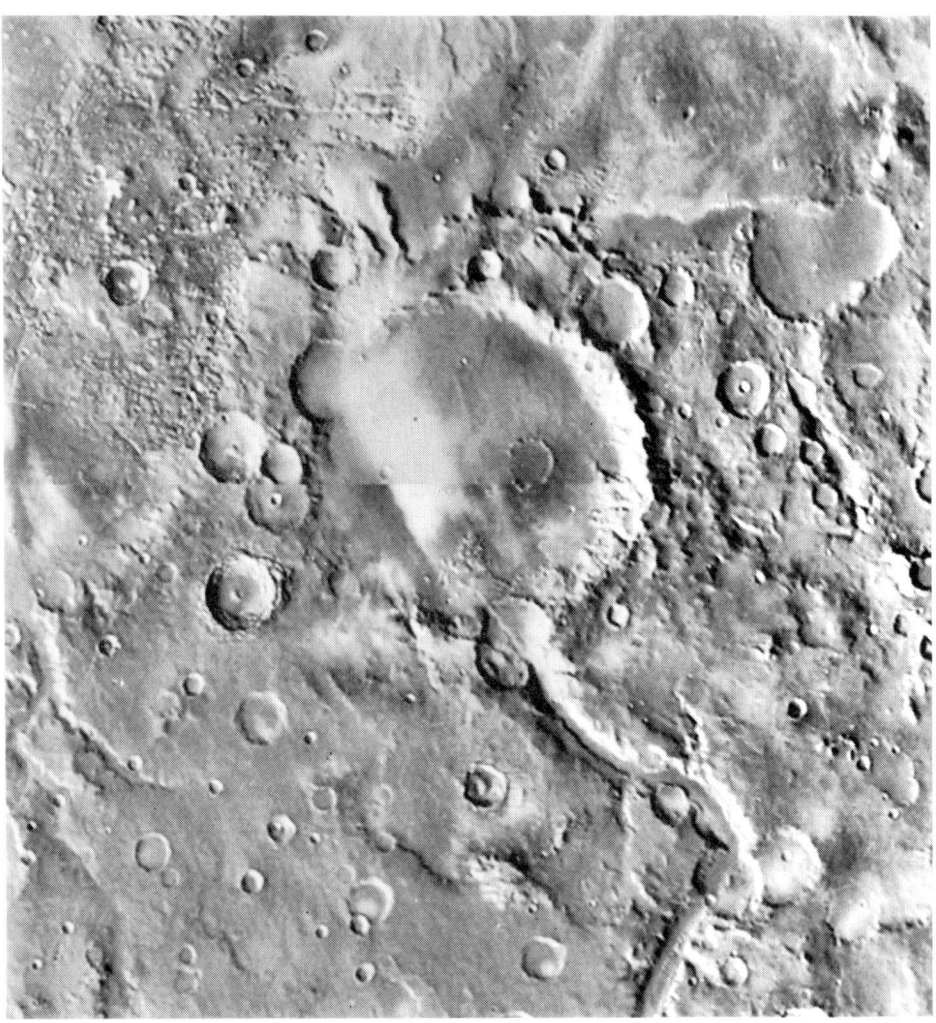

Figure 3 Gusev crater, seen in this Viking photograph, is located at -15 S and 180 W and appears to have been a lake at one time. The river feature flowing into the crater from the south is Maadim Vallis. The crater is about 100 km across. (NASA photo).

These then are the questions that motivate the scientific exploration of Mars' past: Was there life? Can we find fossil evidence of this life in a lake bed? Can we find preserved remains of Martian life in the permafrost? Was this Martian life genetically related to life on Earth? Initially we will conduct this search with robotic missions and with samples returned to Earth. But eventually humans must go to explore this world directly. The question of life is of such scope and complexity that only the most capable of field instruments — the human being — is up to the job.

Human exploration of Mars will probably begin with a small base manned by a temporary crew, a necessary first start. But exploration of the entire planet will require a continued presence on the Martian surface and the development of a self sustaining community in which humans can live and work for very long periods of time. A permanent Mars research base can be compared to the permanent research bases that several nations maintain in Antarctica at the South Pole, the geomagnetic pole, and elsewhere. In the long run, a continued human presence on Mars will be the most economical way to study that planet in detail.

It is possible that at some time in the future we might recreate a habitable climate on Mars, returning it to the life-bearing state it may have enjoyed early in its history? Our studies of Mars are still in a preliminary state, but everything we have learned suggests that it may be possible to restore Mars to a habitable climate. I believe that bringing life to Mars may be a goal worthy of humanity.

Chapter 2
HISTORICAL AND PHILOSOPHICAL SIGNIFICANCE OF MARS

Veteran astronaut John Young has walked on the Moon and is ready for Mars.

CIVILIZATIONS AT THE CROSSROADS: SPAIN AND CHINA IN THE 15th CENTURY*

Wayne H. Bowen[†]

The most successful civilizations have been those who have not abandoned expansion of their sphere of influence, whether military, cultural, or commercial. Conversely, those civilizations who have consciously limited themselves in area, ideas, or markets have atrophied and suffered at the hands of more ambitious societies.

My paper will focus on two civilizations at the crossroads, Spain and China, which had before them the alternatives of exploration or isolation. One of these, medieval China, deliberately chose to abjure further exploration beyond its borders, dismantling its navy and banning trade with the outside world. A counterexample, the Spanish empire, illustrates that nations which expand their geographic, cultural, and commercial boundaries will leave behind a legacy beyond the time of their greatness.

In 1402, during a terrible civil war, Zhu Di seized the Dragon throne from his nephew, becoming Ming emperor of China. This new ruler, taking over through violence less than fifty years after the final expulsion of the Mongols in 1368, was anxious to prove his legitimacy as the Son of Heaven and to reassert the importance of China in diplomacy and trade. Accordingly, in 1403 the new emperor ordered the creation of a massive fleet of seaworthy vessels, placing it under the command of his childhood friend, the court eunuch and Admiral Zheng He. In 1405, Zheng He made the first of what would be seven major expeditions into the Western (what we would call the Indian) Ocean. His fleet consisted of over sixty ships, some over five hundred feet long, manned by over twenty-five thousand sailors, soldiers, diplomats, and scholars. The ships represented the height of maritime technology, including watertight compartments, accurate compasses, stern-post rudders, and an advanced double-masted system of sails. Leaving China in 1405, the fleet sailed south with several objectives. While this was not the first Chinese venture into the oceans, it was the first serious and organized state program of overseas exploration launched by China.[1]

The main intent of the voyage of 1405 was to demonstrate to the world the power and importance of the Chinese empire, bringing areas visited by the fleet under the rule of the Ming dynasty. As they had done with smaller kingdoms elsewhere, the Chinese wanted to receive the respect, submission and tribute of local rulers, kings and princes. At every port at which the Chinese fleet weighed anchor, Zheng He expected local rulers to acknowledge the suzerainty of the Ming emperor. If not, the massive military force which accompanied the admiral might serve to convince the unwilling, or if necessary to install more pliant local rulers.[2]

---

* Panel: The Human Need to Explore, Mars Society Convention, August 16, 1998.

† Assistant Professor of History, Ouachita Baptist University, Arkadelphia, Arkansas 71998-0001.

Admiral Zheng also hoped to open trade routes to Chinese merchants. While China had for centuries conducted trade with Africa and the Middle East, this had previously been conducted via Arab or Persian intermediaries. The fleet provided opportunities for direct trade. Interestingly enough, the Chinese admiral was a Muslim, who had previously made a trip to Mecca, which might have increased his ability to make connections among the followers of Islam in the East Indies, the Indian subcontinent, and the Middle East.[3]

Among the thousands crewing the fleet were also many dozens of doctors and pharmacists, with orders to research new medicines and herbs encountered on the voyage, with hopes of combating the many diseases and epidemics which had struck in China over the centuries. Admiral Zheng also had a more covert mission: to search for the Emperor's deposed nephew, who it was rumored was gathering his forces to attempt an assault to reclaim the Chinese throne.[4]

The first voyage in 1405 was a dramatic success, receiving offerings of tribute and fealty from rulers in what are now Vietnam, the islands of Java and Sumatra in Indonesia, Sri Lanka, and Calcutta, India. Chinese doctors gathered medicinal herbs from practitioners in many ports, Chinese merchants traded silks and spices for local products, and the Admiral was pleased to find no sign of the deposed emperor. On the return voyage, the fleet even defeated a force of pirates who had been attacking merchant ships in the straits of Malacca. On the six subsequent voyages, which left China on 2-3 year missions in 1407, 1409, 1414, 1417, 1421, and 1431, the fleets visited thirty kingdoms in the Persian Gulf, Mecca, the Red Sea, and what are now Somalia, Kenya, the Maldives, and Madagascar.[5]

Although Admiral Zheng He died on the last voyage, his travels accomplished a great deal for the Chinese empire. China established political dominance over the South Pacific and Eastern Indian Ocean, influence which could have been strengthened in following years. Rulers from Siam to Sumatra to southern Africa acknowledged their submission to the Chinese Empire, paying political and economic tribute to the Son of Heaven in return for access to Chinese markets and military protection. Chinese merchants were able to open direct links to markets and trade routes throughout the region, bypassing Indian, Persian and Arab middlemen to acquire ivory, fruits, spices, silver, and exotic animals for the entertainment of the Chinese court.[6] Although now lost to the ravages of time, we can only imagine what improvements to sailing technology the Chinese fleets developed, and subsequently abandoned at the decree of the emperor.[7]

Emperor Zhu Di died in 1424, and his successors only sent one more mission, in 1431. Zheng He died on this mission, but he had not sailed alone: surely out of the thousands of experienced sailors, soldiers and naval officers who had traveled with the eunuch a replacement admiral could be found. After the return of the fleet in 1433, however, no more Chinese expeditions sailed into the Indian Ocean. The ships were mothballed, the crews demobilized, and the hard-won contacts and trade partners abandoned. Why? What convinced the emperors after Zhu Di to end their maritime explorations?[8]

Two possible arguments have been forwarded to explain the Chinese withdrawal from its program of exploration. In the first instance, these expeditions had never been universally popular in the Chinese court. Certain Confucian scholars and officials had never believed in the morality of these expeditions, arguing that commerce was an illegitimate force of economic activity, and that it was degrading to the honor of the throne to sponsor such a program. These Confucians, very influential in the Chinese court and government, also promoted the idea that it was demeaning for representatives of the empire to go forth to seek the submission of foreign rulers; these rulers should have to come to China to show their inferiority to the Emperor, not the other way around. Barbarian nations should come to the Empire of Heaven to be enlightened and to kowtow to the emperor in person. Confucian scholars might also argue that it was selfish and destructive for Zhen He to take his crew on such long expeditions, as this undermined the traditional family, filial piety, and the ritualized forms of respect which sons owed to parents. Perhaps these arguments did win over emperors after Zhu Di, raised as all of them were by Confucian tutors in the Forbidden City.[9]

Another argument explaining the cancellation of Chinese maritime program was the ongoing military threat of Mongol invasion. Despite the expulsion of the Mongols from China in 1368, this enemy did not disappear. China was forced to maintain a standing army of hundreds of thousands of men, in order to man the Great Wall and launch preemptive strikes into Mongolia to prevent the gathering of Mongol forces. Emperor Zhu Di even moved the capital from Nanjing in the south to the northern city of Beijing, in order to coordinate military operations against the Mongols. Faced with the enormous expense of these military operations, and the very real threat of Mongol hordes just a few hundred miles away, perhaps the expeditions of Zhen He seemed frivolous and unnecessary.[10]

Not only did Chinese emperors end their sponsorship of overseas expeditions, but after 1435 gradually forbade all overseas trade, even that conducted by private merchants. Whether for Confucian ideological reasons, hoping to forestall the export of needed capital, concerned about the corruption of foreign ideas, or hoping to charge tariffs on trade conducted on inland waterways, after 1435 China built another Great Wall, this one isolating itself from the outside world. As with the more famous Great Wall, built to shelter China from northern barbarians, this one was also a failure.[11]

In the final analysis, it remains unclear exactly why China chose to end its exploratory expeditions into the oceans of the world. While this will remain an interesting question, the implications for the history of China and the world are even more important. Had China continued with these expeditions, creating a permanent presence in the Indian Ocean, Middle East and East Africa, its relationship with Europe would have developed far differently. Imagine how 16th century Portuguese explorers would have reacted to the discovery of Chinese military bases in India, armed with cannons and superior warships? Imagine the reaction of Ferdinand and Isabella of Spain to a fleet of three hundred massive Chinese ships arriving in Spanish harbors in 1492, just as the tiny flotilla of Columbus was preparing to sail? How might King Henry VIII have reacted to a visit by the Chinese navy, bringing tens of thousands of Asian soldiers and sailors into the port of London in the middle of one of the King's messy divorces? While perhaps a bit fanciful, these scenarios point out the possibilities.[12]

If China had continued with its program of maritime exploration, trade and conquest, it would have met Europeans on equal footing, rather than from an inferior position. Chinese and European emissaries, navies and armies would have vied for status, markets and strategic locations in the lands between them during the 16th and 17th centuries, providing stimulus to both civilizations. Faced with these rivals for influence, Chinese invention and economic development would have continued, with the outcome certainly in doubt. Certainly the nation which invented gunpowder, clocks, paper money, and the printing press might have had a chance to contest European dominance of the world.[13]

Instead, China was unable to resist Western ideas and technology, gradually losing ground to Europe after the 16th century. The first Portuguese traders arrived in southern China in 1516, established a permanent commercial base in 1557 at Macao, and were gradually joined by Spaniards, Italians, Dutch, English, French, and German traders, diplomats, missionaries, and soldiers from then onward. Not just its foreign trade, but even China's domestic policies and economic future became dominated by Europeans after the 18th century, as the Western powers began to carve out spheres of influence from the weakened Chinese body politic. Riven by foreign occupation, wars, civil unrest and economic collapse, China lost its last vestiges of independence in the early 19th century, not becoming again politically unified and independent until 1949, and even then under a harsh communist dictatorship.

The example of Spain is far different. Less than sixty years after the last of China's expeditions, Spain launched a daring expedition West, captained by Christopher Columbus, which led to a global empire, fantastic wealth, and a cultural and linguistic legacy for what was then a small nation on the Iberian peninsula. While it is now difficult to imagine history without Christopher Columbus and a Spanish empire in the Americas, it very nearly did not happen. In 1483, King Joao of Portugal rejected Columbus' plan, as did France and England in 1485, and Spain in 1490, and once in 1492 before finally agreeing to sponsor his expedition.[14]

Indeed, Spain had every reason to reject the scheme of the strange Genoese sailor. Having just completed the final battle of the Reconquista, the Christian reconquest of Spain from Islam, Spain was materially exhausted. If anything, Ferdinand and Isabella hoped to launch expeditions into Africa, to continue the crusade against Islam and to compete with Portuguese gains. Several panels of Spanish scholars, priests and scientists had examined the claims of Columbus and found his ideas lacking in reason and logic. The world simply did not conform to the sailor's nautical and cartographic vision. Everyone agreed that the Earth was a sphere, but Columbus believed the world 25% smaller than was mathematically possible. Finally, Columbus insisted on the right for he and his descendants to rule all discovered lands, taking for themselves the titles of Admiral and Governor General in perpetuity, along with the lion's share of wealth and property.[15]

Nonetheless, thanks to the timely intervene of several important figures in the Spanish court, Ferdinand and Isabella changed their minds and granted the expedition financing and support. Columbus and his fleet of three small ships, which end to end combined were smaller than one of Zheng He's medium-sized vessels, sailed in August 1492 from Spain, and after stopping in the Canary Islands, departed from known waters in September 1492. After only five weeks, on October 12, Columbus reached the Bahamas. His subsequent return to Spain earned him the respect and gratitude of the Spanish monarchs, but even on his additional voyages Columbus never provided many real benefits to Spain. Had Queen Isabella, King Ferdinand and their successors been less patient, this could have been the end of the Spanish presence in the New World. It was not until Cortes and Pizarro conquered Mexico and Peru in 1519 and 1535, respectively, that Spain began to derive substantial benefits from its new colonies in what became known as the Americas.[16]

Spain something indispensable to fuel its continued efforts to maintain contact with the New World: a commitment not just "to explore strange new worlds, to seek out new life and new civilizations," but also to establish permanent settlements, economic connections, and political arrangements with new territories. It was not enough for Spain to show the flag in American harbors: it wanted to plant the banners of the Spanish crown and Catholic Church so deep in the soil that they could never be removed.[17]

While the arrival of Europeans in the Americas exacted a terrible price on the native populations, directly and indirectly causing great loss of life and the destruction of tribes and societies, the Spanish legacy has persisted in the New World to this day. It is not by accident that

Spanish is the fourth most widely spoken language in the world, and that we are speaking today in a state called Colorado.

Beyond Spain, history can provide us with other examples. At the same time as the Spaniards were spreading out into the world, so to were the Portuguese, British, French and Dutch. In competition with each other, these nations developed the technologies, political structures and will to govern much of the planet. In the 18th and 19th century, the Western frontier in the U.S. and Russia's expansion East into Siberia served the same functions: ensuring the survival and strengthening of these societies. Had the United States not expanded beyond the thirteen original colonies, or Russia beyond the territory of Catherine the Great, the survival of these states, and the future of the world, would have developed in ways we can only begin to imagine.

Conquest and military conflict is not indispensable to national or societal survival, however, despite the implicit message imparted by using the Spanish and Chinese examples. The main point here is that civilizations which isolate themselves do not survive, while those that establish permanent contact as far afield as they can muster do, leaving legacies beyond their own eventual decline. The Swiss, for example, have managed to do well for themselves, despite a lack of military adventures or expeditions into the world, but despite their neutrality they have not isolated themselves behind the Alps. Swiss manufactured goods, banking services and trade connections have ensured their place as an important player in the European economy, rather than the periphery.

What difference does any of this make on the question at hand: Should humans go to Mars? The implications of these two examples on today's imperative are tremendous. If we abandon plans to colonize the Red Planet, turning our resources instead to merely terrestrial concerns, our civilization, too, will wither away, staring forever into space, unable to muster the courage to take the first serious step toward becoming an interplanetary species. Turning our eyes to Mars will enable us to direct our aggressiveness from conflict on this planet, so much a part of our history, to a more peaceful endeavor: colonizing our solar system and beyond.

REFERENCE NOTES

1. Gernet, *A History of Chinese Civilization* (Cambridge: Cambridge University Press, 1983), 398-399; John KingJacques Fairbank, *China: A New History* (Cambridge, Massachusetts: Harvard University Press, 1992), 137-138.
2. Dun Li, *The Ageless Chinese: A History* (New York: Charles Scribner's Sons, 1965), 282-283, 285.
3. Li, 285.
4. Li, 283-284.
5. Zhang Ting-yu, *et al.*, "Zheng He's Voyages," in Merry Wiesner, William Wheeler, *et al.*, *Discovering the Global Past: A Look at the Evidence* (Boston: Houghton Mifflin, 1997), 339-340.
6. Li, 285-286.
7. Gernet, 402. Zhang Ting-yu, 338-339.
8. Fairbank, 138. Zhang Ting-yu, 340.
9. Li, 288-289.
10. Li, 287.
11. Gernet, 403. Zhang Ting-yu, 338-340.
12. Rene Grousset, *The Rise and Splendour of the Chine Empire* (Berkeley and Los Angeles: University of California Press, 1958), 264.
13. Li, 287.
14. Peggy Liss, *Isabel the Queen: Life and Times* (New York: Oxford University Press, 1992), 279-291.

15. J. H. Elliott, *Imperial Spain, 1469-1716* (New York: Penguin, 1990), 60-61; John Ramsey, *Spain: The Rise of the First World Power* (Birmingham: University of Alabama Press, 1973), 246-252; Liss, 288-289.
16. Liss, 291-295.
17. Liss, 287.

ON THE LEGITIMIZATION, IMPORTANCE AND DUTY OF COLONIZING MARS

Josef Oehmen*

Fact is: We cannot simply fly to Mars or even colonize it just out of a mood to do so. There does not exist an inherent purpose in doing so that is immediately apparent to everybody. There have to be strong reasons to justify the immense efforts and expenses.

Good afternoon Ladies and Gentlemen, and welcome to this speech at the Mars Society Founding Convention here at Boulder, Colorado.

The question I will address today is the question that should be answered before any technical problems are tackled and before any plans are made, it is the question of the correctness, it is the question of the legitimization of colonizing the Red Planet. – And the answering of that question will make clear why I today consider it so important and maybe even our duty to colonize Mars. Although I am focusing in this speech on the "why", it will also become necessary to briefly address the "how".

When, as an open-minded member of the space-community, one encounters for the first time the idea of sending men to Mars and colonizing the Red Planet, one is instantly absorbed by it.: This bold undertaking, this challenge, this impossibility of establishing a continued and, given a little time, largely independent presence of human beings on another celestial body. Personally, I can hardly imagine anything else being more fascinating and exhilarating. The thought of leaving Earth behind and found a new home for mankind among the stars, an outpost of life, was, and is enough to render me addicted. The whole undertaking of colonizing Mars, all the efforts and expenses, seem to completely justify themselves beyond the shadow of a doubt.

So, what is it what we really want? At this point, we have to be careful: Frankly spoken, sometimes I get the feeling that some people would be perfectly satisfied with a Flags & Footprint mission. To cut a long story short, I mean by this to start some sort of gut-action space euphoria that then, after a short period of admittedly probably marvelous successes, leads into nothingness. We would be successful in convincing a decision-maker or two, we will get one or two manned missions to Mars, but then, sooner or later, funds will be cut and that's it.

But on the other hand, we have this dream of a long-term program of manned missions, ultimately leading to a colonization of the Red Planet. A program with broad public support, a program where all mankind would participate. And then, through all those joined activities, we finally develop into one space-faring civilization, with a healthy population on Mars and already planning the next steps towards the stars. As all mankind is finally united in common goals, wars become fewer and fewer, and it is finally learned how to solve conflicts in a way

---

* Hellerweg 2, D-26524 Hage, Germany. E-mail: j.oehmen@usa.net.

worthy of human beings. Or as Archibald MacLeish put it: "To see Earth as it really is, small and blue and beautiful, is to see ourselves as riders on the earth together."

In plain text: We do not want a Flags & Footprint mission, we want a long-term program!

So far so good, a long-term project then. But why actually? And why should the public pay? Taking a look at the reasons which support these goals, we will see that it is not only a possibility, but also a necessity and duty to conduct the colonization of Mars, as we will see with all mankind in mind, that Mars offers a seldomly encountered chance that must be used!

So let us take a look at what a colonization of Mars has to offer: Fact is: The resources required for the conduct of the research and engineering efforts necessary to establish and continually operate a colony on Mars will be considerable. It could very well turn out that they are, as the subject itself, on a new order of magnitude. Fact is that money and manpower and whatever other resource utilized for this purpose cannot be used to improve the situation of the socially disadvantaged, cannot be used to support kindergartens or schools and cannot be used to save the poorest of the poor on our planet from starving. This shall suffice as a brief glimpse of some of the arguments that have to be won and some of the responsibilities a supporter of the movement to colonize Mars carries. By the way, it should be clear that the problem itself is not solved by discussing some new fancy ways of working with the industry. Without being able to give a legitimization the public can understand, I ask you, what is the point in colonizing Mars after all?

Mankind has in the past years, reached a stage of its development where our technical abilities allow us to eradicate all higher forms of life on our home planet within half an hour. But satellite communication, supersonic passenger airplanes and innumerous other inventions would also allow us to start with a planet-wide cooperation, if we would only choose to do so. The possibilities we have at hand not only make this possible, but they oblige us to use our powers for peaceful purposes, simply because anything else could be fatal.

Never before in the entire history of mankind had we the chance to do anything remotely like crossing the vastness of space to colonize another planet. Never before did we have the chance to unite all people, and I am not only talking of those living in the United States of America, but of the people living in all different countries around the globe, to unite them under one common, peaceful goal. Never before did we have the chance to act as one "civilization", not on a timely strictly limited basis, but to build our Gate to the Stars. With the settlement of Mars we have, for the first time, the unique chance to engage in a worldwide, cooperative challenge. It not only requires a planet-wide coordination, it requires the acceptance of the fact that all men are equal, regardless of the circumstance of birth or upbringing. This is the main point, because I firmly believe that this promotion of peace, freedom and security forms the core of the legitimization, it is this what justifies. Furthermore, it could very well be that the sheer dimension of the effort to support an outpost or a colony on Mars leaves no choice but to be a true international undertaking. Should that, what at first seems to pose the greatest threat, the scale, legitimize the project in the long run?

In addition to this, there are all the other benefits of going into space, "for free" if you like to say so: In the first place, there are the technical spin-offs: The function of space-programs as turbo-pumps in the stream of technical innovation is widely known. Just to name a few: There are the classic examples of the solar cell, the PC or simply the Teflon-coated pan. More recent examples include a LED, which, originally designed to grow crops in spaceships, allows new forms of cancer-treatment. Or geoponic soil, which can help save fertilizer. Or a new medical device which supports the function of the heart, but does not require surgery on the heart itself. Expected leaps in the technical development for us here on Earth through the development of Martian technology can include, for example, new and more efficient machinery which produce less pollutants, or maybe even none at all in the case of purely electric systems, to protect our fragile environment back here on Earth. Or more efficient agricultural techniques which will require a minimum use of artificial substances, because they will probably be quite expensive on Mars. Or new or more advanced materials than we have today.

Other, long-term, justifications include Krafft-Ehricke's Extraterrestrial Imperative. Due to the exhaustion of resources, we simply will have to leave Earth sometime. Maybe we can delay this, but sooner or later our home planet will simply be unable to support any more, or even none at all, of us. At the latest, we can be ready to leave Earth then, or take the alternative and pray that everybody prefers strict birth-control to fighting for food, water and air.

Now I will become almost unbearably subjective and perhaps reach the point where some non-engineers may be of a totally different opinion: In my opinion, there are very, very few other fields of human endeavor where, as is done within astronautics, there is so obviously and consciously defined what is possible or not, where there is in such impressive magnitude created reality. (So much for the engineer's self-esteem!).

In addition to all this, Mars is still something that presents a great challenge to mankind. Without those challenges, we would enter into a slow, but steady decline. At this point, I would like to remind you that Apollo did also pose a great challenge, did also inspire the youth and did also foster technical progress, but went down the drain after a few years, along with almost all connected hopes. Now is a good time to take a closer look at the Apollo Program, the only program that seems comparable, and see what we can learn:

Back in the 1960s, during the R&D phase, during the first launches and of course during the first walks on the Moon, the whole program was popular with the politicians and the public alike, and almost unlimited funds were secure. But after the initial space-euphoria began to cool down, after the primary goal of the people involved had been achieved, namely to demonstrate to the former Soviet Union who is boss in space and who is not, the interest in the program as a whole began to fade away, along with the public and financial support. What remained were separate groups of scientists, engineers, idealists, politicians and, of course, the military. But robbed of a common denominator, which the race with the Soviet Union had so splendidly provided, soon everybody started to pull in a different direction: The scientist, for example, wanted data (fast rockets), the politician wanted prestige (big rockets), the engineer wanted self-esteem (sophisticated rockets) and the public wanted more money to be spend elsewhere (no rockets). Detailed information on the different groups within the space-community can be found in the Parkinson-Classification, which has recently been published in the July Edition of the *Journal of the British Interplanetary Society*. So the Moon-Base quickly and silently died, along with the space-station, along with the trip to Mars. And last but not least, Apollo 18's Saturn-V today makes a marvelous sight at the Lyndon B. Johnson Space Center in Houston.

So we see that it is important for a large-scale program to function to have a common aim, a goal which is supported by all people involved. But today, we are still stuck in the middle of this 'separation'. We can learn that much from Apollo: If a common denominator for all people involved is missing, this will securely put an end to any large-scale program. We need a firm foundation on which to build, a foundation that will, unfortunately in contrast to Apollo, last long enough to support a large-scale and truly long-term program. I do believe that the legitimization given above, the promotion of peace, freedom and security through international cooperation, provides us, in addition to its primary purpose, with such a foundation. Only this kind of common aim makes it possible in the first place to start an undertaking as colonizing Mars. We have to assure that the importance of this common denominator does not get lost in all the benefits this global approach has.

So, what did we get so far? (1) What we want is a large-scale long-term Mars program. (2) This is only *legitimate* as a process that is build around the active attempt to include all mankind. (3) This is only *possible* as an international effort, because this forms an essential part of the foundation on which it is build. (4) A colonization of Mars includes all the "classic" benefits encountered in space-programs.

If we take all this into account, there is only one possible conclusion on how the long-term manned mission to Mars is possible: It prerequisites that the whole undertaking of colonizing Mars is done as a multinational effort, with the honest attempt to conduct it as a mission for all mankind.

We must foster true and long-term international cooperation. We must create a basis for mankind to grow together over time. We must actively support the participation of as many nations as possible. We must create, or better engage in, a challenge that is open for all mankind, in order to promote peace, freedom and security for us and the generations to come.

This way, it not only becomes possible to colonize the Red Planet, but those are the factors that justify all expenses, those are the factors which lift a colonization of Mars from a severe problem to a noble duty.

We must take this unique chance and opportunity Mars offers.

What could this concretely mean for the colonization of Mars? This could, for example, mean that for the colonization of Mars one planet-wide agency is created, which deals with all our Martian affairs. This agency will allow mankind to grow together over time, united by common goals and the effort of bringing and sustaining life among the stars, including our home here on Earth. As more and more people realize that it is better for them to place their attention on the destiny of all mankind, the positive impact in all fields, including the before mentioned kindergartens and schools, and including the before mentioned help for the poorest of the poor, even the impact in these fields will be immense.

Peace and freedom and security through the colonization of Mars, these are the main reasons why it is important. Let us make sure that Mars is not the end of a costly story, but that Mars marks the beginning of a new era for mankind.

For all mankind with all mankind, this is what is important, this is what legitimizes, this is the duty we carry.

MAR 98-006

THE RACE TO SETTLE MARS
EARTH HAVING A BABY
WHAT'S IT WORTH TO THE PARENT CIVILIZATION?

Peter Perrine[*]

Recent technological breakthroughs in Mars mission design, as explained in the book *The Case for Mars*, have vastly reduced the cost of birthing a new human habitable world on the planet Mars. The next President of the United States of America, Congress and leaders of other countries will weigh the value of being the parent civilization that settles Mars. The first habitable Mars biosphere seeded with plants and animals from Earth will be a child of Mother Earth akin to Earth having a baby. Of what worth is this exploration and settlement effort? This question is like asking what benefits do children give their parents?

HUMAN VALUES RESEARCH

During our face to face interviews we chose not to ask the question, "Should we begin birthing a human habitable world on the planet Mars now because of the benefits our civilization will realize?" Such a question is not yet a common thought for most people. Instead we chose to ask an analogous question which citizens immediately understood. From the answers we gain insight useful in getting public support for opening the new frontier on Mars.

WHAT BENEFITS DO CHILDREN GIVE THEIR PARENTS?

We selected 40 roughly random people, young and old, male and female, childless, with children and grandchildren and great grandchildren, and in all economic and social circumstances. We asked them what benefits do children give their parents? The interviews provided us with 239 responses listed in the appendix.

RESULTS

The benefits children give their parents can be roughly summarized into eight categories listed in order of most prevalent to least prevalent.

GROWTH AND KNOWLEDGE
JOY AND LOVE
SECURITY
IMORTALITY
PURPOSE
RESPONSIBILITY
DREAMS
COOPERATION.

---

[*] P.O. Box 41479, Plymouth, Minnesota 55441.

GROWTH AND KNOWLEDGE

The most common response expressed a belief that children cause the parents to gain new knowledge and grow. The main benefits of Mars settlement will also be knowledge and growth for our civilization. The USA is the leading country in the knowledge business. Our leading higher education and high technology industries make a Mars frontier decision a good fit with our strengths. We will also renew many capabilities before they atrophy from only thinking with lack of doing.

JOY AND LOVE

Second most common, people mentioned joy and or love. That children are just fun and children are someone that will love the parent. The second most valuable benefits of Mars settlement to the parent civilization will be the entertainment and cultural value of experiencing this human saga on television, over the internet, on videos and through music, books, toys and clothes. Mars exploration has already stimulated millions of internet hits at the NASA web site. The USA based entertainment industry leads the world in exports and cultural influence. A decision to open the Mars frontier will reinforce and fit with our strong entertainment industries and cultural influence.

SECURITY

A number of people mentioned security as a benefit children give. The emotional and physical security that children give keep one from being alone. A child is someone on whom you can depend in your old age. Children can protect parents in disputes. A valuable benefit of Mars settlement is humans will no longer be alone on Earth in the universe. Ninety five percent or more of all the species that have existed on Earth are extinct. Extinction is the rule on Earth not the exception. Establishing humanity and all Earth life on Mars more than doubles the chance of survival. The USA has the leading defense industries in the world. As they down size a Mars frontier decision will keep our high frontier capabilities preeminent. We cannot only develop new technologies for defense we must build them and use them or loose them. Directing this capability toward saving humanity rather than destroying it fits our strengths.

IMORTALITY

Many people mentioned that children give parents a measure of immortality. They keep the parents young, parents values continue, children do things like their parents. The civilization that makes the decision to settle Mars will be studied by every generation of humans on Mars. Our values, our form of government, our precious rights and freedoms will be taught, our languages will be known. The President and Congress and people of our time who lead us to this frontier will live for ever in the history taught on Earth on Mars and in the history taught to those who leave Mars to settle planets orbiting other stars. Long after all other history of our time is dust this event will be remembered by our children of the stars.

PURPOSE

Many people responded that children give parents a purpose. Participants stated that children are the goal in life and reason for living, that children give you a chance to participate in creation. Children make life real. People who do not have children do not know what they are missing. Children are the reason to keep pushing and striving. Children give a mission. Our civilization stands fundamentally unchallenged after winning all three world wars (two hot and one cold). In the midst of more prosperity than Earth has ever known, our civilization stands without direction or grand purpose. Greatness is fleeting without a challenge and purpose great enough to take our measure. Problems of the environment and of the soul remain on Earth. Problems directly applicable to the environmental and social challenges on Mars. Mars will be Earth's mirror in which to reflect our solutions. It is time for our civilization to have a child. It is time for our civilization to explore, settle, and plant the seeds of our civilization and our life forms on a new world. To do it better, to reflect upon what we have made and make Earth better. The British were the main parent civilization that settled America. The parents reflected upon their children's democratic experiment and America transformed Europe and then most of the Earth. Mars is such a place for our time.

RESPONSIBILITY

Many people said that children caused them to become more responsible. Parents become more aware of safety, take better care of themselves, drink less, and drive more safely, live their life more carefully and thoughtfully. Children cause parents to grow up. Children give the foundation and develop parents' character. Children cause parents to do better at work. A new colony on Mars will need support from Earth just as a young child needs care from its parents. Our civilization will become more responsible because we have a child with needs on Mars. The ancient Egyptians had great wealth from the Nile. Their main problem was keeping the good thing going. So they invented a religion that reinforced stability through a necessity to build and care for the graves of their leaders. Transforming Mars into a human habitable world will be our civilizations' pyramid or cathedral and it will last longer.

DREAMS

People feel that children give them dreams and allow them to realize their unfulfilled dreams. Kids can achieve dreams that parents didn't achieve. Successful children fulfill parents' desire to achieve excellence. Children reflect many of the goals and aspirations their parents had in life. Our parents haven't stopped dreaming and our civilization hasn't stopped dreaming. Mars is a place for great dreams that we can now make real. The civilization that settles Mars will be the one that still dreams great things.

COOPERATION

People mentioned that children cause cooperation. Children cause parental team work. Older kids help with younger kids. Parents become more active in the community through doing things with their kids. The civilization that settles Mars will cooperate to succeed.

CONCLUSION

The civilization that chooses to settle Mars will do it for many of the same reasons the parents in this study chose to raise their children.

APPENDIX: QUESTIONS ASKED

What Benefits do Children give their parents?

ANSWERS

Growth and Knowledge

Opportunity for parents to learn and grow.
Learn from a different perspective, you know how to walk but by watching your child walk you understand walking better.
Have a chance to grow up again with your children and live your unlived life.
Children open parents hearts.
Opportunity for spiritual growth through children's suffering.
A parent child relationship is unique giving opportunity for every kind of growth.
Children help parents to grow.
Challenge.
Parents learn lessons become tolerant and eat a lot of crow.
Teaches parents about life.
Retarded child teaches compassion and tolerance.
Teaches parents how to give.
Learn how to appreciate time children were with you.
New experiences.
New discoveries.
Children give gray hair.
Learn a lot of stuff you would not know otherwise.
Children teach patience.
You get to relive childhood.
Kids retrain their parents.
Children loosen up, up tight moms.
Parents get set in their ways, kids make parents try new things.
Kids make parents become more creative and have more ideas.
Parents relearn with their kids.
With kids you grow up all over again.
Children give parents nervous breakdowns.
Kids and parents learn from each other.
Kids teach parents a lot.
Parents learn what they can handle and what they cannot.
Parents learn a lot more from kids than other adults.
Children make you aware of your surroundings because they see surroundings.
Parents can see through kids eyes and kids see it like it is and also tell it like it is.
Kids teach parents not to take life so seriously.
Children keep your mind active and stimulated.
Grows parents in other facets not possible unless you have children.
Children teach parents tolerance.
Parents become more patient.
Creates a spiritual growth in parents as their center moves from self to others.
Parents have opportunity to relive their childhood through their children.
Kids get parents to try new things.
Parents got saga play station because they had kids.

Kids know things parents do not know for example how to shut of leaking water.
Kids make parents be not so self centered.
Children slow parents down and make them value things they may have missed.
Children lighten up your perspective.
Parents become less self centered.
Sharing of life with children and seeing life through their eyes.
Can be a kid again and live life through child's eyes.
Different perspective of life for child and parent.
Children are like climbing Mount Everest, big challenge.
Children are hardest job on Earth.
Children challenge us to pursue more knowledge ongoing.
Parents learn from children.
Children teach us to learn from our past.
Children put parents through absolute hell and back, builds inner strength.
Children force parents to learn patience.
Children help parents grow up.
Parents gain knowledge from kids.
Parents learn patience.
Fulfillment.
Kids give you nothing and everything.
Children bring back memories of your own childhood.
Children help you see things that you would not see yourself, different perspective clouds look like Winnie the Pooh.
Parents have freedom to act like kids again.
Parents go places they would not go without kids.
Kids can teach or cause their parents to do new things.
Children add a dimension to your life that you can't experience without being a parent.
Take the time to look at things through a child's eyes.
Children allow parent to see a new perspective.

Joy and Love

Children are an opportunity to give love.
Children are fun. Child's spirit.
Fun.
Love.
Good relationship.
Children are people you can count on to love you.
Somebody to love and share with.
Children give unconditional love.
Joy.
Children are just fun.
Enjoyment of watching them grow up.
Watching kids learn new experiences.
Kids give parents joy.
Love, joy.
Kids give benefits more in the emotion sense.
Joy in simple things.
More inner pleasures.

Warms the heart.
Children return love.
Children teach parents how to love more and be more loving.
Kids entertain parents and tell them jokes.
Kids make parents enjoy their marriage more.
Kids make life more fun.
Children intensify the joy in life.
Children make you feel loved..
Children are a blessing.
Children add a lot of fun.
Children make life very interesting and very fulfilling.
Joy in watching them grow up and be successful.
Someone to care for.
Someone that will love the parent.
Enjoyment.
Affection both ways.
Joy.
Children make hobbies more enjoyable.
Unexplainable joy that you cannot get from anything else.
Unconditional love.
Fun to watch kids develop and grow, dance class example.

Security

Children give tax benefits.
Children accept you as you are.
Children can protect parents in domestic disputes.
Can help parents when they are older, car breaks down.
You can depend on kids always being there and always caring.
My wife cooks to feed the kids so husband gets fed to. (Laughing)
Children take care of you in old age.
Kids bury parents.
Kids take care of parents.
Help in lawn mowing.
Children look out for parents welfare in old age.
Children help protect the estate.
Friendship, keep us from being alone.
Gives parents a sense of security.
As parents get older they have family and support.
Keep us company.
Kids can help support old parents.
Children can be there for you.
Another hand to work (farmer)
Company for parents.
Kids cut grass.
Kids clean up rooms.
Kids give parents discounts on theatre tickets.
Kids give parents presents.
Kids give parents company.

Kids hold things for parents and help them.
Kids help prepare food.
Children can make it financially.
Tax deduction.
Day and nights never lonely.
Children could console parents in old age.
Kids help do chores.
Children are people who are concerned about you when you are older.
Knowing someone can take care of you in old age.

Immortality

It's a Immortality future thing.
Children give you opportunity to transfer your values.
Children will manifest parents shadow.
Children give grandchildren.
Children keep you young.
Kids leave.
If you do not have kids you get old quicker.
You can teach kids right from wrong.
Satisfaction is seeing kids make right decisions.
Children fulfill a biological need.
Grandchildren.
Children keep body and soul young.
Children are a window to the future.
Investing in future generations by propagation of the species.
Children keep parents young.
Carries on the family heredity and name.
Because parents have children they care about their image in the community.
Children are part of gods plan because he wishes the human race to continue, and flourish.
Children are a continuation of life our species continues to live.
That's what we are here for on earth, that's why we exist, procreate and continue with the human race.
Seeing yourself in your kids.
To carry another life form is most progressive healthiest contribution to humanity.
To pass on love and care and blessings that we had as kids.
Parents positive input showing in children.
Kids do things like you.
Kids are an opportunity to improve upon yourself.
Parents get to be teachers.
Watching your children parent their own children.
Sharing ideas with the children's minds.
Keeps you young.
Children make parents feel young.

Purpose

Gives you a chance to participate in creation.
Purpose.

People who do not have children do not know what they are missing it is fortunate they do not know because it would hurt them.

Children Keep you busy.

Children make holidays more meaningful, neighbors have no kids and holidays mean nothing.

Parents get to go on trips and do things with kids.

Satisfaction of all the years and raising and hard times and end result and look back and it's worth it.

Gives parents something to do.

Gives parents reason for working.

Children give a sense of accomplishment.

Parents have pride in children if they do good, heart attacks if they do not.

Reason to keep pushing and striving.

Kids give parents a purpose.

Gives people a goal in life that is higher than making money for self and gives reason for living.

Orients one's life purpose.

Children make life real.

Big difference between parents with kids and parents without kids.

Children are the goal in life.

Being able to nurture someone.

Someone that will depend on the parents.

Parents feel successful if children grow up to be successful.

Purpose.

Gives parents a goal.

Children give you a mission.

It all comes back to the kids.

Real world is kids they are a distraction from the other world.

Responsibility

Learn to live your own life more carefully or thoughtfully.

Children keep parents level and responsible and taking less risks.

Children teach parents responsibility.

Children cost a lot.

Children help to mature parents.

Gives foundation.

Gives parents responsibility.

Children keep parents sane or maybe insane.

Children force parents to do things.

Parents live healthier life style because children depend upon them.

Want family to be healthy.

Parents become more safety aware, drive car more carefully.

Children cause parents to grow up.

Children develop parents character.

Parents have incentive to do better at work.

Kids keep parents from doing bad things bars, smoking drinking and driving.

Kids remind parents to do things safely.

Kids get parents motivated and get them out of bed.

Parents gain fear of things of the world.
You drive safer when you have kids.
Parents become more responsible.
Children keep you in line keep you from doing things you should not do.

Dreams

You can get kids to do dreams you didn't do.
Parents can do their best but it is uncertain how they will turn out some times we have to bow down and accept the children even if they do not fulfill parents dreams. Kids will march to the beat of a different drummer.
Kids can do parents dreaming.
Successful children fulfill parents desire to achieve excellence.
Proud of kids gives satisfaction to parents.
If kids turn out good you feel good.
Pride when kid sings well.
Children reflect many of the goals and aspirations you had in life.
Feeling parents get when they see children accomplish a goal.
When you raise kids all the problems that are out there do not seem as bad when you look into a child's eyes.

Cooperation

Causes parent team work.
More things to argue about.
Older kids help with younger kids.
Parents become more active in the community through doing things with kids.
Kids force everyone to eat together.
Children can provide family fellowship relationships and emotional support.
Kids help with other kids.

MARS EXPLORATION: THE SURVIVAL OF OUR CIVILIZATION

E. G. Petrakakis*

WAKE UP CALL
EXPANSION = SURVIVAL

In 1969, when man landed on the Moon, I was listening to the radio. This historic event filled me with hope, vision and excitement. I fully identified with what was happening. I was told that by the year 2000 there would be a permanent base on the Moon, we would have reached Mars, cars would be moving around on cushions of air, and trains would travel on magnetic rails. What has happened to those promises for the future?

RELATING THE REASONS FOR GOING ON TO MARS AND SPACE BY ANALYZING SOCIETY AND THE WORLD TODAY

In order to understand what has happened and how the present day situation relates to Mars expansion, we have to look at what I call the **Three Global Truths**.

GLOBAL TRUTH ONE – Nothing remains still.

GLOBAL TRUTH TWO – Our technological society is contracting.

GLOBAL TRUTH THREE – Space is a channel for expansion

Global Truth One – Nothing Remains Still.

Throughout the history of Mankind, one truth is very clear – civilization survived while it was expanding.

For example, when the Incas and Aztecs, the Romans and Greeks stopped expanding, their cultures started contracting or were taken over by other civilizations, more aggressive and technologically more advanced.

We see a similar phenomenon in southern Africa at the turn of the century, where the Boers pushing northwards clashed with the Zulus. In turn the British arrived and took over from the Boers.

Expansion therefore is directly related to survival and continuity.

Global Truth Two – Our Technological Society is Contracting.

This truth can be seen in several phenomena, such as environmental crime, social collapse, world economic collapse and technological deceleration.

---

* Costa Do Sol Lda, P.O. Box 319, Maputo, Mozambique. E-mail: 101754.1676@compuserve.com.

We can see an example of environmental crime in the global destruction of our forests and Ozone layer. Man-made erosion as a source of environmental destruction is increasing, which is even resulting in the contamination of our food production. This is demonstrated in the advent of Mad Cow Disease and the mass contamination of chickens in Hong Kong recently.

Social collapse is apparent in the increase of the differing aspects of social malaise such as mass migration of whole communities, mass refugee problems, and mass population displacement, which are caused by wars rooted in territorial conflicts.

Urban violence has escalated to the point that projection graphs based on a five-year cycle of violence offer a frightening picture of the future.

Displaced children represent a global wave of humanity that is becoming a commonplace feature of urban society. The future for these children is questionable.

World economic collapse is a very real possibility. The world today is an interconnected global village. What happens in the United States for instance sends shock waves around the world. We are all, in a sense, holding hands.

The United States debt is high and rising. The debt, in turn, of developing countries is also rising. Economic deterioration has reached the point where world financial institutions are forced to reschedule National debts or even writing them off. These countries can't even service their debts, which leads to global debt.

One of the reasons for the increase in debt is that some economies are almost entirely based on an unofficial sector, such as the black market or narcotics. This leads to a break down in Econometrics. It becomes impossible to measure the economic status of a country when the biggest sector is unreported.

As a result, the markets of the Northern Hemisphere (Capital, Equity, etc.) are adjusting and readjusting, and struggling to stay alive while in the Southern Hemisphere the purchasing power is shrinking and poverty is growing.

Technological deceleration is also apparent. At the turn of this century, science was building the bridges that would take us to the end of the century. A few examples include the electric lamp, the diesel engine, airplanes antibiotics, etc. Where are the technological milestones, which will carry us into the next millenium? This time, because of globalization, if society falters it will not be a localized event, such as the collapse of the Roman or Greek civilization—the entire world will go under.

We still judge the comparative power of different countries based on military strength and not on science or other more intelligent parameters. The recent example of the nuclear weapons showdown between India and Pakistan is a good one. Who said the Atomic Age is over? We must be realistic.

Global Truth Three – Space is a Channel for Expansion

The solutions to the problems noted in global truths number one and two lie in the third global truth—space is a channel for expansion. In order to prepare for our future in space, we must engage in preparatory activities in the short term. These include: promoting science, providing a vision to youth, developing strategies relating to the environment and terraforming, focusing on New Millenium objectives, popularizing space, and increasing political awareness.

Promoting science is critical, as science is the first dynamic, which builds bridges for our technological society to develop and grow. It is the father of our technological society.

Because space expansion absorbs real global economic resources, it requires both private and government input. Space expansion involves big numbers, big challenges, and big markets. It also creates new markets and a new economy of scale, possibly the last one.

Vision provides our youth with the hope and direction vital to sustain the development of a healthy environment. The global youngsters of tomorrow will be European, American, African, etc. We need to start the International Young Astronauts School in order to involve these young people in the future.

Terraforming and the environment are integral aspects of space expansion. The process of Terraforming another planet (Mars) will simultaneously teach us to better respect and better understand our own planet and everything here that we take so for granted. It's comparable to giving a child a car versus making the same child work to afford it. In which case would the child most appreciate the vehicle? TERRAFORMING IS THE GREATEST ENVIROMENTAL LESSON.

We should push governments to introduce and adopt serious Environmental and Nuclear bills among the countries; this would be real globalization.

The Moon and the oceans are important areas of study. But they do not possess as many of the necessary qualities to become a genuine third millenium goal, as does Mars. The exploration of Mars addresses the real problems of today's society.

The New Millennium cannot be entered into on the backs of the achievements and goals of the last millennium and space exploration, particularly Mars exploration as a kick off, will be one of the milestones for the 3rd millennium.

This society runs on speed and adrenaline; where better can we expand to?

Political Awareness is critical. We have to help politicians make space exploration "in". Here are a few ideas:

1. Why sell sponsorships for the next launch and paint the next Space Shuttles auxiliary tanks, perhaps one with Coca-Cola and the other with Windows? It would certainly raise a few desperately needed dollars for NASA.

2. Develop a Mars-like mock-up and exhibit it around the world at Trade Fairs.

3. Create Theme Parks around the world in major towns.

Let's push politicians to adopt Bills to convert the military establishment, phase by phase, into a space establishment.

Finally we owe it to our children to give them the same, or an even better, hope and vision at this turn of the millenium, that was given to us at the turn of the previous one. At the turn of 1999, we are entering a new year, a new decade, a new century, and a new millennium.

So, on to Mars, on to Space, onwards forever.

REASSESSING THE HUMAN CONDITION: PHILOSOPHICAL ASPECTS OF MARS EXPLORATION

Richard L. Poss[*]

We who are privileged to be participants in the founding convention of the Mars Society share a vision of exploring and settling the red planet in our time. This paper outlines the strengths and weaknesses of seven "classic" arguments for space exploration in general, and then examines the effect in each of a commitment to Mars as the next human frontier. In addition, several rejected arguments are described, and several counter-arguments are considered. The paper is presented as a set of distinctions and clarifications of practical use to those who will carry the debate to the general public.

ARGUMENTS FOR SPACE EXPLORATION

1. The Advancement of Science and Technology

This is the oldest and purest of the arguments for space exploration and one of the strongest. Indeed it is so strong that we frequently forget how much is left out. While not all space exploration is science, there are some forms of scientific investigation which can only be performed through space exploration. On the simplest level, if we send a probe to one of Saturn's moons, we will acquire information about that moon, which will enlarge our knowledge of the solar system. This knowledge will be interpreted in the light experimental results here on Earth, so that theories of planetary formation, etc., will have a larger data base from which to draw. All this is difficult to refute, if "science" in the limited sense of data collection is our main goal. The two main objections to this are (1) "So what?," and (2) "This can best be performed by unmanned, robotic probes, and therefore human space exploration is unnecessary." The first of these is a serious concern. Funding priorities always involve judgements as to the importance of many different kinds of activities. This "pure" knowledge, if it comes, will come at the expense of flood relief, weapons research, worthy social programs such as Project Head Start, support for the arts and humanities, and unemployment and welfare payments to Americans in genuine need of assistance. The question, "What good is this pure knowledge about Saturn's moons, that we should fund it at the expense of these other needed activities?" is neither small-minded nor short-sighted. Examples from the history of science support the answer to this question, which must take the form of a demonstration that over the long term, basic research in the pursuit of seemingly useless knowledge has shown itself time and again to have life-saving relevance. The lines, however, cannot be drawn clearly at the time the research is conducted, and therefore one must appeal to our historic experience. The argument that the larger good of society (not just American) is served by all forms of basic research has been made persuasively to the American public already, and, one might say, the evidence for its success is everywhere.

The second question, of robotic probes, is also crucial, and goes to the heart of why we want to explore. Much, perhaps most, pure data collection can be performed by machines. The

---

[*] Humanities Program, University of Arizona, Tucson, Arizona 85715. E-mail: rposs@u.arizona.edu.

results of historic spacecraft like the Viking, the Galileo, and the current Cassini mission show the spectacular advances generated by unmanned probes. Without in any way diminishing the accomplishments of those great robotic explorers, we must remember their limitations. Quite simply, they do not report what they have missed. Recent NASA experiments with telepresence in a robotic rover in Arizona show alarming limitations in recognizing quite important features. A rigorous analysis of the *Pathfinder/Sojourner* experience shows that while it constituted a technical triumph and a badly needed public success, the amount of research carried out was very small. Important questions raised by features visible from the lander could not be investigated because they were out of range of the rover. These features could have been explored within hours had there been humans on the site.

Our position should be to try to heal the wounds of the decades of forced competition for funding between the manned and unmanned mission communities. As humans explore the solar system, robotic probes will multiply in number, quality, and importance. Human exploration is unthinkable without a fleet of probes of every conceivable type, doing basic science, targeted experiments, telecommunications and information processing using artificial intelligence, and other marvels we cannot even currently imagine. So the response to the objection that probes can perform exploration must be both "Yes" and "No," for while they can do a great deal, probes cannot replace humans.

The second portion of this argument concerns technology. Briefly stated, it holds that space exploration is worth doing because of the advances in technology which it generates. These arguments have aspects which are ambiguous and problematic, but on the whole the argument is sound. Space exploration requires rockets and other space traveling vehicles, habitats, space-suits, and a range of devices designed to perform tasks in zero-gravity, in low earth orbit, or in the particular environments of the Moon, Mars, or deep space. This "forced" acceleration of technological development then can be applied to terrestrial industrial production, resulting in new manufacturing procedures, new materials, and new products.

This aspect of the technological argument has something in common with wartime. War also spurs development of science, technology, industrial production, new jobs, etc., and yet we do not go to war just to generate jobs. To explore space for the indirect reason of higher technological spin-offs is a disappointing and deflating argument. At the same time the range of technologies which could be expected to benefit from this is quite broad, and includes military and medical applications and considerable meteorological and environmental advances. These benefits are impressive, but not sufficient.

2. The Advancement of Industry and Business

It may seem to us that there is a smooth continuum running from pure science to technology, then to industry and to business. In a capitalist system, when it works, basic science often generates technology which is used by industry to manufacture products which are marketed to consumers. But it need not be that way. A socialist or communist (or any) state could conceivably engage in a major space exploration program and utilize the resulting technologies with no regard for business applications or the enhancing of consumer markets.

It is awkward to argue that we should explore space because of technological spin-offs and in the next moment urge missions to Mars which can be executed "with current technology." It is expected that technological advances are inevitable as exploration proceeds, and they can be expected to have a major impact on industrial processes and materials engineering. Discoveries and applications will extend to medicine, the environment, psychology, agriculture, and human health. The list is endless, but the concrete results in any given area are neither immediate nor assured.

My suggestion is that while these are valuable and important benefits, they do not constitute in themselves a valid reason for space exploration, especially when we consider the enormous dangers this enterprise will entail. Young men and women will die, for if anything is certain it is that deaths will occur in the exploration of space. We should therefore approach the mission with some sense of the solemn character of the enterprise. Humans do not cease to be human when they go into space, and we can expect over the long term to encounter accidents, sabotage, terrorism, crime, even war. Not to acknowledge this is to be naive about the realities of human nature. Too much emphasis on better airplanes or lighter shoes can be misleading with respect to the seriousness of the risks involved.

Of course, space exploration will generate all manner of business and commerce, and it will be of tremendous benefit to the world. Eventually it will involve people building and selling to each other in space, not on Earth. But the whole notion of linear benefits that come back and appear in tangible form on Earth is a limited perspective which in the long run we must grow beyond. Parents do not work, love, sacrifice, and persist for two decades so that their child can write them a check or bring them better shoes. Fundamentally, material pay-off is not what this is about. Like parenting, space exploration is at its core about love and a commitment to the continuity of life beyond our own time.

3. The Argument From Military History—The "High Ground"

While this is the strongest and most immediate argument for the United States and other industrial nations to go into space, as it deals with national security, it relies upon assumptions which are not shared by a significant portion of the community. The "high ground" argument is distasteful (especially at an international conference with world-wide participation) because it accepts the necessity for one group of humans to get the advantage over another group, in anticipation of violent conflict.

The "high ground" argument begins with a review of military conflicts and levels of technology. It proposes that if two countries go to war, the side with the superior technology usually wins. That superior technology can be visualized as the vertical third dimension in a two-dimensional horizontal landscape. If two opposing groups have swords but one group also has the bow and arrow, then in general, the group with the technological "high ground" will eventually overcome its less developed neighbor. If two sides have land armies, but one has a strong navy as well, it will generally be the stronger. If both sides have tanks, but one has air superiority, then it will be the stronger. If both sides have strong air forces, but one has orbital surveillance, then it will have the advantage. This argument suggests that the Gulf War was won when the allies successfully prevented Iraq from obtaining satellite intelligence from a French commercial firm. The argument does not purport to explain the history of warfare, and there are important exceptions to it (Vietnam not the least of those), but as a general principle it can be said to have a limited sphere of applicability.

Thus the "high ground" argues for a nation to go into space because it will lose the next war if it doesn't. However, if there are no more wars, then this argument is null and void. It is the background for much of the history of rocketry, from the V-2 to Sputnik through the Apollo program. President Lyndon Johnson used to ask his advisors, "Do you want to live under a red moon?" This argument assumes and perpetuates competition and lack of trust among groups which have made much progress in cooperation and collaboration. Thus the very strength of this argument is also its greatest weakness.

If the "high ground" argument is valid and recognized as valid, then it implies that when your neighbors go into space, they may be doing so to prepare for hostilities. This was much in the minds of representatives of the member states of the United Nations, who have considered over the years several space treaties limiting exploration, development, and militarization of space. These are legitimate and valid concerns for less-developed nations, which may be placed even further behind in terms of prestige, authority, and wealth. Put another way, any nation which declines to explore space can be expected to fall behind technologically and suffer the resulting losses in the area of national security.

4. The Argument From National Pride

It is possible to see space exploration as sport, and to follow it on TV the way one follows a balloonist or a basketball team. This follows quite naturally from the "high ground" argument, in which the activity of space exploration is evidence of cultural, technological, and military preeminence. Competition in the military arena flows into competition for prestige. The U.S. and Soviet race for space during the Cold War occupies this category, getting to the moon first for the "bragging rights." The Apollo program has been derided for this reason, and it did perform rather little real science. But though the reasons for Apollo were not primarily scientific, they were nonetheless compelling. There are seven arguments for going into space, and science is half of one of them.

The "national pride" element is irrational, insidious, and suspect, but it is a very strong motivator. We see its presence in current negotiations over participation in the space station, and in the dynamics of cooperation in planning future Mars missions. International collaborations for purely scientific research still have major contributors, and the very presence of scientific research indicates political and military prominence.

It has been claimed that during the anti-war tumult of the summer of 1969, when the first human set foot on the Moon, every person in America felt "proud to be American." If it is true, it shows the emotional reach of the national pride phenomenon.

5. The Argument for Planetary Insurance

This argument does not excite the emotions, and is admittedly weak in the short term, but is very strong over the long term. If space exploration occurs on a serious scale, then human civilization will consist of communities living on Earth, on the Moon, on Mars, and on space stations in orbit about the Earth and Mars, and soon out to the asteroid belt and beyond. When this has happened, we will have planetary insurance, which may be thought of as similar to "continental insurance" today. As an example, suppose a major catastrophe occurs and Europe is completely destroyed. What becomes of European civilization? While the human and cultural loss would be great, the truth is we would pick up and keep going in a very short time. What if both Europe and North America were destroyed? The same result. Civilization has spread across the globe and the loss of any country or continent would not halt the living reality of our culture. The ideals of the Greeks and Romans, all European literature, the knowledge of the Gothic cathedrals, the history of all kingdoms and wars, science and technology, music and art, would all survive because it has been transported and transplanted in Japan, Australia, Hong Kong, Africa.

To extend this unpleasant example, suppose an asteroid, or nuclear warfare, or biological or chemical accident or warfare, were to end all human life on Earth. At the current time, it would be the end of everything. But if we lived not just on Earth but in the solar system, then life would go on. This requires us to do a little "growing up." Many feel that "if life on Earth were destroyed, I wouldn't care about it continuing on Mars or the Moon." This provincial childishness is like the Englishman who figures that if England sinks, the rest of humanity is not worth considering, or the Frenchman who thinks if Paris is destroyed, the rest of humanity is not worth bothering about. We must realize that humanity and civilization are larger than the city-states of ancient Greece, larger than western Europe, larger than the countries which currently rule the Earth, and larger than the small collection of planets and moons we are soon to inhabit. What we do here matters, and our actions have consequences beyond our own lifetimes. We are at the beginning of human history, not at its end, and there is a greater world out there in which we will play a important role.

6. The New World Argument

If space exploration is more than doing science and making better machines for a more convenient life on Earth, then what is its real essence? For the answer we must refer to the first and greatest of explorers, Odysseus, whose voyage through great pains and marvelous wonders, recounted in Homer's *Odyssey*, was all in search of his home, the isle of Ithaca. The ultimate goal in his wanderings was to get back home where he belonged, as king of the island, husband to his wife, son to his elderly father, and father to his own son. Later, when Virgil writes the *Aeneid* under the patronage of the Emperor Augustus, he rewrites Odysseus' voyage, transforming it into a different kind of journey. Aeneas' world, Troy, is destroyed, and the gods coerce him into leading his people to a new land. He encounters the same dangers and marvels as Odysseus, but for a different kind of overriding purpose, which is to create a new home and a new civilization. He wages war on the Italian mainland, then marries the Princess Lavinia. "Lavinia" is a woman but is also the name of the land near Rome. His followers intermarry with Italians and settle on Italian soil, which becomes their home. Thus the goal of his adventures is to join with the new land and plant the seeds of what will become the civilization we know as Rome.

This narrative has a long and problematic history. It is the essential myth of colonialism, but it is also the essential trajectory of space exploration. We will go into space, not to bring back things which will make us richer or more comfortable here on Earth. We will go there to live and make it our home. This is the "New World" argument, and just as few Americans today want to return permanently to their country of origin, in two hundred years few residents of Mars or the space stations will want to move back to Earth. This argument raises the hopes for a new start, for a new landscape where new kinds of cities, universities, arts and letters will arise. This kind of exploration is the precursor to migration, and it is backed by a biological imperative which is more basic and more powerful than the languages or ideologies we generate to describe it.

7. The "Substitute for War" Argument

This is my favorite argument but one we hear only rarely. It has a sociological dimension and is not a strong motivator in the short term. If we recall the personal accounts of the members of any community in serious crisis, but most particularly the personal accounts from England and America during World War II, we repeatedly hear remarks like the following: "It was very hard to get by, but there was a feeling of togetherness, of common humanity and community that was so strong, I have never felt anything quite like it." From any number of quarters but most classically from London during the "blitz," one encounters these testimonials. Of course the ones who gave these accounts are the ones who survived, and no one ever recommended going to war as a way to generate feelings of togetherness, but it is a fact that the social phenomenon of mobilization for war had many positive features.

Would it be possible for a society to commit itself to so radical a process as mobilization without the threat of war as a motivation? This argument echoes the famous phrase, "the moral equivalent of war," and it suggests that instead of several nations mobilizing to attack each other, that they all mobilize to explore, settle, and spread civilization throughout the solar system.

This is a social argument, which prompts the hearer to look around at the "way things are going" in the world. Is society becoming more rational, more humane, more virtuous? Or is it degenerating? This argument makes use of examples like the following: The youth of the inner cities who join gangs do so for many reasons, many of which are in themselves good. The opportunity to be part of a team, to experience trust, to test one's skill and courage, to encounter real dangers, to depend on one's companions and have them depend on you for their survival—all of these are healthy motives. There is something in youth and in growing up that needs these challenges, and a society at peace often invents substitutes for combat so that these challenges can still be met. If space exploration were to become a large-scale societal imperative, as military and technological competition was during the Cold War, then the young would know exactly where to turn for the most awesome and dangerous challenges, to the strict meritocracy of an astronaut corps, open to all.

DISCARDED ARGUMENTS

1. Overpopulation

The argument to explore space because of overpopulation on Earth is fraught with flaws and difficulties, and for this reason I have rejected it. First, population growth continues to evade accurate quantification and predictions of its increase or decrease continue to miss the mark. Secondly, and more critically, this argument is open to the charge that space exploration is an attempt to avoid responsibility for coming to terms with social problems here on earth. Any arguments for space exploration which fall into this category should be discarded. For instance, one could argue that we have polluted the atmosphere, and therefore we should move into space. Or, we have failed to deal with violent crime or to care for the poor, and therefore we should create "suburbs" in space. Space flight cannot be flight from the consequences of failure to deal with any social problem, first because it is cowardly, second because it invites easy and devastating rhetorical attack.

COUNTER-ARGUMENTS: ARGUMENTS AGAINST SPACE EXPLORATION

1. The Argument From Hubris

"We should not presume to conquer other worlds. It is not for us to disturb the pristine state of other planets." This argument invokes guilt over past genocides, over manifest destiny, over pollution, and ridicules the effort to explore space as a vain attempt to reclaim past imperialist glory. This is essentially a cultural argument, and can only be met on cultural terms. If we rip off the mask of piety, we expose this argument for what it is, fear and cowardice. It accuses space exploration of being a flight from responsibility when in fact it is the opposite. A commitment to space exploration involves nothing less than taking responsibility for the solar system the way we now take responsibility for the Earth. Space exploration is dangerous, expensive, and not for the weak of heart. It will eventually entail "environmental management" not just of one forest or of the entire Earth, but of the entire solar system. As such, it involves a profound turn outward, into territories some feel are "alien" and "inhuman." It will always be a threat to those who fear the wide vistas revealed by modern science, who prefer to huddle here in endless contemplation of the past, in comfortable ignorance of the reality of other landscapes than our own.

This argument has curious blind-spots. It has no problem with the current state of high technology, for example. It is presumably not "hubristic" to hop around on a commercial jet from New York to Paris to Tokyo, to lecture audiences on the evils of science and technology, but it is only the gravity well of Earth which is a sacred taboo.

2. The Argument From Misplaced Perfection

This argument takes a current problem, portrays it in dramatic terms as a pressing need, and then asks the question, "With such urgent responsibilities here at home, shouldn't we wait until we have solved the problem of (fill in the blank) before we go shooting rockets off to outer space?" So, before we expend resources on space exploration, we should put an end to the problems of world hunger, poverty, war, AIDS, pollution and environmental damage, global warming, education, domestic violence, racism and ethnocentrism, homophobia, and el niño. In short, each and every legitimate social concern can be placed in front of space exploration and shaped into a rhetorically effective attack.

This particular formulation is effective in a number of ways. It makes the partisans of space appear insensitive, elitist, "haves" over the "have-nots." It pushes the audience toward the comfortable populist left, toward arguments for social welfare, thereby pushing the space partisans toward the right, lining them up (again) with the military, the aeronautics industry, and big-budget science programs. A crucial part of our task is not to allow this demarcation to take place. We must get there first, and clarify the benefits and opportunities of space exploration, and infect every audience with the wonder, the dream, the idealism of this enterprise. We must also resist the political polarization which renders space exploration the prisoner of one side of the ideological landscape. In space we will experience greater diversity than on Earth, not less. So far we have been on the losing end of this debate.

The answer to this counter-argument has two parts. First, some of the social ills are permanent features of human society, so that "solving" them would be comparable to solving the problem of anger. There is appropriate use of anger and inappropriate use of anger, but anger is a part of life and is not going away. Second, both for those social ills which are permanent and for those which can be eliminated, our moral obligation to do all we can about them still stands, with or without space exploration. But why is it that it is always space exploration which is "put on hold" until all these ills can be permanently cured? Why is it that physics, chemistry, engineering, and biology laboratories are not told to close their doors and wait until hunger is "solved" before they can resume their activities? It cannot be that they are useful in the effort to solve hunger and space exploration is not, because that is not true. More likely, space exploration is a victim of its own visual imagery, because it wants to go "out there" and seems far removed from the daily necessities of human life, when nothing could be farther from the truth. It has taken five hundred years for European civilization to expand onto the American continent. Has this resulted in any benefits to Europe in the areas of medicine, physics, technology, or culture? What a ludicrous question! But it is a question for which five hundred years are required to produce a definitive answer.

3. The "New Age" Argument

This is more a prevalent attitude than a coherent argument, but if it had to be stated plainly it would read like this: "Science and technology are bad, and anything that takes as much science and technology as space exploration does must be bad too." This is the popular mind-set which relaxes in a fuzzy subjectivism about the possibility of truth or reality, thereby avoiding responsibility for judgement. Instead, it luxuriates in a variety of might-be's: astrology, crystals, channeling, pyramids, and the diverse popular forms of belief and ritual which go in and out of fashion. To engage in the long-term program of space exploration is to accept enormous risks and to show enormous faith in the future. These are not currently popular, and the debate (more a conflict of mythic images) between these two mind-sets must be joined before real progress will be made in the area of large-scale societal support for exploration.

4. Practical Arguments

These are the concrete short-term arguments concerning all the specific aspects of exploration. Frequently these take the form of "too much": too much cost, weight, time, coordination. Each of these arguments must be met on the particular turf of technical specifications. We are on good ground with this kind of argument, partly because in the years since Apollo, during which little exploration has been attempted, technology has progressed significantly. The result is that several technologies are available for exploitation in space that were unanticipated by the Apollo planners.

THE ROLE OF MARS

What happens to these general arguments when we introduce the commitment to Mars as the next human frontier?

1. Science and Technology

Mars is a gold mine for geological and meteorological investigation, and for inquiry into the origins of life. Having a base for large-scale information gathering on another planet provides several sciences with a doubling of the available data set. It will expand the scope of our understanding of Earth processes in ways we cannot now anticipate. In terms of technology, the task of creating a livable environment on Mars, dealing with the challenges of Martian dust, the generation of water, breathable atmosphere and sufficient biomass, will all by necessity force the growth of our technological prowess. All these mechanisms will benefit the Earth.

2. Industry and Business

The importance of Mars is that it will be a permanent place, a country, in which a culture and a way of living will emerge. People on Earth will soon have the feeling that things are getting "out of our hands." Indeed, much of the commerce which develops in space will be between different groups in space, rather than between space and Earth. But products and processes generated on Mars will wind up on Earth, just as products made today for the American market also find their way around the world.

3. The "High Ground"

At first glance it is difficult to imagine the presence of a human community on Mars having any singular effect on military conflict on Earth. The "high ground" argument applies very well to Earth orbit, and secondarily to the Moon. In the early stages, wars on Earth will produce shock waves on off-world communities, as European wars had their effects on the American colonies. It was difficult back then to foresee that the young American colonies would return to Europe and resolve a global conflict not once but twice in the 20th Century. We should not underestimate the importance of Mars to the future health of the Earth, even though it will happen beyond the horizon of our current understanding.

4. National Pride

Mars provides a wider playing field for matters of national pride. Instead of launching men and women in tin cans out into empty space, or to several destinations, we can now send whole communities to another landscape, another country. Mars will provide a permanent location for expansion beyond the Earth, offering more possibilities for the ephemeral enthusiasms of the "sport" and "soap opera" aspects of public consciousness. Exploration will continue on the surface of Mars, with challenge upon challenge to play out on the nightly news as an edifying example in contrast to the latest governmental scandal.

5. Planetary Insurance

Mars is the first realistic step in the direction of a truly planetary civilization. A minor outpost composed of military and scientific technicians does not constitute a true "settlement." For example, on Earth, if a catastrophe destroyed everything but the research stations on Antarctica, that would not be enough. Similarly, Mars will not be enough unless there is large scale settlement of the red planet. Large scale migration can more likely be achieved there than anywhere else in the solar system. We are more likely to survive in large numbers, and to create real cities, real communities, on Mars, than anywhere else. It is the first step beyond the Earth.

6. New World

Mars is the crux of the New World argument—if we only had the Moon or an L5 space station, the New World argument would not be valid. But with Mars so large, so vast, and so adaptable to human presence, this is the strongest of arguments. Survival on the red planet will not be easy, nor will it be impossible, rather, it will be just difficult enough. Mars is a cosmically fallow field, waiting for the human presence to disturb it into a flourishing majesty.

7. Substitute for War

It is difficult to say if the human community is willing to accept this sort of mobilization, nor is the settlement of Mars dependent on its acceptance. Mars can be settled by any one of several countries now active in space. It will even be possible soon to achieve its settlement through private consortia. Like the East India Company, these multinational concerns will have their own motivations for establishing communities around the solar system. So the sociological scenario of the "moral equivalent of war" may or may not come about, but the presence of Mars even here gives a concrete focus, a palatable goal, where "outer space" becomes not a vacuum but a territory and a place.

CONCLUSIONS

Arguments for space exploration in general take place within several modalities, each with its own assumptions, its own inherent values, its own rhetorical strengths and weaknesses. Some are valid but rest on assumptions not shared by the general public; some are unfashionable; some are rhetorically ineffective. The commitment to Mars as the next human frontier helps all these arguments. Generally, it brings them down from the level of vague "futuristic" predictions to the world of the practical and the doable.

"Outer space" is a vast emptiness, but Mars is a place, and we can go live there. The Martian frontier gathers all the impulses of bravery, adventure, science, technology, idealism, and the yearning for freedom to move, and localizes them in a palpable, actual, reachable landscape. Martian life will be hard and unforgiving at first, but there will also be new arts, new literary forms, new religions, new dreams, and greater glory. The cumulative effect of all these arguments is to urge a reconsideration of the role of humanity in the cosmos. Earth alone is not our destiny. There is a larger context to all our decisions, and that larger context into which we are now beginning to grow is that of the solar system as our home. A commitment to Mars is the first great step towards our maturity as a civilization.

Chapter 3
HISTORICAL LESSONS

The Rev. James Heiser explains the need for a Shining City on a higher hill.

THE OUTWARD COURSE OF EMPIRE:
THE HARD, COLD LESSONS FROM EURO-AMERICAN INVOLVEMENT IN THE TERRESTRIAL POLAR REGIONS

Marilyn Dudley-Rowley*

In the late 1800s and the early part of the 20th century, American explorers had a vision of the polar regions as a logical extension of Manifest Destiny. Vilhjalmur Stefansson, however, had a different viewpoint, referring to the entrée into the Arctic by Euro-Americans as "the northward course of empire." Popular history would have it seem as if this vision came true. But, approximately 100 years later, we substantially fall short of these explorers' dreams. Most American claims in the Arctic fell through, not from lack of interest by average Americans, but the lack of government sponsorship, backing, and going back on promises made. Even the purchase of Alaska from Imperial Russia was a transaction that was almost not made. In expedition after expedition, men, women, and children died in the field waiting for pick-up from ships that would never come.

Ironically, what interest there was for the Arctic eclipsed a promising beginning of interest in the Antarctic. American explorers either had to pass themselves off as foreign nationals to join the expeditions of other countries or use their own money to launch expeditions to the southern continent. Interest in aviation caused the government to establish the United States Antarctic Service (USAS) and bases were established to protect territorial claims. However, the onset of World War II drew resources away and the bases were closed, and when the United States returned to Antarctica, it was with a different strategy of scientific investigation. In 1959, twelve nations signed the Antarctic Treaty, agreed to use Antarctica for peaceful purposes, "froze" territorial claims, and forbade new ones. The Antarctic Treaty set the tone for similar agreements among nations which dictated similar use of the entire Cosmos.

This presentation points out that the two polar regions represent two separate kinds of human progress, and reviews lessons from Euro-American polar exploration useful to the private Mars initiative, making recommendations for the public outreach and financing of the venture.

The polar regions have had a way of luring people out of a temperate-zone mentality and bringing nations into conflict and cooperation.

In the late 1800s and the early part of the 20th century, American explorers and their supporters had a vision of the polar regions as a logical extension of Manifest Destiny. Broadly speaking, Manifest Destiny meant that Americans were a chosen people ordained by God to create a model society. Specifically, it referred to a conceptualization of American expansionists to extend the boundaries of the United States from the Atlantic to the Pacific; and later, it was used to annex a number of territories and to justify involvement in Mexico, Cuba, and the Philippines (*Encyclopedia Britannica* 1979, p. 567). So confident that Manifest Destiny would legitimize their activities, Euro-American gold prospectors, trappers, and hunters entered Siberia from Alaska in search of life, liberty, and the pursuit of happiness. Following their own ex-

---
* OPS-Alaska, 2664 Montana Road, Fairbanks, Alaska 99709.

pansionist policies, Russians moving eastward stumbled over these Yankees and their influence on many occasions. Russian explorer Innokenty P. Tolmachoff came upon an American prospector and trader, a Mr. Wall, living at Serdze-Kamen Cape on the northern shore of Anadyr Bay, married to a Chukchi woman, during his Siberian expedition of 1909. A number of Chukchi in this area could speak English fairly well, too (Tolmachoff 1949, pp. 208-209).

Awareness of lands to the south of Magellan Strait seems to have emerged in the 1500s, and by the 19th century, Dutch, Portuguese, French, British, Russian, American, Belgian, German, and Norwegian ships were making exploratory, scientific, whaling, fishing, and sealing voyages to Antarctica. The early 20th century brought a number of heroic interior and aviation explorations of the continent. Territorial claims by several nations were made or implied. Although Germany never made any claim over Antarctic territories, it mapped Antarctica extensively by air in 1938-1939 and used the subantarctic Iles de Kerguelen as bases for resupply during World War II. German ships based there destroyed approximately 193,000 tons of allied ships (Mericq 1987, p. 68). The prospects of aviation and the military significance of Antarctica caused the American government to establish the United States Antarctic Service (USAS) and bases were established to protect territorial claims. The military significance of Antarctica is this: "Control of Antarctica allows a tie between the Atlantic and Pacific oceans and improves the potential logistic support by air and maritime forces (p. 45)."

Ironically, what interest there was for the Arctic eclipsed a promising beginning of interest in the Antarctic. American explorers either had to pass themselves off as foreign nationals to join the expeditions of other countries or use their own money to launch expeditions to the southern continent. American James P. Shetland was on the 1821 voyage of Fabian Gottlieb von Bellingshausen who served Czar Alexander I of Russia and who made it south to the islands which bear his name (p. 7). In 1823, the American Nathaniel B. Palmer in the company of a British crew hunting sea lions discovered the South Orkney Islands (p. 7).

Using the Arctic as an example, Vilhjalmur Stefansson called the Euro-American expansion into the circumpolar world, "the northward course of empire" (Stefansson 1922). The concept of this northward course of empire is part of a larger idea. This larger idea conceives of human progress continually moving away from tropical and temperate-zoned lands. (It is also called "the coldward course of progress".) Stefansson saw the Arctic as one of the new cradles of civilization for humankind. Other authors voiced similar-sounding sentiments in the titles of their books about northern lands: *The Path of Empire* (George Lynch 1903), *Lost Empire* (Hector Chevigny 1944), *Quest for Empire* (Kyva Petrovskaya Wayne 1986). But, they were only depicting Manifest Destiny or the Russian version of it. Stefansson and his antecedents were talking about something else: the breachment of humanity from its lands of origin, and with that, the common use of the Arctic for humanity.

Euro-American efforts in the Arctic seem to have taken after the model of Manifest Destiny, however, than after the "coldward or northward course of progress." Manifest Destiny operated in spite of lack of national commitment to foresight and the future and the politics of the moment, which often translated out into the lack of government sponsorship and backing. Where sponsorship and backing were evident, promises and effort often did not come through. For example, even when the United States had international competition and motives to acquire polar lands, it tended to put its best interests on the back burner. The Alaska Purchase was a case in point. Alaska is an enormously rich and strategic country which was appreciated by a number of 19th century American statesmen and a good portion of the public. A number of presidents prior to the Andrew Johnson administration had considered the purchase of Alaska from Imperial Russia. The stated reasons were many: (1) the profitability of the fisheries of the

Pacific Coast; (2) resistance to the occupation of the Northwest Coast by another nation, such as Britain; (3) the establishment of the United States as a Pacific power; (4) to annex British Columbia; (5) to secure unlimited American commerce with China and Japan; and (6) a feeling of friendship for Russia by the United States (Shiels 1967, pp.1-3). The issue went back and forth till the Johnson Administration. When the principals in the second largest land deal in history finally transacted the Treaty of Cession on March 30, 1867, there was much difficulty with the American payment of $7,200,000. The 19th century version of insufficient funds occurred – in other words, the American check bounced twice. Following the October 18, 1867 formal ceremony turning over Alaska to America, undisciplined American troops abused the Russian citizenry so badly that most returned home (Cohen 1996, p. 48). All those who stayed or came into Alaska from then on out "complained bitterly about the inefficient and often corrupt Army rule (p. 52)." In 1877, troops in Alaska were pulled out to deal with a rebellion of Nez Perces in Idaho and Montana (p. 51). Although Alaskans were happy about that, they were left unprotected and with no police to enforce any law and order. The street fighting in Sitka, the colonial capitol, and strife between whites and Tlingit got so bad that the British warship *Osprey* steamed into Sitka harbor on March 1, 1879 and turned its guns on the Tlingit village near Sitka. The captain announced that he was staying until the U.S. government did something about protecting its own citizens (pp. 52-53).

The Lady Franklin Bay Expedition in the Eastern Arctic was organized in response to an international polar science project. Officially mandated, it was shabbily put in motion because of Congressional and military mis-coordination. When the expedition got in trouble owing to the bumbling of the pick-up ships under the command of the U.S. Navy, the government simply abandoned 25 healthy expeditioners. As a result, only seven men survived, and Commander Adolphus Greeley has gone down in popular history with the "bum rap" of being an incompetent leader. Digging into the facts of the expedition, however, one finds that it was the efforts of Greeley's wife, who through much networking and private means, got a bounty imposed among international naval, whaling, and trading vessels for the rescue of the party. Without her effort, there would have been no survivors. Fort Conger, the scientific and military post established by Greeley in the Eastern Arctic, does not seem to be ever again used by Americans until Peary and Henson used it for an emergency stopover in the early 20th century.

The Wrangel Island debacle is similarly tragic. And, it is a matter whose issues extended into the latter half of the 20th century. Stefansson had already embraced the concept of "northward course of empire" by the time he organized the Wrangel Island colonizing expedition of 1921. Wrangel is a sizeable island lying about 90 miles off of Cape Jakan off the Chukchi Peninsula and not far from the northwest coast of Alaska. Stefansson launched the Wrangel Island Expedition for several reasons: 1) as part of a continuing campaign to debunk the extreme image of uniform hostility of Arctic regions which was popularly held; 2) because at the time a million square miles of the north polar region was still unexplored; and 3) for a stopover for the fledgling air transport industry (Stefansson 1925, pp. 69-72). However, to attract funding for the expedition, he advanced Manifest Destiny arguments rather than the more intellectual and social evolutionary "northward course of empire" credo.

Wrangel Island was known by Native peoples of the Arctic and was a waypoint on the proto-historic route to Point Hope, Alaska, an early center of Arctic civilization. Native informants told explorer Lieutenant Ferdinand Wrangel the approximate location of the island. Wrangel, in the employ of Imperial Russia, testing the theory of a high northern continent, never actually saw the island. It was the British who found it looking for Sir John Franklin's lost expedition in 1849. At that time, it was taken possession of in the name of Queen Victoria

(Stefansson 1925, p. 18), but promptly neglected by the British for the next 32 years. Then, in 1881, an American expeditionary team landed on the island (p. 12), constituting an American claim, particularly in light of the Treaty of Cession which fixed one boundary of the Purchase of Alaska as "to the Frozen Sea." (Wrangel Island lies within the semi-permanent pack ice.) However, by the end of 1916, the Russian ambassador in London had given notice to the United States that territories and islands situated in the Arctic Ocean and discovered by a 1911 visit by Russian icebreakers were being claimed by Russia. Wrangel Island, however, was not specifically named (p. 22).

Stefansson sought funds from Britain and Canada on the basis of their claim of the island. Funds and support promised to him by those countries fell through because of interminable political dickering, the competing claims of the four nations involved, and by the momentum of world events. The poorly stocked expedition failed in the face of a period of scarce game on the island. By the time a proper rescue could be mounted, only one survivor remained of the Wrangel Island Expedition. It was only through a private network that Stefansson raised enough funds to retrieve the lone survivor (pp. 157-169) and send replacements to the colony. Two families and three bachelors from Alaska were recruited to occupy Wrangel Island (Webb 1981, pp. 85-87). Great Britain and Canada remained disinterested in pursuing their claim on the island in the meantime. Financially at the end of his rope, Stefansson sold his property and improvements on Wrangel Island to Carl Lomen of Nome, who headed the largest reindeer industry in the United States (pp. 89-90). Lomen traveled to Washington, D.C. to consult with the Secretary of State on his plan to reinforce the United States' claim on the island. Secretary of State Charles Evans Hughes essentially told him to "go and hold it" (p. 91). Shortly thereafter, the British renounced their claim on Wrangel Island and the Soviets sent an armed vessel, *Red October*, to the island "to imprison all inhabitants, confiscate all goods and furs, and establish Russian ownership of the island (p. 92)." Once there, while the officers of the *Red October* confiscated the colonists' four gunny sacks of skins and ivory for the Soviet Union, they failed to mention to them that they were under arrest. In fact, they were treated courteously and told they would be returned to Alaska (pp. 94-95). That was not the case, however. They were taken to Vladivostok in the middle of November 1924. The Eskimo members of the colony were permitted to work. The health of the one white man, bachelor Charles Wells, however, began to fail and he had to be supported by a friendly Soviet official and an American trader with a permit to do business in Siberia (pp. 96-97). One of the children died while the Americans and Soviets dickered over the Russia-America boundary referred to in the Alaska Purchase's Treaty of Cession (p. 97). Hampering negotiations was the fact that the U.S. did not then recognize the Soviet Union and had no direct diplomatic relations with them (p. 99).

When it was found that the cost of transporting the colonists from Vladivostok to Seattle would cost $1,600, Secretary of State Hughes told Carl Lomen that the State Department had no funds available for their relief or transportation (p. 98). When Lomen pointed out that the Eskimos were wards of the United States, the State Department backpedaled from Secretary Hughes' earlier stand, saying that the United States "at no time asserted a claim to Wrangel Island" and that the Lomens should have foreseen complications (p. 98). Both Lomen and Hughes appealed to the American Red Cross for help (p. 99). The Soviets next ordered the colonists to leave Siberia for Harbin. By this time, Charles Wells was ill with pneumonia and could not travel. In crossing the border into China, the American Consulate Officer there told Chinese officials that the Eskimos were not American citizens which led to their being detained in a hotel (p. 99). Another child died in detention. Back in Vladivostok, the Soviets sent word to the Americans that Wells would not be allowed to leave until the United States apologized for a boundary marker placed on the Siberian coast by a United States Coast and Geodetic Survey ship. No apology needed tendering because Wells died three days later on 8 January 1925 (p. 100). On January 10th, the United States Secretary of the Interior acknowledged that the Eskimos were wards of the American government, but also voiced the claim that there were no funds available for their transportation. Finally, the American Red Cross broke down and advanced the money and the Eskimos were permitted passage on a Japanese steamer. In Seattle, waiting for transportation to Alaska, another child drowned (pp. 101-102). Carl Lomen and his family and friends continued to fight for the American claim of Wrangel Island and for reparations. The American and Russian dust-up over Wrangel Island did not abate until 1974 through the Mutual Protocol Agreement which permitted exchange of scientists, which allowed one American ornithologist to do fieldwork on Wrangel.

In Antarctica, World War II would draw resources away and the American bases were closed. However, the United States did not back away from the southern polar continent:

> The United States is the chief architect of law and policy for the Antarctic. For over three decades the United States has exerted the political and diplomatic clout necessary to enhance and expand the legal regime governing the Antarctic continent and Southern Ocean. In fact, the legal basis for U.S. involvement in Antarctica stems from the 1959 Antarctic Treaty, an agreement originally called for and substantially forged by the United States (Joyner and Theis 1997, p. 1).

The Antarctic Treaty of 1959 stemmed from an American initiative (p. 1). Its central theme is the international preservation of the continent for scientific and peaceful purposes. It works along the line of consensus led by the duopoly of the United States and Russia (Peterson 1988, p. 212; Joyner and Theis 1997; Vicuna 1988; Suter 1991; Stokke and Vidas 1996; Jorgensen-Dahl and Ostreng 1991; Francioni and Scovazzi 1996; and Klotz 1990). The twelve nations which signed the Antarctic Treaty acknowledged that there were competing or potential claims among signators but that the Treaty would not settle these, but rather forbid any new ones. As one legal scholar has noted, "…it is no more than the grandest internationally agreed upon 'question begging' that diplomacy has devised (Haley 1963, p. 122)."

The International Geophysical Year of 1957-1958, a year of scientific research and cooperation, led to the Antarctic Treaty of 1959. It also led to the development of the law in space (Haley 1963). The model for the handling of sovereignty over celestial bodies is Antarctica (p. 121). On December 13, 1963, the General Assembly of the U.N. adopted an "Declaration of Legal Principles Governing the Activities of States in the Exploration and Use of Outer Space" which advocates exploration and use of outer space for the benefit and interests of all mankind; and declares that outer space and celestial bodies are not subject to national appropriation (Jenks 1965, pp. 317-318). When the United Nations Moon Treaty was up for ratification in the early 1980s, the United States backpedaled away from it and its language about lunar resources in terms of "the common heritage of mankind", a phrase found in Antarctic Treaty System (ATS) documents (Simpson 1982, p. 12). A stumblingblock appears to have been the issue of mineral resources, the same issue which has come up in international law concerning Antarctica in recent times (Vicuna 1988).

This notwithstanding, the ATS is a model of progress in human cooperation. Antarctica, its treaty system, and its precedent for the use of the Cosmos appears to be more in line with the model of the northward or coldward course of progress, in spite of its sticking points. The Manifest Destiny model of expansion in Antarctica was effectively nipped in the bud with the expense of World War II and the evolution of the Antarctic Treaty System itself.

Let us now return to the Arctic in its more modern incarnation. The futurist R. Buckminster Fuller's vision of the Arctic had the tenor of Stefansson's "northward course" ideas. Fuller saw the Arctic regions as a kind of world heartland where Arctic rivers could provide hydroelectric power for all of the planet (Fuller 1979). One has only to remember geopolitical theorist Halford MacKinder's maxim: "He who controls the heartland controls the world" to understand how Manifest Destiny proponents would view this. The Fuller vision has not come to pass.

In actual fact, the irony has been, the Manifest Destiny model accounts for the permanent human presence in the Arctic, and its lack of usage in the Antarctic goes a long way in explaining why there are no permanent Antarcticans. (This is not forgetting the relative proximity of overland routes to the Arctic from temperate-zoned centers of population and the lack of a thick ice cover except in Greenland, in contrast to Antarctica.) Everything being equal, there are a number of human features of the Arctic landscape missing from the Antarctic. A few are as follow. An overriding feature found in the Arctic is the presence of people who live and work there for other than scientific reasons. In Antarctica, with perhaps the exception of a couple of Argentine and Chilean bases, families do not make Antarctica their permanent address.

In the Arctic, military bases, installations, and submarine activity which have emerged during World War II and the Cold War have kept a third world war at bay. The Antarctic Treaty restricts a military presence in southern polar lands.

The Trans-Alaska Pipeline (TAPS), the largest construction project since the Panama Canal, keeps the United States from being held hostage by other oil-producing nations. Other countries having Arctic lands also exploit oil and gas there. In fact, Russian experts were consulted in the construction of the TAPS in Alaska and a lot of technology was invented on the spot as a result. Environmental activists also played a role by forcing pipeline builders into upper levels of technology to protect the Alaskan environment. In Antarctica, oil and gas exploration has just begun and is stymied by the ATS, as any mineral exploitation has been. Most authors writing about mineral exploitation in Antarctica cite technological problems anyway. But, the first hurdle is the ATS itself. It has been pointed out that a "third stage" of Antarctic development has dawned (Mericq 1987, p. 31), but no one seems to know how to proceed.

Highway systems crisscross many of the lands of the northern polar regions. The Alaskan-Canadian Highway (Al-Can) was built in response to Japanese encroachment in the Aleutians during World War II. In an effort to bring their transportation network into the 21st century, Alaskans have even been discussing building a bridge or tunnel over the Bering Strait, not an insurmountable piece of engineering which would bring economic relief to the people of Siberia, cause economic growth in Alaska, and change the world in a landmark way. Regular transportation networks are instrumental to human permanency. Antarctica is a land where no conventional roads are possible. Antarctic workers must depend on aircraft and over-the-ice vehicles for travel from station to station or fieldwork. Perhaps Antarctic tourism will introduce some kind of regular transportation network on the continent, but stauncher supporters of the ATS think that what tourism exists should be curtailed.

It is clear there are two disparate models of human progress operating in the two polar regions respectively. This dichotomy more or less tracks with the dichotomy of *res nullius* and *res communis*. *Res nullius* is a legal and political position which means that lands with no owner can be appropriated and subjected to national sovereignty and jurisdiction. *Res communis* is a legal and political term meaning "belonging to all", and not subject to appropriation and national sovereignty for any purpose, but subject to exploitation for the benefit of all humanity (Mericq, p. 58). I have identified, for the purposes of this paper, the dichotomy of Manifest Destiny vs. the northward or coldward course of empire. In the Arctic, Manifest Destiny seems to have played a role in bringing population from all parts of the world there and setting up a human infrastructure for them there. The drawback of the Manifest Destiny model is that it was expensive in terms of lives and fortunes lost. The model wavered in whatever political winds happened to be blowing in the world and not all Arctic lands reached the potential of their explorers' dreams.

The northward or coldward course model represented by the Antarctic, exalted science and technology as tools of diplomacy and served as a laboratory for international cooperation. The drawback of this model is that the environmental exploitation necessary to ensure a permanent residence for human populations has not been achieved.

The exploration and settlement of Mars can not tolerate a loss of too many lives and fortunes. It can not be held hostage to the winds of world politics. And, it must rely on the exploitation of a fragile and pristine environment for humans to make a go of it there.

1. The first lesson to be learned from these examples is that Mars settlement will require an amalgam of the Manifest Destiny and northward course models. This hybrid model will require international cooperation to set up the human infrastructure on Mars. But, once ensconced, the settlement will need a certain amount of autonomy to exploit the planet for their immediate needs and also for the long-term success of the settlement. Their multinational sponsors must be prepared to allow self-government among Mars settlers and to permit them stewardship over the planet. It will be a frontier society in a very large territory a long way from Earth.

What other lessons can we learn from Euro-American exploration?

2. For any one government to keep a Mars exploration and settlement agenda, a large and serious enough threat must be perceived and Mars must be viewed as a solution to that threat. Several such threats can be imagined. The *first* is the threat of world-wide nuclear war. While the threat of nuclear war between Cold War superpowers have abated, other nations pose risks with their nuclear brinkmanship. Casting about for a moral equivalent of war, the scientific enterprise is a good candidate. What greater scientific enterprise than something so big that many nations, former combatants, have to work toward meeting the goal? Mars is such a goal. Antarctica, but more to the point, the International Space Station project are good exercises in such cooperation necessary to attain Mars. The *second* threat which makes Mars an attractive solution is world overpopulation. One demographic school of thought warns that world population is growing out of control; another model demonstrates that as nations industrialize, they bring their populations down. The truer picture is probably made up of both predictive models. Industrial nations do bring their populations down, but there are plenty of non-industrial countries whose birth rates are many times that of industrial and post-industrial nations and whose death rates do not cancel out this effect. They enjoy to greater or lesser degree the benefits of the one-world economy. Birth rates are not tempered by old-time horticultural or agrarian agricultural death rates owing to modern medicine, nutrition, and sanitation. In addition, human longevity is on the rise in the world as well. World population stands at about 6 billion persons at the moment, but soon there will be many more billions. Mars can be terraformed in almost the same amount of time that petroleum explorers take to ready an oilfield for production. In a few generations, the fourth planet out could be ready to receive the human overflow. The *third* and largest threat to which Mars is a solution is getting "our eggs out of one basket." That phrase has come to be associated with the overwhelming threat of

asteroid and cometary impact to Earth. As Gene Shoemaker said, the Earth moves through the Cosmos as in a hail of bullets (National Geographic Films). Implanting a human population on Mars would increase our chances of not becoming extinct following an impact event. Developing the technology for the seeding of Mars with a human genome would in turn produce better technology to protect the Earth. Related to this vein of thought is a *fourth* more exotic threat which is of no immediate relevance, but I will mention it anyway: passive extinction. All species which have ever lived on planet Earth run the risk of losing out over the course of time in the constant adjustments and accommodations made between their genomes and changing planetary environments. The human species is just as vulnerable as any species. Sending a human population to Mars might increase our genetic vigor by allowing this group to differentiate in subtle ways, away from Earth humans for the most part, with little gene flow between.

3. A third lesson from the polar record is that investments will flow into exploration and settlement of Mars if it is presented as a marketable resource. The dawning of the 22nd century could hold in store at least a two-world economy, requiring a re-engineering of systems of supply, demand, capital, and production. Whenever exploration is presented as a business solution, it has gone forward typically. The collective influence of multinational corporations far outweigh the efforts of national governments. We must not bind Mars up too tightly under some kind of treaty system like the Antarctic Treaty System or through some kind of legal lands lock-up such that business opportunities are stymied. Nature will take its course and Mars-grown eco-politics will prevail as a regulatory mechanism against over-exploitation. Concern over the resources of Mars, which are renewable and which are not renewable, will bring the Green Movement to the Red Planet. Green concerns will feed back into the exploitation and terraforming effort and make for better technology and products.

4. Another obvious lesson which can be gleaned from the polar record is to expect a lot of attempted government intervention in the agenda of exploration and settlement, but to expect very little government help, even in terms of rescue of explorers and colonists.

5. A fifth lesson from the polar record shows that the more overlapping and interconnected networks acting in the interests of the expedition are, the better off it is. Mars exploration and settlement should be more than just a grassroots effort. It needs to be a virtual, if not actual, joint venture with the grassroots organization, with the public, with governments, and with major corporations, with the grassroots consensus taking a proactive, guiding role.

6. In a related vein, the sixth lesson which can be learned from human activity in terrestrial polar regions is the worth of cultivating private backer networks from among the public. Rather than wait to solicit private backers through a public appeal when the going gets tough, devise and depend on a private resource pool from the beginning, while at the same time cultivating all the other support networks. This is along the line of what Robert Zubrin has in mind with these proceedings.

The media must be harnessed to accomplish this task, with the Internet providing a personal interface with backers, again, as Zubrin has suggested. Virtual communities make individuals in large populations powerful, creating a strong collective influence. The polar record shows that the public's imagination is seized by exploration. The task is to convert that imagination into a motivation to provide the resources to explore and settle Mars. There has always been a "reserve army of explorers" in this century. This is nowhere better illustrated than with the Cold War Baby Boomers who thought they would live on the moon, beneath the seas, and travel to Mars. The *Weekly Reader* told us so. However, few structures were in place to sustain the Boomer interest. As Han Mark has said, "...the nation's space program has only one important constituent: the President of the United States." He maintains that only John Kennedy and Ronald Reagan care much about space exploration. He says that he sees no evidence of a voting bloc of Boomers to support a strong space program (Mark 1997). He has a point. Perhaps no voting bloc has existed because no one really presented the Boomers with an opportunity to vote on issues directly relating to space. A private Mars initiative must rely on the consensus of backers. It bucks the old presidential consensus-of-one model. And, this is important, the initiative must rely not only on the new generations, but must rekindle the feeling of the Baby Boomers for space exploration. It can be done. I recently gave a lecture on Mars in a Russian commercial fishing camp on a rugged bit of coast along the Sea of Okhotsk. The fishermen were all Baby Boomers. I gave about a 20-minute after-dinner lecture, but they asked enthusiastic questions well into the wee hours of the morning. The captain of the crew said it made him want to leave fishing to go back to school. The Baby Boomers, where ever they are, remember their lost legacy of space very well. They should not be discounted as a voting bloc, either. The time that all political demographers have warned about is upon us: the Baby Boomers are aging. Aging people are an important voting bloc, and there's an enormous number of Baby Boomers in the bloc which is forming. They have longevity, individual autonomy, and discretionary income. They are a force to be reckoned with. The Boomers can put the "Boom" into the rockets to Mars.

REFERENCES CITES

Chevigny, Hector. 1944. *Lost Empire: The Life and Adventures of Nikolai Petrovich Rezanov*. New York: The MacMillan Company.

Cohen, Daniel. 1996. *The Alaska Purchase*. Brookfield, Connecticut: The Millbrook Press.

Encyclopedia Britannica, Micropaedia, Vol. VI. 1979. Chicago: University of Chicago Press.

Francioni, Francesco and Tullio Scovazzi, Eds. 1996. *International Law for Antarctica*, Second Edition. The Hague: Kluwer Law International.

Fuller, R. Buckminster. 1979. Personal Communication. Future Frontiers of Alaska Conference, Brookings Institution, Anchorage, Alaska.

GilFillan, S. Columb. Sep 1920. "The Coldward Course of Progress." Publication Unknown (cited incompletely in Stefansson's *Northward Course of Empire*, pp. Iv-vi).

Haley, Andrew G. 1963. *Space Law and Government*. New York: Meredith Publishing Company.

Huntington, Ellsworth. 1915. *Civilization and Climate*. New Haven: Yale University Press.

Jenks, Clarence Wilfred. 1965. *Space Law*. New York: F.A. Praeger.

Jorgensen-Dahl, Arnfinn and Willy Ostreng. 1991. *The Antarctic Treaty System in World Politics*. London: MacMillan Academic and Professional Ltd.

Joyner, Christopher C. and Ethel R. Theis. 1997. *Eagle Over the Ice: the U.S. in the Antarctic*. Hanover: University Press of New England.

Klotz, Frank G. 1990. *America on the Ice: Antarctic Policy Issues*. Washington, D.C.: National Defense University Press.

Lynch, George. 1903. *The Path of Empire*. London: Duckworth and Company.

Mark, Hans. Sep 1997. Presentation. Medicine on Mars Conference, Center for Advanced Space Studies, Houston, Texas.

Mark, Hans. 24 Nov 1997. Personal Communication.

Mericq, Luis H. 1987. *Antarctica: Chile's Claim*. Washington, D.C.: National Defense University.

National Geographic Films. *Deadly Impact*.

Peterson, M.J. 1988. *Managing the Frozen South: the Creation and Evolution of the Antarctic Treaty System*. Berkeley: University of California Press.

Simpson, Christopher. 1982. "Battle for the Moon," *Science Digest*, June, Vol. 90, No. 6.

Stefansson, Vilhjalmur. 1922. *The Northward Course of Empire*. New York: Harcourt, Brace and Company.

Stefansson, Vilhjalmur. 1925. *The Adventure of Wrangel Island*. New York: The MacMillan Company.

Stokke, Olav Schram and Davor Vidas. 1996. *Governing the Antarctic: The Effectiveness and Legitimacy of the Antarctic Treaty System*. New York: Cambridge University Press.

Suter, Keith. 1991. *Antarctica: Private Property or Public Heritage?* London: Pluto Press Australia.

Tolmachoff, Innokenty P. 1949. *Siberian Passage: An Explorer's Search Into the Russian Arctic*. New Brunswick, NJ: Rutgers University Press.

Vicuna, Francisco Orrego. 1988. *Antarctic Mineral Exploitation: the Emerging Legal Framework*. Cambridge: Cambridge University Press.

Wayne, Kyra Petrovskaya. 1986. *Quest for Empire: The Saga of Russian America*. Blaine, WA: Hancock House Publishers.

Webb, Melody. 1981. *Chronicles of a Cold, Cold War: The Paperwork Battle for Wrangel Island*. Occasional Paper No. 28. Fairbanks, Alaska: Anthropology and Historical Preservation, Cooperative Park Studies Unit, University of Alaska, Fairbanks.

A SHINING CITY ON A HIGHER HILL: LESSONS FROM THE LAST COLONIZATION OF A 'NEW WORLD'

James D. Heiser*

> The colonization of the New World provides examples of three possible motivations for colonization of Mars: (1) military expansion or competition between colonizing nations, (2) economic exploitation of the natural resources of the colony, and (3) pursuit of political and religious freedom. The first motivation, international competition, played a crucial role in the early development of the Soviet and American space programs, particularly in the race for human exploration of the Moon, a situation roughly analogous to the competition between Spanish, Portuguese, French and English colonization efforts in the Americas. However, the end of the U.S.-Soviet 'Cold War' eliminates this motivation for colonization of Mars, although some experts hope that cooperation between the two nations might yield exploration of Mars, a crucial preliminary step to colonization. The second potential motivation, economic benefit, is even more tenuous, since the success of such an approach rests on profits from hypothetical scientific advances or economically feasible exploitation and/or exportation of Martian natural resources, while relying on a transient population of workers motivated by a desire for a quick profit, not permanent settlement. A similar situation can be found in the early Virginia colonies that were financially disastrous until the beginning of the exportation of tobacco and the importation of slaves. The third motivation—freedom—is the most fruitful motivation for colonization in terms of stability, steady growth, and cultural cohesion. The desire for religious self-determination was the guiding motivation for successful English settlement in New England, efforts that were initially quite meager in terms of personnel and financial resources. The pursuit of freedom—particularly religious freedom—is a motivation powerful enough to move men and women to leave behind land, home, family—even a world—to build a new life.

For we must consider that we shall be as a City upon a Hill.
The eyes of all people are upon us...
—John Winthrop, 1630.

INTRODUCTION

Nearly 400 years ago, on December 11, 1620, one hundred and one men, women and children landed at New Plymouth, an event which, as one historian describes it, was "the single most important formative event in early American history, which would ultimately have an important bearing on the crisis of the American Republic."[1] Of course, these settlers, a mix of Pilgrim Separatists and non-Pilgrim "Strangers", were not the first colonizers of the 'New World'; indeed, they were not even the first English settlers. Two key factors distinguished these settlers from those who came before them: (1) their unshakable self-identity as a community, and (2) their commitment to a world view which shaped their motivation for colonization.

It is the purpose of this brief paper to set forth—albeit quite broadly—lessons which can be learned from the colonization of the 'New World' which could have much to teach us as we

---

* The Rev. James D. Heiser, M.Div., Pastor, Salem Lutheran Church; Publisher, Repristination Press, Route 1 Box 285, Malone, Texas 76660.

contemplate colonizing Mars. We will then examine the specifically religious motivations which led to the landing at Plymouth which have reappeared in our age.

POSSIBLE MOTIVATIONS FOR COLONIZATION OF MARS.

For purposes of our study, we will consider three possible motivations for colonization of Mars: (1) military expansion or competition between colonizing nations, (2) economic exploitation of the natural resources of the colony, and (3) pursuit of religious freedom. Such an analysis does not discount the possibility of other motivations, or that individuals or nations might be influenced by more than one motivation. Still, it is our understanding that these were the most significant motivations.

1. Military Expansion or Competition

Time does not permit a recounting of all of the struggles between the colonial powers in the New World; rather, our concern is with the early colonial division of the New World. This division first occurred between Spain and Portugal, but later included Sweden, England, France and the Netherlands.

Following Columbus' discovery of the New World, little time was lost in dividing up the 'spoils' between the European powers. Through a series of four papal bulls, Pope Alexander VI 'gave' the New World to Spain, setting down a line of demarcation running between the poles, "one hundred leagues towards the west and south from any of the islands commonly known as the Azores and Cape Verdes."[2] However, this action did not amuse King John II of Portugal, whose rather substantial navy placed him in a good bargaining position with Ferdinand and Isabella; this led to a more equitable line in 1494.[3]

However, although Spain and Portugal recognized this 'border', other Europeans did not. Indeed, one of the incidental 'fruit' of the Reformation was that it freed England from the ecclesio-political concerns connected with papal decrees. "In 1561, Queen Elizabeth I's Secretary of State, Sir William Cecil, carried out an investigation into the international law of the Atlantic, and firmly told the Spanish ambassador that the pope had no authority for his award."[4]

Sweden and the Dutch Republic (having secured its independence from Spain in 1581) also established a 'presence' in North America in 1638 and 1624, respectively.[5] Although the Swedish colonies were quite small (never numbering more than 400 inhabitants), the Dutch made no effort to encroach on their claims until 1655. "The reason for this toleration was that in the Thirty Years' War—which was then being fought in Europe, Sweden, and the Netherlands—Protestants were allied against Roman Catholic powers"[6]—a war which ended in 1648.

French Huguenot seamen automatically dismissed the Roman Catholic Church's claims and maintained that "the normal rules of peace and war were suspended beyond a certain imaginary line running down the mid-Atlantic. ... [T]he theory, and indeed the practice, of 'No Peace Beyond the Line' was a 16th-century fact of life."[7] Already in 1564-5 French Huguenots established a colony (Fort Caroline) on the northeastern coast of Florida. "Less than a month later, Florida being considered Spanish territory, the Spanish captain-general Peter Menendez attacked the settlement and massacred all the Huguenots. He was reported to have explained, 'I do this not as to Frenchmen but as to Lutherans [*Luteranos*].'"[8] The Protestant Huguenots did not fair much better at the hands of the Portuguese, either. In 1555-6, Gaspard de Coligny, Admiral of France, sent two groups of colonists to settle in the harbor of Rio de Janeiro; many of the more than 300 colonists were chosen by John Calvin himself. However, the Portuguese struck the colony in 1560, hanging all of the colony's inhabitants.[9] While one should not underestimate the religious nature of these attacks, it is impossible to dismiss the political motivations for both settlement and repression. International competition was no small motivation; the choice to settle (or destroy a settlement) expressed European tensions and conflicts.

In our present situation, however, a parallel motivation for colonization seems unlikely. The end of the U.S.-Soviet 'Cold War' eliminated much of the military competition motivation for colonization of Mars. This was not always the case, of course; it was the 'Cold War' which propelled the early 'space race', including the moon landings.[10] (Still, the late Carl Sagan, among others, hoped that cooperation between the two nations might yield *exploration* of Mars, a crucial preliminary step to colonization.[11]) This does not mean it is impossible such a motivation could arise again in the future, although such a 'solution' seems worse than the 'problem' of motivating Mars colonization.

2. Economic Exploitation

The second potential motivation, economic exploitation by a transient worker population, is possibly even more tenuous, since the success of such an approach rests on profits from hypothetical scientific advances or economically feasible exploitation and exportation of Martian natural resources, when relying on a population of workers motivated by a desire for a quick profit, not permanent colonization. Economic motivations require more than simply proving tremendous profits *could* be gained by going to Mars; it is necessary to convince multinational corporations and international banks that there is a guaranteed profit which equals (or exceeds) the profits of a comparably risky investment on Earth.[12] Such an endeavor will also be hampered by a lack of personal 'stake' in the development of Mars: For those drawn to Mars only by massive salaries, the new world will simply be a particularly unpleasant worksite with higher pay than usual—a place to be survived and *left as soon as possible*. The emphasis will be on stripping Mars of its resources, not developing a new civilization.

Again, the colonization of the 'New World'—particularly in the English colonies of Virginia—offer interesting historical lessons. First, consider the debacle of Sir Walter Raleigh's Roanoke colony. Certainly the settlement did not lack for resources: seven ships, over 100 men, troops led by experienced officers, an able scientist (Hariot) with instructions for scientific investigation, and knowledge of the local Indian language. Nevertheless, the assessment after the first year of the settlement was *not* encouraging: Ralph Lane (who commanded the settlement's troops) declared that the colonists "thought they would find treasure and 'after gold and silver were not to be found, as it was by them looked for, had little or no care for any other thing but to pamper their bellies.'"[13] Nevertheless, 150 colonists sailed for Roanoke in 1587 and 114 (including sixteen women and ten children) remained behind when the fleet sailed for England. The results were, of course disastrous: the colony was apparently wiped out by Native Americans.

Roughly thirty years later (1607), Jamestown suffered nearly as disastrous a fate. Again, the problem was centered in colonists who were more interested in adventure (i.e., fast profits and a trip home) than colonization; "... lacking a family unit basis, the colony was fortunate to survive at all. Half died by the end of 1608, leaving a mere fifty-three emaciated survivors."[14] John Smith's firm leadership (beginning a year after the establishment of the colony) slowed the colony's collapse, but he returned to England in 1609. Once again, of 400 new settlers arriving in 1609, by May 1610, "scarcely sixty settlers were still alive. All the food was eaten, there was a suspicion of cannibalism, and the buildings were in ruins."[15] It took martial law under Thomas Dale to whip some order into the Virginia colony—and it took the establishment of *families* in the colony and a profitable export. In 1619, ninety unmarried women were brought to Virginia; "Any of the bachelor colonists could purchase one as a wife simply by paying her cost of transportation, set at 125 pounds of tobacco."[16] By 1616 Virginia tobacco had developed to the point where it was an exportable crop and in 1619 the first twenty African slaves arrived to grow tobacco.[17] The future course was set. Virginia survived in no small part because: (1) families were established in the colonies, moving away from disastrous experiments with 'adventurers', (2) tobacco was developed to give the colony a valuable export, (3) indentured servants and slaves were imported to shore up the economy.[18] Certainly the English settlers did not 'invent' modern slavery,[19] but this development has had a profoundly harmful influence on our society to this day.

In his essay, "On Plantations" (1625), Sir Francis Bacon examined the reasons for the failure of the earlier colonies. "He pointed out that any counting on quick profits was fatal, that there was a need for expert personnel of all kinds, strongly motivated in their commitment to a long-term venture, and not least, that it was hopeless to try to win over the Indians with trifles 'instead of treating them justly and graciously.' ..."[20] All of this does not automatically doom the usefulness of short-term 'adventurers' in a future Martian colony—a certain number of them in certain specializations may be indispensable. But certainly the Virginia colony provides powerful warnings against relying heavily or exclusively on temporary workers; if nothing else, the cost of rotating hundreds of colonists 'home' every few years seems an unnecessary burden on the financial viability of a young Martian colony. We must keep in mind that initially Mars, like seventeenth century America, will not be an extremely wealthy place. As one historian observed:

> Even after the whole of the seaboard from stony Maine down to the swampy Florida frontier had been occupied by the British, the North American colonies would be of small value to Britain except as suppliers of tobacco, furs, naval stores, some dyestuffs and foodstuffs, and a few other products of no great consequence. Barbados or Jamaica alone—or, for that matter, some other islands in the West Indies—seemed worth more to London than all the thirteen mainland colonies put together.
>
> True, the North American colonies served in some degree to hold in check England's French and Spanish rivals in the New World, and therefore they were worth defending in time of crisis. In general, nevertheless, those North American territories were disappointing to the British trading companies that made the early settlements; they were disappointing to the great English proprietors—some, noblemen; others, commoners—who for decades controlled the vast empty lands; they were sufficiently disappointing even to the English kings who eventually asserted direct sovereignty over the Thirteen colonies.[21]

3. Pursuit of Religious Freedom

We turn, therefore, to the third motivation under consideration: religious freedom. The first two motivations sought to exploit the New World for advantage in the Old World. The motivation of religious freedom, however, is only incidentally interested in the concerns of the Old World; it is, if anything, a flight from the Old World.

It is our contention that religious freedom is the most fruitful motivation for colonization in terms of stability, steady growth, and cultural cohesion. The desire for religious freedom was the guiding motivation for successful English settlement in New England, efforts that were initially quite meager in terms of personnel and financial resources. The pursuit of religious freedom is a motivation powerful enough to move men and women to leave behind land, home, family—even a world—to build a new life.

It has been observed that: "Between the abortive Armada in 1588 and conclusion of the Thirty Years' War in 1648, Europe was a fascinating but fearful place. Transitional paroxysms of the later Renaissance and uncontrolled tremors of several sorts were felt everywhere in western Christendom. In England especially, for more than sixty years, there were profound crises of mind and spirit, reflected in religion, literature, and political thought."[22] In 1588, the Spanish Armada attacked England with 130 ships and 30,000 men, confident that their forces (and a hoped for, but unrealized, Catholic uprising) would return Protestant England to the Roman Catholic fold.[23] Thirty years later, the Thirty Years' War began, and again Spanish troops (and those of other Roman Catholic territories) were on the move against Protestant forces, this time on the Continent. Indeed, the fall of the Bohemian forces in the Battle of White Mountain (November 8, 1620) was nearly simultaneous with the Pilgrims' landing.[24] Fear that European Protestantism would fall utterly to Roman Catholicism was a major concern motivating Puritan immigrants, particularly during the Great Migration of the 1630s. As Thomas Hooker declared in a 1630 sermon:

> Will you have England destroyed? Will you put the aged to trouble, and your young men to the sword? Will you have your young women widows, and your virgins defiled? Will you have your dear and tender little ones tossed upon the pikes and dashed against the stones? Or will you have them brought up in Popery, in idolatry, under a necessity of perishing their souls forever, which is worst of all?
>
> ... God begins to ship away his Noahs, which prophesied and foretold that destruction was near; and God makes account that New England shall be a refuge for his Noahs and his Lots, a rock and a shelter for his righteous ones to run unto; and those that were vexed to see the ungodly lives of the people in this wicked land, shall there be safe.[25]

Even more than the foreign threat, however, Pilgrims (Separatist) and Puritans (non-Separatist)[26] were appalled by the decay of English civilization and religion. The division within the English church between Anglican and Puritan is well known: differences over clerical attire and worship forms, the form of church government, communalism and discipline were all areas of disagreement.[27] However, the spread of immorality and the burden of (perceived) overpopulation also motivated Puritans to emigrate from England. Writing in 1629, John Winthrop gave the following reasons for "the Intended Plantation in New England":

> First, it will be a service to the church of great consequence to carry the gospel into those parts of the world, to help on the coming in of fullness of the Gentiles and to raise a bulwark against the kingdom of Antichrist, which the Jesuits labor to rear up on those parts.
>
> 2. All other churches of Europe are brought to desolation, and our sins, for which the Lord begins already to frown upon us, do threaten us fearfully ...
>
> 3. This land [England] grows weary of her inhabitants, so as man who is the most precious of all creatures is here more vile and base than the earth we tread upon, and of less price among us than a horse or a sheep...
>
> 4. The whole earth is the Lord's garden and he hath given it to the sons of men, with a general condition, Genesis 1:28, *Increase and multiply, replenish the earth and subdue it*, which was again renewed to Noah. ...
>
> 5. We are grown in that height of intemperance in all excess of riot, as no man's estate almost will suffice to keep sail with his equals, and he who fails herein must live in scorn and contempt; hence it comes that all arts and trades are carried in that deceitful and unrighteous course, as it is almost impossible for a good and upright man to maintain his charge and live comfortably in any of them.
>
> 6. The fountains of learning and religion are so corrupted (as besides the unsupportable charge of education) most children (even the best wits and fairest hopes) are perverted, cor-

rupted, and utterly overthrown by the multitude of evil examples of the licentious government of those seminaries...

7. What can be a better work and more honorable and worthy a Christian than to help raise and support a particular church while it is in the infancy?

8. If any such who are known to be godly and live in wealth and prosperity here shall forsake all this to join themselves to this church and to run a hazard with them of a hard and mean condition, it will be an example of great use both for removing the scandal of worldly and sinister respects which is cast upon the adventurers to give more life to the faith of God's people in their prayers for the plantation...

9. It appears to be a work of God for the good of his church in that he hath disposed the hearts of so many of his wise and faithful servants (both ministers and others) not only to approve of the enterprise but to interest themselves in it...[28]

The Pilgrims (and the Puritans who followed) were, therefore, different from those who came before: They did not come as individuals, but as a community. They did not come as adventurers, but as 'planters' (colonists). They came with a specific *vision* motivating their settlement—a revitalization of the Christian faith—and understood themselves bound up in a covenant with God in this task. Again, as Winthrop told his shipmates in 1630 on the way to the New World:

> Thus stands the cause between God and us. We are entered into covenant with him for this work. We have taken out a commission, the Lord hath given us leave to draw our own articles. We have professed to enterprise these actions, upon these and those ends, we have hereupon besought him of favor and blessing. Now if the Lord shall please to hear us, and bring us in peace to the place we desire, then hath he ratified this covenant and sealed our commission, [and] will expect a strict performance of the articles contained in it. But if we shall neglect the observation of these articles which are the ends we have propounded and, dissembling with our God, shall fall to embrace this present world and prosecute our carnal intentions, seeking great things for ourselves and our posterity, the Lord will surely break out in wrath against us, be revenged of such a perjured people, and make us know the price of the breach of such a covenant.

> Now the only way to avoid this shipwreck, and to provide for our posterity, is to follow the counsel of Micah, to do justly, to love mercy, to walk humbly with our God. For this end, we must be knit together in this work as one man. We must entertain each other in brotherly affection, we must be willing to abridge ourselves of our superfluities, for the supply of others' necessities. We must uphold a familiar commerce together in all meekness, gentleness, patience and liberality. We must delight in each other, make others' conditions our own, rejoice together, mourn together, labor and suffer together, always having before our eyes our commission and community in the work, our community as members of the same body. So shall we keep the unity of the spirit in the bond of peace. The Lord will be our God, and delight to dwell among us as his own people, and will command a blessing upon us in all our ways, so that we shall see much more of his wisdom, power, goodness, and truth, than formerly we have been acquainted with. We shall find that the God of Israel is among us, when ten of us shall be able to resist a thousand of our enemies; when he shall make us a praise and glory that men shall say of succeeding plantations [i.e., colonies], "the Lord make it like that of New England." For we must consider that we shall be as a city upon a hill. The eyes of all people are upon us, so that if we shall deal falsely with our God in this work we have undertaken, and so cause him to withdraw his present help from us, we shall be made a story and a by-word through the world. We shall open the mouths of enemies to speak evil of the ways of God, and all professors for God's sake. We shall shame the faces of many of God's worthy servants, and cause their prayers to be turned into curses upon us till we be consumed out of the good land whither we are agoing.[29]

The Puritan decision to aggressively pursue a path of emigration was not accidental. Sir Robert Rich, Earl of Warwick, became a member of the Virginia Company in 1612 and worked with other like-minded Puritan gentry to reform England. However, should that effort prove impossible,

> [H]e wanted the alternative option of a reformed colony in the Americas. Throughout the 1620s he was busy organizing groups of religious settlers, mainly from the West Country, East Anglia and Essex, and London—where strict Protestantism was strongest—to undertake the American adventure. In 1623 he encouraged a group of Dorset men and women to voyage to New England, landing at Cape Ann and eventually, in 1626, colonizing Naumkeag. John White, a Dorset clergyman who helped to organize the expedition, insisted that religion was the biggest single motive in getting people to hazard all on the adventure: 'The most eminent and desirable end of planting colonies is the propagation of Religion,' he wrote. ... He admitted: 'Necessity may oppress some: novelty draw on others: hopes of gain in time to come may prevail with a third sort: but that the most sincere and Godly part have the advancement of the Gospel for their main scope I am confident.'[30]

CONCLUSION—RELIGIOUS MOTIVATIONS FOR EMIGRATION TO MARS

While our present study has focused on the influential Puritan immigrations, much could also be said regarding the influence religious freedom had on the formation of other colonies. Lord Baltimore's interest in chartering the Maryland colony was to provide refuge for fellow English Roman Catholics.[31] Pennsylvania was founded on the principle of religious freedom; William Penn was concerned with providing a refuge for fellow Quakers.[32] In addition, much could be said concerning the later immigration of Christian communities into the colonies and the American Republic. Please indulge a few Lutheran examples: The Salzburg Lutherans were forced out of Austria because they would not convert to Roman Catholicism; these refugees eventually formed the Ebenezer colony in Georgia in 1734.[33] In 1843, a group of 1,600 Prussian Lutherans immigrated to New York and Wisconsin because they would not be part of a attempt by King Friedrich Wilhelm III to force a union of the Reformed and Lutheran churches.[34] In 1839, 602 Saxon Lutherans arrived in St. Louis, having fled their homeland because of the rationalism and secularization they believed were destroying the church; the Lutheran Church—Missouri Synod they formed now numbers 2.5 million members in 6,000 congregations.[35]

Granted that religious freedom was a significant motivation for the successful colonization of the New World, is it possible this motivation would be operative in a future colonization of Mars? The stark reality is that Western Christians are probably experiencing a greater sense of cultural isolation and alienation than their seventeenth century forefathers. Then the division was between groups which still shared many points in common; even Deists still upheld the existence of God and were, in fact, outspoken defenders of the existence of natural law. Turner's groundbreaking study of the roots of Western unbelief, *Without God Without Creed*, contains the following conclusion concerning the present cultural division:

> Unbelief has transformed the hopes, aspirations, purposes, and behavior of millions of unbelievers. It has affected believers almost as remarkably. Wrangling over prayer in public schools is only one minor eruption, pointing to a major shift of the tectonic plates on which our culture moves. The option of godlessness has dis-integrated our common intellectual life, both in formal disciplines like philosophy, science, and literature and in those informal habits of mind by which we, as a culture, experience and order our world. God used to function as a central explanatory concept. As cause and purpose, the idea of God shaped and unified natural science, morality, social theories, psychology, political thought into one vaguely coherent (though very loosely assembled) approach to understanding humankind and the cosmos.

> ... [A]t the most fundamental level, God provided the frame of an agreed-upon universe in which to argue. Our web of shared assumptions has not unraveled altogether—without some unity, a culture collapses. But the traditional linchpin is missing; our culture, in this sense, now lacks a center.[36]

The division between the Christian world view and that of the secular elite culture has grown increasingly distinct. In the words of Stephen Carter, "Our culture seems to take the position that believing deeply in the tenets of one's faith represents a kind of mystical irrationality."[37] From a Christian perspective, many of the currents in Enlightenment, Modern, and Post-Modern thought have been as progressively de-humanizing as they have been increasingly agnostic or atheistic. Charles Darwin, Karl Marx and Sigmund Freud (and their intellectual heirs) are all understood to have simultaneously attempted to tear down God and those who were made in His image; both reason and faith become essentially meaningless terms when man's thoughts and actions are dictated by biology or economics. (Such determinism could be considered the agnostic's equivalent of John Calvin's double predestination.) For a Christian perspective, it has become an age in which, in the words of William Butler Yeats,

> Things fall apart; the centre cannot hold;
> Mere anarchy is loosed upon the world,
> The blood-dimmed tide is loosed, and everywhere
> The ceremony of innocence is drowned;
> The best lack all conviction, while the worst
> Are full of passionate intensity.[38]

In summary, then, it is our contention that a significant number of Christians would embrace having 'somewhere to go.' As John Lewis observes in *Mining the Sky*, "It was the search for freedom of religion that brought most of our ancestors here, and it will be the search for freedom from religious, political, and ethnic persecution that will send the first colonists forth into space."[39] As with the settling of the New World, Christians colonists setting out for Mars would do so as a community seeking the freedom to worship the Triune God who created heaven and earth, redeemed us from our sin, and sanctifies us through His Word and Sacraments. This unity of faith will, Lord-willing, provide the motivation to endure the hardships of opening up a second new world, perhaps on the four hundredth anniversary of that historic landing at Plymouth.

Robert Cushman, numbered among the Pilgrims reaching our shores in November, 1620, wrote a defense of immigration to the 'New World' which was included in the work known as "Mourt's Relation" (1622), the first published account of the Pilgrims' arrival. His concluding words are so *apropos* that they will be our last word, as well.

> To conclude, without all partiality, the present consumption which groweth upon us here [in England], whilst the land groaneth under so many close-fisted and unmerciful men, being compared with the easiness, plainness and plentifulness in living in those remote places, may quickly persuade any man to a liking of this course, and to practise a removal, which being done by honest, godly and industrious men, they shall there be right heartily welcome, but for other of dissolute and profane life, theirs rooms are better than their companies. For if here, where the Gospel hath been so long and plentifully taught, they are yet frequent in such vices as the heathen would shame to speak of, what will they be when there is less restraint in word and deed? My only suit to all men is, that whether they live there or here, they would learn to use this world as they used it not, keeping faith and a good conscience, both with God and men, that when the day of account shall come, they may come forth as good and fruitful servants, and freely be received, and enter into the joy of their Master.[40]

Soli Deo Gloria

REFERENCE NOTES

1. Paul Johnson, *A History of the American People*, (New York: HarperCollins, 1998) p. 28.
2. quoted in *The Discoverers*, by Daniel J. Boorstin, (New York: Vintage Books, 1985) p. 248.
3. Boorstin, p. 248-249.
4. Johnson, p. 9. As early as 1530, the Lutheran Augsburg Confession declared, "If bishops have any power of the sword, that power they have, not as bishops, by the commission of the Gospel, but by human law, having received it of kings and emperors for the civil administration of what is theirs. This, however, is another office than the ministry of the Gospel." (*Concordia, or Book of Concord*, [St. Louis: Concordia Publishing House, 1922] p. 23.) Here we see a return to the earlier medieval church's distinction of the three estates of mankind: those who pray, those who fight, and those who work.
5. E. Clifford Nelson, *The Lutherans in North America*, (Philadelphia: Fortress Press, 1980) p. 4-5.
6. Nelson, p. 5. In 1664, the Dutch, in turn, surrendered to English troops without firing a shot. (Nelson, p. 6)
7. Johnson, p. 9-10.
8. Nelson, p. 3.
9. Johnson, p. 10.
10. John S. Lewis, *Mining the Sky*, (Reading, Massachusetts, *et al.*: Helix Books, 1996) p. 5.
11. Robert Zubrin with Richard Wagner, *The Case for Mars*, (New York: The Free Press, 1996) p. 279-282.
12. This is, of course, a marked weakness of this motivation in comparison to military competition, since nations rarely allow cost considerations to interfere with programs deemed necessary for national security, as can be amply proved from the arms build-up in the final years of the U.S.-Soviet 'Cold War'.
13. Johnson, p. 16.
14. Johnson, p. 25.
15. Johnston, p. 26.
16. Johnson, p. 26.
17. Although austensibly only 'indentured' servants, it is doubtful any of this first group ever became free. (Johnson, p. 27)
18. The increase in the number of slaves rose sharply after the 1670s: "The number of blacks in the South rose after the 1670s from 6 to 21 percent of the population, an influx concentrated in Maryland, Virginia, and above all in South Carolina, which in 1700 counted 43 percent of its people black. By that year over 80 percent of blacks in America lived in the South. Meanwhile, their numbers in the North remained fairly static—perhaps 7 percent of the population." (David Freeman Hawke, *Everyday Life in Early America*, [New York: Harper & Row, 1989] p. 128.) Concerning the economics of slavery, see Jeffrey Rogers Hummel, *Emancipating Slaves, Enslaving Free Men*, (Chicago and La Salle: Open Court, 1996) p. 38-52.
19. Indeed, the Portuguese immeshed themselves in the economics of slavery a full forty years before Columbus' famous voyage. "The Portuguese entered the slave trade in the mid-15th century, took it over and, in the process, transformed it into something more impersonal, and horrible, than it had been either in antiquity or medieval Africa. The new Portuguese colony of Madeira became the center of a sugar industry, which soon made itself the largest supplier for western Europe. The first sugar-mill, worked by slaves, was erected in Madeira in 1452." (Johnson, p. 5)
20. Johnson, p. 18.
21. Russell Kirk, *The Roots of American Order*, (Washington: Regnery Gateway, 1991) p. 303.
22. Michael Kammen, *People of Paradox*, (Ithaca, New York: Cornell University Press, 1980) p. 117.
23. Williston Walker, *et al.*, *A History of the Christian Church*, (New York: Charles Scribner's Sons, 1985) p. 524.
24. Walker, p. 530.
25. *The Puritans in America, A Narrative Anthology*, ed. by Alan Heimert and Andrew Delbanco, (Cambridge: Harvard University Press, 1985) p. 68-69.
26. 'Separatist' in terms of consciously severing all ties with the state church of England.
27. Kammen, p. 125.
28. *The Puritans in America: A Narrative Anthology*, p. 71-72.
29. *The Puritans in America, A Narrative Anthology*, p. 90-91.
30. Johnson, p. 30.

31. Walker, p. 574.
32. Walker, p. 562, 577.
33. William J. Finck, *Lutheran Landmarks and Pioneers in America*, (Philadelphia: The United Lutheran Publication House, 1913) p. 121 ff.
34. "From a German Jail" by Alfred H. Ewald, in *Church Roots*, ed. by Charles P. Lutz, (Minneapolis: Augsburg Publishing House, 1985), p. 41-59.
35. See "The Saxon Migration to Missouri," by Eugene W. Camann, in *Confessional Lutheran Migration to America*, (Eastern District, LC—MS, 1988), p. 13-20.
36. (Baltimore and London: The John Hopkins University Press, 1985) p. 263-4.
37. quoted in *Their Blood Cries Out—The Worldwide Tragedy of Modern Christians Who Are Dying for Their Faith*, by Paul Marshall, (Dallas: Word Publishing, 1997) p. 189.
38. quoted in *The Roots of American Order*, by Russell Kirk, (Washington: Regnery Gateway, 1991) p. 7.
39. p. 240.
40. *Mourt's Relation, A Journal of the Pilgrims at Plymouth*, intro. by Dwight B. Heath, (Bedford, Massachusetts: Applewood Books, 1963) p. 96.

Chapter 4
MOBILIZING THE PUBLIC

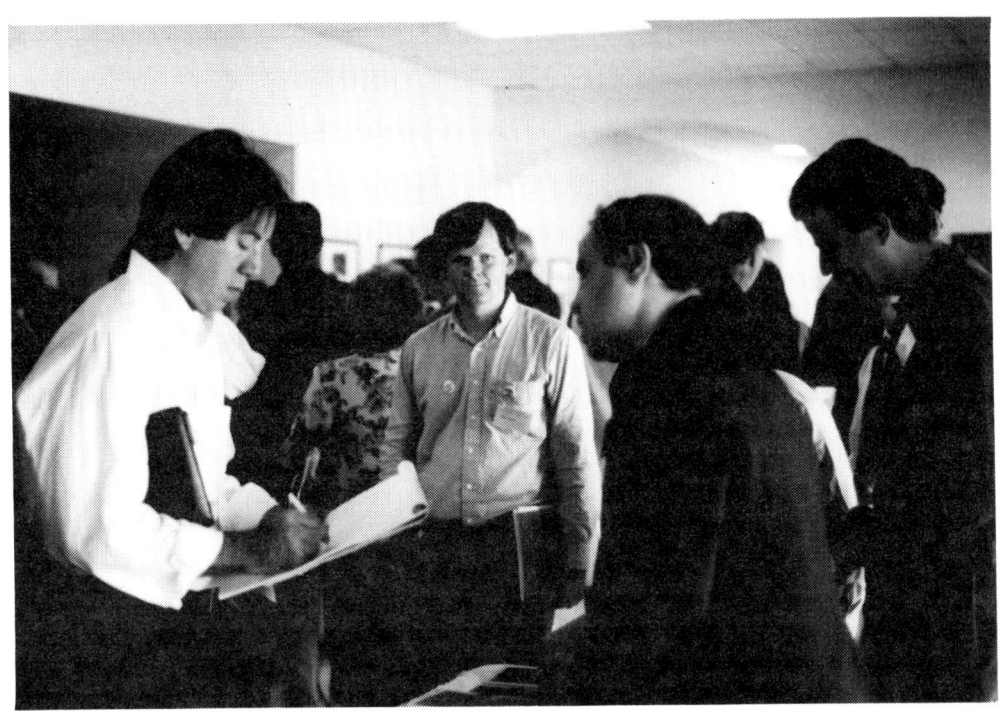
Washington Post reporter Joe Achenback gets his story. The conference was also covered in the New York Times and the Boston Globe.

Sam Burbank proposes methods of public outreach.

IS THERE A SHORT-TERM ECONOMIC AND SOCIAL JUSTIFICATION FOR HUMAN EXPLORATION AND SETTLEMENT OF MARS?

Robert E. Becker[*]

The Space Exploration Initiative died stillborn years ago, largely unfunded, widely viewed as unfounded. While a stream of smaller, robotic missions to Mars has started, human exploration seems to have been removed from any official timeline!

This, despite the robust economy, possible evidence of ancient life in a Martian rock, millions of hits on the Mars Rover Web Site, availability of relatively inexpensive missions like Mars Direct, and energetic advocacy by luminaries like Carl Sagan. Traditional arguments—the human exploratory spirit, intellectual excitement, and economic benefits down the road—suffice for exploration enthusiasts, but are not sufficient for the general public, and especially not for their elected representatives.

Our nation has a notoriously short attention span. All these arguments pale in comparison with immediate economic and social problems. Proponents of exploration often take a "not my job" attitude towards these issues. But until we embrace them, our arguments will remain largely unpersuasive to the public.

Human beings can not be "Faster, Cheaper, Bettered" out of the process if Mars Exploration is to gain broad support. It is the very ambition of full-fledged Mars Exploration, which can provide that justification, starting in the SHORT-TERM, during the DEVELOPMENT phases. It is this scope—living and working on another planet—which can bring space closest to the person in the street, just as on the frontier of yore.

It is the responsibility of the exploration community to apply its collective CREATIVITY and formulate cogent arguments RELEVANT to those who are NOT enthusiasts. Answering the title question affirmatively will do that. The Mars Society is the perfect multidisciplinary organization to launch a CONCERTED effort to answer it. If we succeed, a new exploration initiative, Mars Holistic, would be announced, not with a bureaucratic whimper, but with trumpets blaring.

WHAT HAPPENED TO THE DREAM?

Only eight years ago, at the same venue as this Founding Convention of the Mars Society: The 1990 Case for Mars IV Conference. The atmosphere was enthusiastic and optimistic. Besides the "usual suspects" of aerospace engineers, planetary scientists and NASA managers in attendance, there were a number of educators and social scientists, people not normally associated with space.

Even more notable was the participation of many non-aerospace companies. Heavy construction companies, marine drilling firms, and heavy equipment manufacturers were all represented. There was a contingent of such companies from Japan. Indeed, one of them established a kind of Mars infrastructure testbed in Florida.

---

[*] 131 Old County Road, #168, Windsor Locks, Connecticut 06096. E-mail: Roberte.becker@mcione.com.

What drew this diverse group of people and companies to this Conference? An obvious answer is the unveiling of the NASA's Space Exploration Initiative (SEI) not long before. SEI was a serious, aggressive, multi-decade, plan for human exploration of Mars. While it began with an assortment of robotic exploration missions, SEI culminated in human missions to Mars. Potential commitment of major funding for human exploration and Martian resource exploitation caught the attention of these non-aerospace firms because, for perhaps the first time, their products became relevant to the space program.

Now go forward three years, to 1993 and the Case for Mars V Conference (where this author presented an identically titled Paper). This time, a totally different atmosphere, almost depressing. Most of the non-aerospace companies in attendance in 1990 are not present. The quick demise of SEI in these three short years accounts for this reversal of fortune. Faced with estimated costs of hundreds of billions of dollars, and a traditional strategy of selling the overall plan by getting small-scale precursor missions through the budgetary door first, Congress quickly administered the coup de grace. Unfortunately, much lower cost mission architectures, such as Robert Zubrin's Mars Direct, were not incorporated into the plans. As government lost interest in Mars Exploration, so did the non-aerospace firms.

Five years later we find ourselves in a curiously dichotomous situation, with cause for both optimism and pessimism. On the positive side, by now Mars Direct is well known and has achieved wide acceptance. The days of estimating 12-figure budgets to just get humans to Mars and back are over. Discovery of possible evidence of ancient life in a Martian meteorite elicited great interest on the part of the general public, as did new robotic Mars missions, as evidenced by millions of hits on the Mars Rover Web Site.

In fact, this was just the first of a planned steady stream of smaller, "Faster, Cheaper, Better" (FCB) robotic missions to Mars. The space industry itself has undergone substantial changes in recent years. Telecommunications space missions are burgeoning to the point that non-aerospace companies like Motorola are now full-fledged satellite manufacturers, while aerospace companies like Lockheed-Martin and Boeing have expanded into product domains from their traditional space mission infrastructure roles. Space is increasingly privatized, as space industry sectors from imaging to launch vehicles pass from government sponsorship to commercial industry. The U.S. economy has flourished for almost a decade; the budget deficit is decreasing.

Finally, the very formation of the Mars Society demonstrates that exploration enthusiasts remain as committed as ever.

Overshadowing all this, however, is one salient fact: There are no official plans for human missions to Mars. The robotic missions will be steady and relatively frequent, but not fast. The first samples will not be recovered until about a decade from now. And even these FCB missions have recently run into stumbling blocks. Meanwhile, the Space Station continues its cost overruns and schedule slips. To this point, the commercialization of space has not extended to major private investment in Mars missions.

Anyone approaching Congress with an idea to fund a major Mars Exploration and Settlement program would surely have his or her sanity questioned. Even though costs have been lowered and there is some public interest in Mars exploration, there is no official interest or sympathy for the idea. The costs are viewed as still too high, and the returns intangible or too long long-term to be justifiable.

In short, the Title question has not been answered in the affirmative. Even worse, it hasn't even really been posed. All the arguments so persuasive to exploration enthusiasts—human destiny; the challenge of the next obstacle; visceral excitement of the chase; long-term economic benefits—even when annunciated with elegance by people of the stature of Carl Sagan—remain unpersuasive to the public and to their elected representatives.

The arguments that ring clear to the Mars Exploration community do not necessarily have relevance to the country at large. Our country has a short attention span, and with it a short planning and investment returns horizon. Quotidian social and economic problems are much more tangible to most of our citizenry than the adventure of Mars exploration or the "because it's there" rationale.

Unless the Title question is answered affirmatively, addressing the concerns of the majority of people in our country, human missions to Mars, let alone exploration and settlement, will be a very long-time coming.

HOW TO KICKSTART MARS EXPLORATION

Several ingredients are required to revive Mars exploration. These are not so much specific technology solutions or mission strategies, though plenty of these must still be delineated. Rather, they are general philosophies and methodologies, which should guide the entire project. Call this project Mars Holistic, because it addresses issues and encompasses disciplines that affect all aspects of our lives.

First, full-scale exploration and settlement must be a HUMAN endeavor. On the surface, this is tritely obvious. What else could characterize human Mars exploration? Yet, in practice, it is an element that is often lost or de-emphasized. If Mars exploration is to draw popular support, and be most useful to the economy and social fabric of our nation, it can not "Faster, Cheaper, Better" humans out of the equation.

FCB usually is cheaper; is often faster; and definitely can be better, depending upon the eyes of the beholder. We can never return to the days of wasting billions of dollars due to a lack of creativity and an insistence on doing things the "traditional way" for that sake alone. Nor can it be a bureaucratic refuge for hordes of seemingly functionless workers.

Mars Direct, though originally not explicitly part of FCB, is perhaps the best example of what FCB can do: eliminate hundreds of billions of dollars; safely compress the mission timeline; and actually enhance exploration capabilities while on the surface.

But Mars exploration also can not be just a parochial playground for an elite cadre of aerospace engineers and planetary scientists. It must be recognized that the management optimizations and streamlining associated with FCB may work very well on small robotic missions, but shoestring budgets, extremely compressed schedules, and stripped-down mission capabilities are not always appropriate for long-term missions of the magnitude of Mars Holistic.

FCB, when translated into the language of modern business—"lean and mean", "best-in-world competition"—often does mean eliminating the human element to the greatest extent possible. Variously spurred on by the need for survival, new technology utilization, or the desire to improve profitability, the end result is usually human dislocation.

Perhaps for the first time in economic history, business not only takes advantage of labor saving technology innovations—as it has since time immemorial—but actively seeks them as a means to an end. One result of the confluence of all these trends has been massive dislocations in traditional labor intensive industries. New information-intensive technologies are supposed to be the refuge for the dislocated, but not all are made out for this kind of work. Furthermore, even this software work is increasingly outsourced to other countries, consistent with the natural evolution of the global economy.

By its very nature, Mars Exploration and Settlement must be labor intensive in much the same way as the colonial or western frontiers were rigorous. While living and working on Mars should not be as onerous as the lives of pioneer sodbusters, we are also not going to Mars to open Internet software boutiques. Frontiers require human participation to a degree not always entailed by civilized lands permeated with technology.

Likewise, the development of the infrastructure necessary for living, working, and playing on Mars will in itself require an augmented distribution of industries and talents than heretofore has been utilized in space.

Thus, in both the development of what will be needed, and then its actual usage by first, explorers, and then, "ordinary" settlers, the opening of this new may offer a chance to reverse macroeconomic trends which are eroding human participation in economic development, without retreating to Luddite extremes.

Using the very size of a Mars Exploration and Settlement program to make it accessible and useful to the general public is a bit heretical, and will require considerable creativity to actually implement. It is surprising that an agency like NASA, under the leadership of a man who justifiably has won renown as an exemplar and proponent of creativity, seems to have simply dropped the subject. The creativity that went into devising FCB has not been applied to this larger problem; and the principles of FCB alone will not suffice. Mars Exploration and Settlement presents an opportunity to apply out-of-box creativity on a scale appropriate to this problem. Mars Direct is an excellent start in the technological and mission planning realms. Similar creativity must now be extended to the social and economic disciplines, which in the past have not been a focus of space development. It has not been the job of the space program to solve broad social and economic ills. Now, to do its own job, it must undertake non-traditional roles.

And that creativity must bear fruit early in the process. It is somewhat antithetical to grander vision inherent in Mars exploration and settlement, that the returns from this creativity must be realized so quickly. Yet, we are a people with a short time horizon. It is not just business that is driven by the need for a quick Return on Investment (ROI), but the nation as a whole. It is unrealistic to expect allocation of major funding for Mars exploration over a multi-decade period, unless tangible economic and social benefits result in the near-term. Perhaps the situation would be different in Japan (pre-financial crisis), but not here.

This means another requirement for Mars Holistic is that it be so structured that the social and economic ROI occur in the early development phases of this project, not after completion, forty years down the road. The way to do this is to use the scope and size of Mars Holistic to broaden the base of the space industry, and to bring both its work and benefits directly to the industries most related to daily living on earth. It is not only opening a new frontier that should benefit our nation, and humanity as a whole, but the very steps of preparing for that opening.

SHOULD THE GOVERNMENT LEAD?

One of the saddest aspects of the present state of space travel is the transition from an image of remarkable NASA competence and leadership during the earliest period of space travel through the glory days of the Apollo lunar landings, to today's atmosphere of aversion to government involvement and management of space exploration.

That aversion is hardly unjustified. From the singular lack of any overarching vision for space exploration over the last quarter century; to a perceived complacency with the status quo in space transportation systems, which never fulfilled their initial promise; to wasteful and glacial bureaucracies characterized by continual fighting between different NASA centers; to the biggest albatross of them all: the Space Station.

Through endless redesigns of the initial Space Station concept; an inability to estimate budgets and schedules with any assurance; creation of a space structure which for many people has no overriding raison d'être, a Station that exists merely to continue industry contracts and comply with foreign commitments.

Nowadays, ask someone how to build something in space, get back the advice to go anywhere but to NASA to get it done! This frustration appears endemic, and is certainly true of human Mars exploration—apart from the fact NASA seems to have abdicated that role in any case. Must this be so? Despite the triumphs of the Hubble Telescope (once repaired), a long and memorable string of robotic planetary missions, and Apollo itself, is the situation now irredeemable?

Perhaps not. One substantial difference between the genesis of, say, the Space Station, and that of Mars exploration, will be the existence of an overarching objective. This time there will be one! We have seen in the past—in domestic mobilization for WWII, in the Manhattan Project, in Apollo itself—that such a higher, ennobling objective motivates people, breaks down bureaucratic and disciplinary boundaries, encourages innovation, and, in the end, is manageable.

Politics and gamesmanship will always be part of any human activity. Another advantage of a large-scale program is that there are enough "goodies" to go around so that the often irrational and petty allocations of resources in space programs can be rationally ramified from the beginning, while still assuring constituencies that they will not be neglected so long as they let the process proceed logically. This still will not be easy to accomplish and very strong leadership will be required, but the scope and nature of this project does hold out the reasonable promise that it can be done.

Even if government could again be entrusted with this leadership role, should it be? From a historical perspective, the answer is yes. Zubrin has correctly pointed out that in the end, successful polar expeditions, and oceanic exploration, were not conducted under the aegis of a government authority.

But this occurred well after the initial stages of the great exploratory and colonial age, which were largely sponsored by sovereign nations. It should also be noted that no major Arctic or submarine settlements resulted from these private initiatives. Only nations had the means to make the investment to promote exploration and settlement. Each nation took a different path. Northern European powers like Britain enlisted and encouraged quasi-private-government entities such as the Hudson's Bay Company in North America and its equivalents in south and East Asia. Spain kept colonial exploitation under tight royal control. In either case, the ultimate fallback for protection and a logistics lifeline, even for the most independent of settlements, was the Mother County.

Of course, the royal houses of Europe did not have to be concerned about angry voters or a recalcitrant Congress when they approved these voyages. Which is why the case for Mars exploration and settlement must be made to a much broader constituency than Columbus ever had to worry about.

It all depends on the goals: If the objective of the Mars Society is to get a small party to Mars and bring them back safely, that is probably within the realm of financing from private industry or even voluntary contributions.

But if the real objective is a sustained, protracted period of human Mars exploration and settlement, then the only realistic source of long-term funding is the government. Thus, the case must be made to them, and to the public which elects them.

THE COMMERCIALIZATION ALTERNATIVE

If the government can and should lead, what role does that leave for commercial business? Given governmental stumbling, why not instead rely on entrepreneurialism to boost Mars exploration and settlement?

The short answer is that it will! Private industry will indeed play a major role, and as the aerospace base is broadened to include non-traditional technologies and infrastructure, the role of purely commercial companies should increase drastically. Part and parcel of that should be sponsorship of smaller, entrepreneurial firms. This is already encouraged in Small Business Innovative Research (SBIR) and Advanced Technology Program (ATP) programs, and is a natural starting point for dispersing innovations to the rest of the economy.

The best model for what will likely happen, is what did happen in 1990. Once SEI was announced, and it appeared that Mars exploration would finally receive full government backing, private industry was poised to jump in. Government investment, validating the market, was going to pave the way for significant private investment.

Furthermore, just because the ultimate sponsor and igniter of sustained Mars exploration is the government, does not mean that all activities that occur on the surface, especially during the settlement phase, need be controlled by a government agency. Quotidian affairs should be conducted as close to that of life on earth as possible. The government doesn't run the barbershop on Main Street and there is no reason for it to do so on Mars Street. Nor need it run Martian mines or Martian atmosphere fixation plants.

If government is dismissed because of its decades-long dawdling, is private industry ready to take the lead? In the short run, there is no evidence of this. In the forty years of the space age, only the telecommunications sector is thoroughly commercialized. Other sectors are starting to take off: Imaging; a more diversified commercial launch service industry; space tourism; even planetary science data acquisition. But these are embryonic, or at best, still in infancy. Some would argue, with considerable justice, that one reason for the slow spread of commercialization is the too mutually comfortable relationship between large aerospace companies and the federal funding spigot. The other way to look at the historical evidence is that without that spigot, what commercialization that did occur, likely would not have. Once the market door is open, and a critical mass of activity has been established, it is much easier to bring the private sector into the process. This is especially true of a new market that requires massive upfront investment in technologies and infrastructure stretching across many economic sectors.

Clearly, there are many aspects of the Martian environment, and its ultimate socio-economic fabric that can be exploited for profit. But profit is not sufficient! What is required is sufficient ROI in a fast enough time frame. And ROI is most definitely in the eyes of the beholder. Both government and private investors will require ROI to launch Mars exploration. However, there is an important distinction. Government may be satisfied with the social, cultural and economic advantages that will accrue to the whole nation, and with assurances that sufficient tax revenues will flow from the economic activity fueled by this initial investment.

Private investors work from a much narrower criterion: A certain amount of money must be returned within a certain time period. That period is not long in private industry. No matter how socially conscious private investors might choose to be, their investment was made for only one purpose: to make money. More money, made faster, than in competing investment opportunities. Presently, there is no solid evidence that there is a reasonable ROI, by Wall Street criteria, for the massive investments required to land humans on Mars, explore and settle Mars, and then launch profitable resource exploitation.

The bar to private investment is not the cost of Martian real estate—which costs nothing now—but getting there to exploit it. Relatively much greater investment is required for Martian resource exploitation then was needed to claim a parcel of land in Oklahoma, or stake a claim in California, a century or more ago.

Ironically, if the government were to offer direct funding to private entities, this would certainly spur private settlement of Mars, but would put the government directly in charge of determining winners and losers in this competition, much as the FCC regulates the frequency spectrum.

The same legal precedent—the Outer Space Treaty—that appears to forbid outright national ownership of solar system real estate does seem to imbue national governments with some responsibility for oversight of the private entities which can own such property. This is all for the good, since it is hard to imagine that totally unfettered exploitation of Mars resources along the lines of the worst colonial excesses, or even the California Gold Rush and Oklahoma Land Rush, is desirable. But this oversight is a far cry from direct allocation of property grants.

Until there is very solid evidence of immense, relatively short-term economic opportunities on Mars, there is no reason for Wall Street to invest vast sums in these missions any time in the near future. Such massive investment might finally come when terrestrial space markets or resources have matured, and the commercial space infrastructure is well established. Then the endless search for new markets will drive commercial exploitation of Mars and other solar system bodies, lifting off from a strong terrestrial base. But this may take decades.

On the other hand, if the first few stages, the landing and exploration, and early settlement, are begun under government auspices, the boundaries to such investment start toppling and Wall Street can quickly move in. Then the benefits of this commercial investment, which must otherwise accrue primarily to risk-taking private investors, could start spreading through the wider society and economy immediately. The risk to private investors would be reduced, while the benefits would be distributed to society at large.

THE VOLUNTARY ALTERNATIVE

Perhaps the most astonishing development in recent Mars exploration advocacy is the serious possibility of voluntary funding for such missions. If carried out, it would be one of the most remarkable accomplishments in human history, not only for what was done, but for how it was done. Yet, its realism is open to question, and even if possible, its desirability is also open to question.

Advocates of this route to Mars exploration find encouragement in the hundreds of millions of hits on the Web during the Mars Rover peregrinations. However, this link to Mars was free. A simple way to test sustained public support for Mars exploration would be to set up a Pay-Per-Hit (PPH) site during the next NASA Mars landing. Requiring payment for each hit, or even making it voluntary, say at $10/hit, would either quickly raise the necessary funds, or demonstrate once again that people prefer to get something for free.

One also suspects that if tangible support were this widespread, older, established space advocacy organizations such as the Planetary Society would have raised similar sums already. Although its purview encompasses much more than just Mars advocacy, the Planetary Society already co-sponsors some mission and technology development projects, though not on the scale contemplated here. And the Planetary Society is just one of perhaps dozens of space advocacy organizations staking out different positions and priorities. The end result is that not only is funding dissipated amidst these sometimes cooperating, sometimes competing organizations, but so is advocacy itself.

Sponsorship of smaller robotic missions, alternative technologies, or smaller facets of Mars Holistic by the Mars Society is eminently feasible and desirable. But funding entire human missions is a different matter.

First, even if it were accomplished, there is the risk that one would have another Apollo lunar debacle—a short series (perhaps just one) of missions which deliver humans to Mars, but without the follow-up needed to bring broader visions of human exploration to fruition. One assumption buttressing arguments for this voluntary route is that once an organization like the Mars Society places humans on Mars, the media reaction, and public enthusiasm and demand for support of further missions, will so overwhelming, that government funding spigots will open wide. This is entirely possible. But it is just as possible that people will conclude that an endeavor of primary appeal to a narrow interest group should be funded only by that interest group. Our delivering humans to Mars may simply prove that not only should this be funded solely by voluntary means, it can be!

Again, the question comes down to goals. If the goal of the Mars Society is to land humans on Mars at least once, as fast as possible—any which way we can—then pursuit of voluntary funding, in the absence of government interest or readily available private financing, becomes mandatory.

But if the goal of the Mars Society is to advocate for, and help bring about, the most expansive vision of launching human exploration and settlement of the solar system so that it is sustainable over the longest term, then this requires more patience. Single missions, or even a few missions, will not accomplish this, and as viscerally exciting as they may be for participants, may actually delay that grand vision, much as the Apollo landings proved to be a temporary dead end.

WHEN BIGGER NEED NOT MEAN BADDER

In recent years, virtually anything tarred with the adjective "big" is assumed to be bloated, wastefully expensive, sluggish, and numbingly bureaucratic and uncreative. Granted, with that aforementioned albatross in mind, this is understandable. Yet, some (relatively) large-scale projects have worked: Once fixed, Hubble has performed magnificently; so did the other members of the first generation of Great Observatories. The larger planetary missions demonstrated great resilience in delivering high scientific value.

All things start small, but incorporating the grandest vision of human exploration and settlement of Mars requires "big". The question is whether the benefits of this size are confined to the nuts and bolts of reaching and settling Mars, or whether they can be dispersed more widely.

Dispersal is very different for each size scale. For FCB robotic science missions, the cost is indeed cheap, but the benefits are strictly confined to the science returns, and any eventual technology spin-offs.

Medium scale—a string of Apollo-like human missions to Mars—may present the worst of both worlds in some sense. While accomplishing its goal of landing humans on Mars, the cost of these medium-scale missions is only small when compared to that of the old style SEI mission architectures or that of full-scale Mars exploration and settlement. The benefits, which will derive from solving the problem of sustaining a small, trained astronaut crew on Mars, for a limited period of time, are likewise constrained. Exchange Mars for the moon, or even the Space Station, and the scope of the problem is similar. Communities are not being established; widespread exploration might not even be attempted. The expense will not break this country's economy, but any funding pumped in will also not make much of a positive difference either, and might confined to the traditional aerospace industry. Justification for funding these missions (assuming they are not privately or voluntarily financed) must then come from the intrinsic merits of sending humans to Mars. And that, of course, has been precisely the problem: Thus far, those merits do not constitute sufficient justification

That leaves Mars Holistic, with its expansive objective identified and planned from the beginning. The obvious objection is cost, which could run as high as those early SEI estimates—albeit, for far more than just delivering an expedition to Mars, thanks to innovations like Mars Direct. This cost is high enough to damage the national economy. So to justify this cost there must indeed be very tangible benefits. Fortunately, the scope of the project also means the door is opened for benefits and justifications beyond the intrinsic, "philosophical" merits of sending humans to Mars. Justifications whom those who are not exploration enthusiasts may now find persuasive because it starts addressing their needs and aspirations.

Table 1 lists some of the advantages of this largest scale: Advantages that will assist in the development of the exploration and settlement program itself, and those that should disperse to the economy and society as a whole. They are grouped according to the general factors that must contribute to a justification for this program. Models describe historical precedents upon which these advantages can be based; current practices leading to the same conclusion; or even fictional analogies, which function along the same lines. (For the most part, traditional tenets of aerospace advocacy are not covered here.)

The crucial factor is the first, since it goes to the heart of the issue of the relevancy of Mars exploration. Exploration and settlement are written into the plan from the beginning, and full-scale development of settlement infrastructure is started early; the economic sectors most closely associated with living and working on earth are now drawn into the space program. Naturally, some of this already goes on in designing a structure like the Space Station. Building a village on Mars for "ordinary" inhabitants is a step beyond this. The economic, social, and technological sinews, which bind towns on earth, must be established on Mars. It must now play in New Peoria, Mars. Funding to do this should be dispersed to those elements of the economy most closely related to this infrastructure, sectors that traditionally have not been closely associated with the space program, including the consumer products sector. Thus, expenditures ultimately on the order of tens or hundreds of billions of dollars will not be confined one small segment of the economy.

Table 1
ADVANTAGES OF MARS HOLISTIC

Factor	Advantage	Model
Economics	Investing in Settlement Infrastructure Early in Development Spreads Rapidly to Broader, Non-Aerospace Economy	- Age of Exploration - Settlement of Frontier West - Financial Modeling of RLV
Economics	Investment by Commercial Firms, Independent of Government Capital, is Encouraged	- Evolution of Space Industry - Initial Response to SEI - NIST ATP
Socio-Economics	Development of Human-Intensive Infra-Structure with Human-Intensive Goal Help Reverse Trend to "De"-Humanize Work	- Frontier Settlement - Hollywood Portrayals - "Space Flatware"
Social	Opportunity for Socioeconomic "Underclass" in Human-intensive Industry And as Settlers	- Demographics of Colonial and Settler Populations - Artemis Project
Politics & PR	Overarching Theme Provides National Vision and Direction. Offers Something for Others to Jump Aboard	- Democrats: "Bridge to the 21st Century" TO WHAT? - Republicans: ???
Science	Demand for Planetary Science Increases, but Insularity is Broken as Other Fields Must be Integrated and Fused into Project	- Origins & Astrobiology Proj - Columbia Un Earth Institute - SSC
Engineer. & R&D	"Faster-Cheaper-Better" Continues in Toolbox for :Advanced Scout" Missions. Identify Spin-off Applications in Parallel with Technology Development	- Robotic Lunar and Mars Missions - SBIR Programs - Space Sensorium
Education	Opportunity for Outreach and Participatory Programs at all Educational Levels, Including, Adult Education	- JPL Pluto Outreach - Un Colorado Mission CTRL - Artemis Project
Politics	Large Enough that RATIONAL Resource Allocation can be Done Slighting Certain Regions or Constituencies. Elicits Further International Cooperation, with "Big Stick".	- Apollo & Wartime Missions - Space Station - Economic Sanctions on Pariah States

Historically, once past the initial exploratory voyages, expeditions had to bring much of their infrastructure with them. More importantly, they usually brought the people who could sustain that infrastructure with them; expeditions intended to settle and exploit a region were usually not small military parties. Once past the phase of penetration by solitary trappers and traders, settlement of the Western Frontier proceeded similarly. Overshadowed by the gunfire at the OK Corral, were the farmers, blacksmiths, innkeepers, dentist-barber etc., who comprised the real woof and weave of Western society. This must now be transplanted to Mars, and likely will mobilize our entire economy.

At the moment, these are suggestive, but still vague ideas. To be convincing, they must be buttressed with econometric analysis. One way to do this may be to extend current financial models analyzing returns from government and private investment in the new Reusable Launch Vehicle industry.

While infusion of government funding into Mars infrastructure development should positively affect broad segments of the economy, the other side of the coin is that it should encourage private investors to jump in. The evolution of the space industry supports this belief, as does the interest engendered in the private sector during the early days of SEI. We already have a smaller-scale model of how an industry-government partnership can be conducted: the National Institute of Standards and Technology ATP. ATP infuses government support for R&D on problems with great practical importance, but which cross industry boundaries and is beyond the ability of individual companies or industry sectors to solve. (On the smallest scale, an SBIR-type program would pull in small businesses.)

Recent decades have witnessed what might be called a gradual "de-humanization" of the workplace and the economy, particularly in the U.S. Manufacturing operations are now routinely shifted overseas to cut costs. At the first sign of profitability troubles—or even when there is no such evidence—many companies transition to a downsizing mode. In general, manufacturing, skilled trades, even office infrastructure, jobs, have been eliminated and opportunities have been diverted to highly skilled technical positions or the service sector, or out of the country entirely. Even the supposed employment refuge in the software and computer industries may eventually prove illusory as many companies already outsource much of their software work to India, Russia, and China.

This is a natural, even favorable, evolution of the global economy, as the rest of world develops and catches up to the U.S. But it does raise the question of whither we go? One answer is into the Cyberworld for sure. Yet, the only way to ensure gainful employment for our own population well into the next century may be to open another frontier, one in which pure efficiency, automation, and elimination of human participation need not take center stage. This is the model of Frontier Settlement.

By making the endeavor most relevant to the person in the street, it holds the greatest potential for mobilizing their interest and support. One can envision a worker who had no previous contact with the space program, sitting down for lunch in his consumer products company cafeteria, reading in his company newsletter about new contracts his employer has won for "sticky space flatware". Maybe he'll be working on this project. The word spreads.

Hollywood depictions of life in space, like "Armageddon" and "Total Recall", draw some of their popularity from portrayals of "ordinary" people living in space, whether they be heroic asteroid busters, transplanted to space, or the plight of mutant Martian miners. A few years ago, this author proposed a Mars Roadshow, (format conceived by Stephanie Bishop), to try to communicate to an audience how Mars settlement might affect "ordinary" people. "Interviews" with a Martian sanitation man, explaining how such a person came to live on Mars, may strike more of a cord than a similar interview with a planetologist.

Another lesson from the Age of Exploration is that the participants in these voyages of discovery and settlement were not usually from the upper crust of society. Even the leaders were often second sons of nobility, disenfranchised from any inheritance of wealth by feudal laws of primogeniture. And the crews and settlers themselves, more often than not, came from what used to be called the "dregs" of society, seeking a better life.

The space program, despite opening a new frontier, can not solve all social ills. But beyond the usual mechanisms of Enterprise Zones and the like, it can offer hope to our own underprivileged classes, where success need not ride quite so heavily on superior education and conformity to prevailing social standards. Indeed, one can imagine that the cohesiveness, mutual support, and brotherhood of urban youth gangs—if it could be redirected to positive ends—might be precisely some of the qualities necessary for survival through the rigors of an alien planet.

Though not directed specifically at social problems or an underclass, the Artemis Project for lunar settlement and exploitation offers a useful example of how people who are not denizens of the aerospace industry can be brought into the fold.

It has often been noted that our nation lacks an overarching vision, theme, or driving impetus since the end of the Cold War. We are all being propelled into a Cyberworld, and bound ever more tightly together by these cyberlinks. But this is not a great vision that we set for ourselves; it "happened". Many had hoped that human space exploration would be that grand vision. It has not turned out that way. Instead, we have seen that the vision by itself can not justify the price.

But once justified by other rationale, such as that suggested here, this overarching vision can take its rightful place. Politicians have a natural inclination for this sort of thing. The Democratic Party already claims to have "Built a Bridge to the 21st Century". What does one see after crossing that bridge? That has not really been defined yet. When coupled with solutions to domestic problems (a redefined "Mission to Planet Earth"?), Mars Holistic might be that Valhalla (erected with greater wisdom than Odin's home!). To Republicans, who do not seem to have settled on a similar catchy concept, perhaps the same offer could be made.

Many scientists object to shifting money to human space programs on the grounds that it is both unnecessary to the conduct of good science—the traditional van Allen argument—and, in fact, ends up severely diluting science funding. Nowadays, the Space Station is an example par excellence of scientist's worst fears.

On the other hand, Mars exploration and settlement requires science as an absolute prerequisite. NASA already established the Origins Program and Astrobiology Centers around the country. Robotic missions have long proven their worth, and will continue to do so, as they can best draw from the discipline of FCB. Nonetheless, human geologists, able to make on-the-spot decisions, apply human insight rapidly, or even perform laboratory analysis on-site, have value, pace Prof. van Allen.

As integral as planetary science (and biomedical science) will be to human exploration and settlement, these can be fused even more with the human aspects of the endeavor. An excellent example of groundbreaking work in this direction is the Columbia University Earth Institute. Shattering traditional barriers between scientific fields has not proven easy, but the Earth Institute has started down this path. Biologists, geologists, and social scientists are joining forces on problems affecting humanity on earth.

Conversely, availability of potential new observation platforms on the surface of Mars or its vicinity, for instance, helps ensure that Mars exploration is not a zero sum game for other scientific disciplines. The fate of the Superconducting Supercollider cautions us as to what happens when one discipline, perforce or by choice, becomes too insular.

Engineering is already proceeding in innovative directions as the recent robotic Mars landing missions have shown. FCB works very well for these missions, though even they can benefit from stable budgets. Technologies inherent in Mars Direct, when fully developed, also offer benefits on Earth. Renewed emphasis on human involvement in exploration should make commercial technologies and R&D much more applicable to space. The old dream of transparent technology transfer, not only from the defense-aerospace sector to the commercial, but the more recent goal of smoothly going in the reverse direction, will come much closer to realization. True dual use technologies become an essential ingredient of both space exploration and life on earth.

As technology is propelled forward ever more swiftly, applications are increasingly aggressively sought very early in the development phases, rather than waiting for a technology to mature and applications to fortuitously stumble upon it. Formalization of this process, as proposed by the author on a private project called the Space Sensorium, would hasten the spin-offs and benefits of Martian exploration and settlement.

Like technology transfer, education outreach has long played an important role in the space program. The Internet brings us the wonders of space with remarkable immediacy. In of itself, this promotes education, but the experience is still more absorptive than participatory. If Mars exploration and settlement is to garner significant public support over generations, education must play a key role. And to do that, it should become increasingly participatory..

Telecontrol via the Internet will assist in this process. Participation in the development process will help even more. The JPL Pluto Flyby program is a good example of how student efforts can be integrated into a real design program. The new University of Colorado Mission Control center is a good counterpart on the mission operations side. AMSAT might be archetype for all others.

Continuing education for adults is crucial to modern life. The Artemis Project points the way—utilizing the Internet—to involving, and instructing, adults with no prior experience in aerospace disciplines.

Perhaps the biggest hurdle to the success of this venture is quintessentially human: Politics. Even assuming the case can be made to the satisfaction of our politicians, will the conduct of these programs be allowed to proceed with a fair degree of rationality? Not all delays and cost overruns can be laid at the feet of NASA, or even industry. Appropriation decisions, scheduling, and resource allocation are often made not on any objective assessment of merits, but on the basis of partisan politics, geographical parochialism, or personality conflicts. (Appropriations for aircraft, which the Air Force neither needs nor wants, exemplify this.)

There is no way to eliminate these defects in our system. Two things may ameliorate them somewhat: First, the very existence of a goal of some grandeur tends to create a spirit of unanimity of purpose, counteracting the tendency to engage in the worst of these practices. Second, this endeavor—perhaps the most ambitious project in human history—will send out tendrils into all corners of our economy and society, touching enough geographical areas, that it may satisfy the competing needs and ambitions of our elected representatives, without having to compromise the integrity of the program.

This discussion has focused on the role of the United States, in the belief that we should and will take the lead in Mars exploration and settlement. Ultimately, the rest of the world will also become involved. Of course, leading space powers like Russia, Europe, and Japan, already have their own robotic planetary missions. Despite the travails of multinational financing of the Space Station, the benefits of international cooperation before and during its construction are pretty clear. This should extend into the era of Mars exploration.

Beyond this, participation in this great international venture can be as used leverage against the behavior of pariah nations, a form of economic sanctions, leveled not against the present, but against the future. Even the most brutal of dictators may pause before mortgaging their countries' future, for decades or even centuries, as the rest of the world pulls away and leaves them behind.

THE ROLE OF THE MARS SOCIETY

If the membership of Mars Society believes private investors will finance a Mars exploration program in the next few years, there is no need to answer the title question.

If the membership of the Mars Society believes voluntary funding for permanent Mars exploration and settlement is realistic, there is no need to answer the title question.

If the membership of the Mars Society desires to simply demonstrate that it is possible to bring humans to Mars without use of government funding, there is no need to answer the title question.

But if the Mars Society recognizes that, ultimately, human exploration and settlement will require public funding and support, the first priority for this newly constituted body will be to answer this question. For the ideas and suggestions in this Paper are just that, concepts, not convincing demonstrations. The priority is to muster what creativity we have to not only answer this question, but answer it in the affirmative.

Fortunately, there could not be a better organization to tackle this problem. The Mars Society draws together people from a very wide variety of specialties: science and medicine, administration, engineering, business, legal, education, and the social sciences, to name a few. What is not available in our organization can probably be found through the good offices of our members.

Studies are the very last thing people involved with space want to see! They freeze and deter action, risk, and accomplishment! One last time, though, we must be patient, and apply our resources to a Study to achieve the larger goal. Without this Study, there is virtually no chance to achieve the larger objective of Mars exploration and settlement.

Plan on completing this Study in one to two years. The whole idea is to push Mars Holistic in the short-term, and the same should be true of the Study. Funding—and there will be plenty of expense for something of this nature—should come from part of membership fees and perhaps from voluntary contributions.

If a concerted, creative effort ends up answering the question in the negative, then what is lost is one or two years. Not good, but not devastating, since the voluntary route can be resumed at that point, and private investment by business is independent of this Study anyway.

How should the Study be conducted? Start with Integrated Product Development (IPD) teams, formed under the aegis of a Mars Society Steering Committee. IPD methods have spread from its origins in aerospace to just about every corner of the business world. IPDs are designed to tackle interdisciplinary problems where specialists from multiple disciplines must work together from the start of a development process, lest expensive mistakes be made from ignorance of factors outside an individual's expertise.

Now the scope of IPDs and their Concurrent Engineering techniques must be widened even further, encompassing disciplines not usually brought together. This requires a willingness on the part of disciplinary experts to accept reasoned input from specialists outside their ken, and conversely, seek ways in which their own ideas can benefit these "alien" fields. For example, systems engineers may have to add new criteria for broad social and economic benefits to their systems configuration space when selecting mission architectures and infrastructure development technologies.

Once Congress approves settlement initiative, the Mars Society can transition into another critical role: Independent, objective advisors and counsel to Mars Holistic, who like Robert Jordan's fictional Aiel warrior societies, can cut across institutional boundaries when the inevitable conflicts between organizations arise, and keep the lines of communication open. This way, there is always a unified, reasoned voice supporting Mars exploration and settlement. Crucial issues, like cost estimates, which have typically been subjected to gamesmanship amongst constituencies, now have a home where they can receive (relatively) unbiased evaluation.

This project will require unprecedented cooperation between government, industry, and academia. For perhaps the first time, voluntary organizations like the Mars Society will also be added to the mix. The Mars Society may sponsor its own R&D, perhaps even its own missions once the Martian frontier is penetrated. Its members will surely serve in many leadership capacities in their home organizations.

Nonetheless, the most important role of the Mars Society will be as a resource, even a refuge, for problem-solving advocacy of Mars exploration and settlement over the centuries. The challenge will be to not become atrophied in the process.

MARS HOLISTIC

To earn the funding and popular support needed to initiate and sustain full-fledged Mars exploration and settlement requires answering the question "Is There a SHORT_TERM Social and Economic Justification for Human Exploration and Settlement of Mars?" with a resounding YES.

To attain significant short-term benefits requires adoption of an unusually holistic approach to space travel. Unless the great size of Mars Holistic is used, and unless the definition of who participates in, and who benefits from, it is extended well beyond what have been the traditional constituencies of the space program, it will not succeed.

The welcome news is that this Holistic vision, this all encompassing viewpoint, is precisely what most enthusiasts always foresaw as the essence of Mars exploration and settlement: Something at once wonderfully exciting and magnificently beneficial to all of humanity. One would think that even if there were other feasible alternatives, Mars Holistic would be selected because it is indeed most aligned with this grandiose, yet condign, vision of the role of Mars in human destiny.

If the answer is yes, and Mars Holistic is affirmed, then we can announce a rebirth of the space age, this time not with a bureaucratic whimper, but with TRUMPETS BLARING!!

MAR 98-012

MARS: FOSTERING PUBLIC SUPPORT

Sam Burbank*

Many obstacles block any plan to send humans to Mars. In conversations on the subject one often feels the discussion has ended before it has begun. Supporters of a human mission to Mars often end up preaching to the choir rather than facing questions of an educated and adamant friend or colleague, let alone a stranger. We need to become comfortable, each of us, with discussing and arguing for a human exploration of Mars.

Mars is a world waiting for life. It is tantalizingly close, always recognizable, and yet completely new. Rooted in the folklore of our oldest cultures as well as in our own popular fiction, Mars is a beautiful and harsh old world waiting to be seen for the first time.

There was a time when a description of Mars was easy, because knowledge of our neighbor was so limited. We could imagine the blurry, red place, but hadn't really seen it. The eye on the telescope saw many things; how people wonder. While America and Russia raced to the Moon, they both sent probes to Mars, each success bringing more knowledge home. And during the three and a half years when America landed six crews on the Moon exploration of Mars continued. Those scientists at NASA knew what the real goal was.

NASA's Mariner 9 was placed in Martian orbit in 1971. We had known little of Mars before its arrival, but hadn't imagined enough. Mars was amazing, diverse, fulfilling fantasies, and shocking skeptics who had thought it to be a larger version of our cratered Moon. Research continued until the stunning successes of Vikings 1 and 2 a very long term (four year) mission to Mars. The Vikings' two orbiters and two landers, arriving in 1976, would define a successful robotic mission. Their data are still being studied.

The images sent back by the Viking spacecraft invigorated my childhood. They stirred my imagination and sense of wonder like nothing since. Luckily my godfather was at Stanford at during those years. He would come by with computer printed images and hand them to me, knowing full well what he had was gold to the awkward eleven year old. I was ready to pack my bags. I, like the scientists at NASA, knew we would be there soon.

---

* E-mail: Birds@wenet.net.
Sam Burbank is a gaffer and lighting director in the entertainment industry. He recently completed work on a four part PBS documentary about microbial life. His Hollywood contributions include *The Game* with Michael Douglas and *The Rock* with Sean Connery. Some of his corporate clients are: Sun Microsystems, Apple, and Silicon Graphics. Sam's musical group The Birdwatchers sells records around the world. *The San Francisco Bay Guardian* recently wrote "Mastermind Sam Burbank's songwriting vacillates between wistful storytelling and buoyant pop gems." Britain's *Bucketful* magazine described the CD as "A masterpiece of inventiveness. One of the strongest albums of the year." Sam studied creative writing at San Francisco State University from 1990 to 1994. He lives in San Francisco with his wife Linda and two very large cats. Since childhood he's been fascinated by Mars.

What Happened?

For those of us who live and breathe space that question is mystifying. What slowed things down, how could the energy after Apollo 11 have shifted so quickly? Frankly, it's not something we like to think about. We'd better start. We could have been on Mars ten years ago—and could be ten from now if the word goes out that invigorates not just the adventurers, but everyone footing the bill. We're not selling one single idea; this is an entire planet of possibility. Selling it as anything less, or just to a specific group of people, is selling Mars short.

Apollo Program - 1960s Counterculture. A Tough Mix.

It's unfortunate that such unique moments in human history as the Apollo program and the 1960s counterculture/revolution occurred at the same time. Apollo was the pursuit of a dream of the ages. It was realized by a clean, efficient group of crewcut fellows whose hard work was cut out for them. Conversely, American popular culture was quickly heading away from the conservative mode of the 1950s, the decade that saw the birth of the American rocket program. People were experimenting, breaking from the past, trying hectically to create a new world and end an unpopular war. A war that many saw NASA aligned with because of it pilots and its contractors. It was a time of change with factions competing for the energy of the country. After Apollo 11 there was little crossover between those enthusiastic about space exploration and those who saw themselves exploring other realities. Some even suggested that since Apollo 10 had proven that a manned lunar landing was possible there was no need to go ahead with an actual landing.[1] We visited the Moon but it soon became clear which side had the momentum. Apollo 18 was built but never flew.

Ironically the most recognizable image from the Apollo program is the now famous "Earthrise", a photo of our jewel-like Earth rising from the horizon of the Moon, first taken during Apollo 8. "Earthrise" has been a beacon for one of the main spin-offs of 1960s counterculture: environmentalism. That picture alone, because of the good work it helped inspire, justifies some of the costs of the very expensive Apollo program. It helped galvanize a generation of concerned citizens towards a common goal. Its powerful effect on so many was unforeseen.

So in the end there was a communication between camps, through a photograph. But by then human space exploration was on hold. Some of the blame rests on NASA's shoulders, for not building on the momentum created by "Earthrise" or finding more flexibility to bring their good work into the focus of such a rapidly changing society. And, more than anything else, for not including women or minorities sooner than they finally did. Those white male engineers in their button up shirts and pencil holders didn't represent much of an America that was changing quickly.

Right now America is a measurably better country for most people than it was during the late 1960s. It may not be as exciting, but because of the end of the cold war it is an undeniably saner and safer place. This is a time to celebrate, a time to create our own challenge, not against an old enemy but against the limitations we always find for ourselves. Going to Mars is one way to do that. We will be going there someday. Going now, in this relatively peaceful time, we can be sure it's for the right reasons, and that it remains a long term goal.

I've talked to people around the world about my hopes for Mars and space exploration in general. I'm thankful to those who were kind enough to tell this overzealous thirty-one-year-old what they really thought of my ideas; those who put the dreamer in his place and made me find them a reason to go. Their debate has helped me to understand some of the basic and serious misgivings people have about the idea of space travel, and what value, if any, it might have for them. It has also shown me that beneath the seemingly apathetic exterior of many a skeptic is a dreamer who, when their values can be included, is easily excited about a Mars mission.

Below are samples of common arguments against a human mission to Mars and some possible arguments I've found to be useful in return:

"Socially what would be different there? How would society benefit from the trip?"

It has been said that capitalism is the worst possible system except when compared to all other political systems. America, for all its faults, has been an amazing place for social and political change. The freedom of our time for the majority of its citizens is unparalleled in history. It is the great experiment on this planet, always in flux, and yet constantly imitated. America's founders didn't plan on it looking as it does. Their language was broad and a new culture evolved. Its outcome couldn't and cannot be assured. Equal rights for its citizens, social freedom, freedom of the press, only now are these things starting to be taken for granted here. But the idea of them as something to strive for is taking root everywhere.

Most importantly America is where women took their power. Our example, constantly sent around the world, is being imitated everywhere. A strong resistance to this can be seen in some cultures but it seems unlikely they will succeed. It is not in the interest of any culture at all connected with the rest of the world to keep half of its population silenced or bound. For those bothered by the idea of women sharing power with men this is not a good time. But to me there could be no better sign for humanity.

Who knows what will change in us on Mars, what we will become? Much of humanity is moving in the right direction; most no longer live under the rule of kings, robber barons, slavemasters. Individuals have more freedom and wealth now than in any other time in history. And the power of the Internet to organize people around the world has yet to be measured. This progression will continue on Mars. Just as America's founders couldn't have imagined the changes that would occur here, neither will we know who the Martians will become.

"Technology has ruined so much here, yet that's what it would take to get there. Wouldn't you be basing a society on it?"

Another obstacle on the road to Mars remains the genuine and absolutely reasonable view that humans tend to ruin environments in their path. There are few untouched places left on Earth. Yet the technology needed to survive on Mars may change the way people perceive the Earth's environment, and could spark a revolution of sustainability on both worlds.

Mars is desolate. There is little there that most people would consider delicate. There are no forests to destroy or oceans to pollute. The prime directive of anyone living there will be to create and maintain a stable environment in which to grow food and make air. It will remain that way for a long time. No one will be building the newest computer processor or cell phone on Mars any time soon. They will need to survive, reminding all of us watching from Earth that we take a the luxury of our natural environment for granted.

Golden Gate Park runs half the length of San Francisco, from the foggy ocean shore to busy Haight Street. Anyone who has spent time in the park remembers their shock of hearing that it is entirely man-made. I couldn't believe it, having visited there many times as a break from the madness of city life, a trip into nature. This park had represented to me a city before its development.

Golden Gate Park was built on sand dunes and it was no small project. It took years, and were it not for the singular vision of John McLaren, who devoted the better part of his life to this 1,000 acres it would have never been completed. (The future Martian might study its construction as an example of what can be built from sand. But this is not why I bring this up.) The scientific thinking and technology that built the park from a harsh environment to a green wilderness is the same kind of technology that will be most needed on Mars. Over time Mars will slowly be transformed from its present dry, desert state into one capable of, in small pockets, independently supporting life. It may not seem appropriate to call that kind of work technology, but the botanists who built Golden Gate Park would disagree.

There is nothing unnatural feeling about Golden Gate Park. It feels good. It's an example here of what can be created from sand. It can be used as an example of the kind of technology that will be needed on Mars. And as more natural environments are used here on earth, it will need to be imitated. Not as a substitute for environmentalism, but because, for better or worse, we humans will continue to grow and consume and will need to recreate some of what has been used carelessly. Those living on Mars, trying to find stability there will show us how possible that is and how to do it.

Creating sustainable environments on Mars will be a great challenge. The initial explorers' greenhouses and then, in Mars' not too distant future, covered craters or valleys will serve as examples to Earth of the kind of care needed to maintain stability on any scale. I can imagine Mars' red landscape dotted with green communities, possibly visible from space, small worlds working hard to find an equilibrium, microcosms of sustainability.

"I've had it with the frontier spirit, with white men ruining everything in their path. If that's the plan count me out."

If there is one thing to be done to garner support for a human study of Mars it is to change the image of the militaristic, white man explorer. Without a doubt this is the main stumbling block I've encountered in conversations about Mars, especially with women. The conquering of the west brought riches to the old world and the unique experiment that is America might not have happened otherwise. But the trade of lives and culture for America was harsh and ruthless. No one is proud if the destruction of our giant redwoods or the pollution of the Great Lakes. For many Americans the image of the cowboy frontiersman represents a better time, but for most now he represents destruction and callousness.

It's important that the Mars Society's objectives remain open. That should be easy. That's a whole planet next door. Yet we sometimes hear people discussing plans for Mars as if it were only interesting for a specific group. Yes, the initial years will be tough and in some ways analogous to the American West, but as a selling point that idea is not only uninteresting for many, it's a showstopper for some. I don't try to sell a human mission to Mars to a microbiologist the same way I do to a sculptor. They have different hopes for the future, yet neither dream is wrong or less realizable than the other. Anyone willing to dream about a human presence on Mars, about that next step, should be given a wide latitude for their thoughts of what Mars might become. The only agreement needed is that it is possible and worthwhile to start now. That's the goal.

During the cold war we sent some heroic cowboys into space. During that time, such as it was, they represented us well. One, John Glenn, will be returning to space soon but thankfully he now represents something new. He and Shannon Lucid inspire more people than the rest of the corps combined. I don't think there is a woman alive who heard about Mrs. Lucid's long stay on Mir and didn't think 'My God, that mother of three made it up there. That could be me!' I am invigorated by this transformation. It allows the adventure to be all of ours, it represents us honestly.

Whether heartfelt or not, try to think about a human presence on Mars from the perspective of the average person. Why write off a potential supporter? Find them a reason to go. What are some tangible benefits that can be brought home from it? What are subtler benefits? A supporter need not be a frontiersman or scientist. If they are asking questions, even confrontational ones, a door is open.

"We should concentrate on the Earth's problems before going anywhere."

Going to Mars and fixing the Earth are not mutually exclusive ideals. This statement is analogous to "We should concentrate on the Earth's problems before building a telescope, or before doing deep-sea research, or before studying Antarctica." It operates on the premise that (a) human exploration of Mars is not good science, or (b) that no science is worth doing until the Earth is peaceful and healthy. The continued study of Mars tells us about our solar system, about its history, about how planets and environments can change, and possibly about a different strand of life.

Of course the Earth needs to be cared for. And a human study of Mars needs to be efficient like any other science project. But the money spent for the work done there will be well used. Not only in terms of the knowledge that will be gained, but also the excitement infused into an apathetic world and the potential for ultimately a new branch of humanity, not disconnected from Earth, but separate from it.

Anyone who reads the newspaper or tunes in to the nightly news might not have noticed that crime is way down, unemployment is very low and America is on top of the heap economically again. The world is free of large scale conflict.[2] Humans always have problems, so of course we do now. But compared with most of history this is a remarkably peaceful and prosperous time. One way of celebrating the amazing good luck of having gotten through the cold war in one piece and the exciting new economy that has bloomed in the past couple of years is with a scientific human study of Mars.

"NASA is part of the military. Remember the Vietnam War? Their goals are suspect."

I heard a story earlier this year from a man who ran Stanford's hypersonic wind tunnel during some of the Vietnam years. His work was space related. One day he came to his office to find that the office door had been broken in with an ax and the lab was destroyed. He didn't understand it. This was an anti-war statement, probably by students, but at the time, for him, the connection was vague. His work was space, not military. As time went on though the connection between the war and his space science became less easy to distinguish. It was when one of his co-workers, a kind friend, a father and husband, casually left the wind tunnel to pursue making napalm burn more efficiently that his perspective changed.

Many of those conflicts are in the past. But even now the military still shrouds its space work in secrecy. NASA recently considered replacing an overdue Russian component of the international space station with a "previously secret Navy spacecraft".[3] When space science and military work are aligned, when that connection is vague, supporters are lost. I believe this is a reason why human space exploration stopped after we beat the Russians to the Moon.

If the study of Mars is funded privately this shouldn't be an issue. Also, to many NASA is looking more credible and frugal. Its image after the excellent Pathfinder mission, Global Surveyor, Shannon Lucid's stay on Mir, and John Glenn's soon to be space shuttle flight is looking up.

"Who pays? How much?"

It's possible that a small scale human exploration of Mars could be paid for within NASA's existing budget. It would probably require scaling back the shuttle as a cost cutting measure. Burton Edelson described the shuttle as "an absolute failure as a launch vehicle. It's expensive and has made too many demands on its payloads."[4] It may mean getting the Russians involved with their big, cheap bulletproof rockets. But as a selling point this is hard to beat. You're already paying for the budget, now you'll have something to show for your money. In this regard, getting support for a human exploration of Mars has less to do with asking the public to pay more, rather it is asking NASA to do something worthy of such a well funded government agency.

The wealth exists privately as well. Bill Gates could pay for the entire initial series of flights and still feel okay about himself at a dinner party. The price of a simple mission would not be staggering. The two most important assets, the needed technology and human enthusiasm and capability, already exist, and public support seems to be growing. It's hard to believe we've waited so long.

Get out and talk to people about Mars, hear their concerns, seek out their arguments against going so they can be addressed. If you can't find them a reason to go we're not going! It's not the fault of the average person that they know next to nothing about Mars. We need to get the word out. Every supporter has felt the frustration of hearing a well educated adult ask "Haven't we sent astronauts to Mars?" or "It's about twice the size of Earth, right?" Most people don't know the first thing about Mars, its size or climate, its unique place as the most hospitable neighbor in our solar system. They know about *War of the Worlds,* not Olympus Mons, about *Mars Attacks*, not Valles Marineris. A little information would go a long way towards helping people understand why it's Mars and not Venus or the Moon that has captured our attention. Below is a list, a cheatsheet if you will, that everyone you know should know about Mars. You'll be surprised who doesn't.

1. Mars is half the size if the Earth, about 4,000 miles in diameter. The Moon is about half the size of Mars.

2. But because Mars has no oceans it has roughly the same land surface area as Earth.

3. Its gravity is about 1/3 that of Earth.

4. Mars' atmosphere is thin, about 1% of the Earth's. It's mostly unbreathable carbon dioxide.

5. Mars' year is 687 days long. Like the Earth it has seasons, which are longer, like Mars' year.

6. Its day is 24 hours and 37 minutes.

7. Mars has surface features that dwarf anything on Earth, a canyon (Vallis Marinaris) that would stretch across the U.S. and a volcano (Olympus Mons) two and a half times as high as Mt. Everest.

8. Mars has two small moons, each around the size of San Francisco. They are captured asteroids.

9. Mars has water in its polar caps and trapped in the soil as permafrost.

10. The trip from Earth to Mars can take less than a year, with present technology.

Ask a submariner if that's possible. They're doing that right now, deep underwater. The line of talented, adventurous souls ready to go will be long.

Lets let them know what they have to look forward to. On to Mars.

REFERENCE NOTES

1. *The Soviet Manned Space Program*, Phillip Clark, Orion Books. p 84.
2. *Wired Magazine*, Julian Simon, January 1998, p. 66.
3. *Aviation Week & Space Technology* "Agenda for Congress," February 17, 1997, p. 44.
4. *Aviation Week & Space Technology* "The Shuttle at 15," April 8, 1996, p. 52.

MAR 98-013

PLANET MARS HOME PAGE©
SOME RESULTS AFTER THREE YEARS OPERATION*

Thomas A. Gunn†

As we embark on a new era of the Mars Society, it is helpful to take a brief look at the experiences of day to day Mars interest interaction with the world public. The World Wide Web site "Planet Mars Home Page" has provided outreach to a public hungry for participation in the Mars human settlement adventure for three years.

The site experience has included persons and companies from all walks of life in all age groups. The format is not only informative, but encourages actual input of content from everyone. In the email questions and answers are to be found the depth and breadth of substantive issue discussions and proposed solutions. Although aimed primarily at private sector rocket programs, the implications gleaned from this public interactivity contains important insights for all who are involved directly or indirectly in the quest for Mars human settlement.

Three activities of the Planet Mars Home Page have yielded significant results during the past three years. They are: Outreach Expansion, One-Way To Stay Mindset, and the Mars Rocket Survey. This report discusses those areas, and for some, their surprising outcomes.

OUTREACH EXPANSION

Analysis of email responses of email responses revealed the following breakdown of interest levels and support sectors for permanent settlement of Mars:

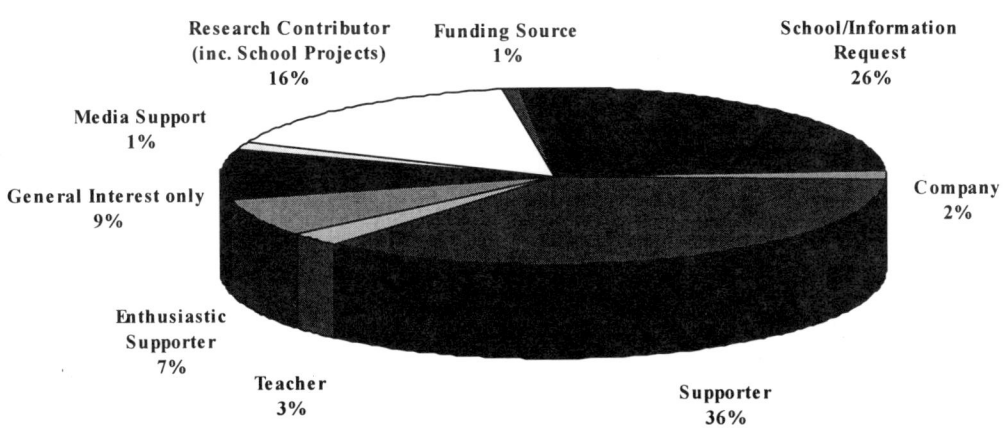

---

* Copyright © 1998 by Thomas A. Gunn, All Rights Reserved.
Search for "planet mars" on Excite and is listed as the "Planet Mars Home Page." Mirror Site is: http://personalwebs.myriad.net/tgunn. http://www.marshome.net is the new site currently under construction.

† 414 West Brookside, Bryan, Texas 77801. E-mail: tgunn@myriad.net.

This is our current market. All segments are potential resources to contribute funding, equipment, facilities, research solutions and manpower. All segments need coordination, participation, and encouragement.

Government people and programs were noticeably missing altogether which should be reason enough to rethink current mission support planning. This not only reinforces a general comment that "space science is dead," but also strongly suggests that the correct meaning is that "*government* space science is dead" from a public market viewpoint. The support has simply shifted to a host of other sector interests - and the support is stronger than ever - just misunderstood by media, government, and government-dependent space business.

Also noticeably was interest in robotics missions. The general feeling expressed in the email survey was robotics that is needed to gather information, but public interest was short-term unless each mission clearly provided key information that solves problems directly needed to get people on Mars.

The hot button key factor of sustained interest and support is almost exclusively centered on **permanent Mars settlement.**

Furthermore, one-on-one response to the public through web participation can mold and solidify opinion, understanding, and support to a larger extent than other mass media.

Lessons learned about the Web's essential role in funding, design, education and construction of solutions for Mars exploration and settlement are important to the success of the Mars Society, not to become just another planetary society, but to become a working structure for outreach and participation beyond that of conventional groups. Several groups historically overlooked by government dominated planetary mission proponents have participated in the Planet Mars Web page.. These sectors are:

- Students & Schools: 7-12, University, Community College Retraining
- Non-space related (NSR) Businesses
- Adult supporters from all walks of life
- Institutes, Societies & non-space related Agencies
- Wealthy Individuals
- State government (and their equivalents in other Countries) including State Commissions, and Municipalities.

Any ONE of these groups could, by themselves, produce the funding, manpower, technology and effort to design, build and launch a permanent Mars human colony.

The size, scope, and cost-effectiveness of their efforts might vary, falling short of or exceeding the slower, costlier tourist approach built into a totally government run Mars effort. Pressures to organize these NGR groups are increasing. More of their constituents want to participate in this effort but find themselves outside the select few running the government dominated effort. It is therefore important to expand the effort to include, promote, coordinate, and foster cooperation between all available resources. The narrow path of the past should not be continued.

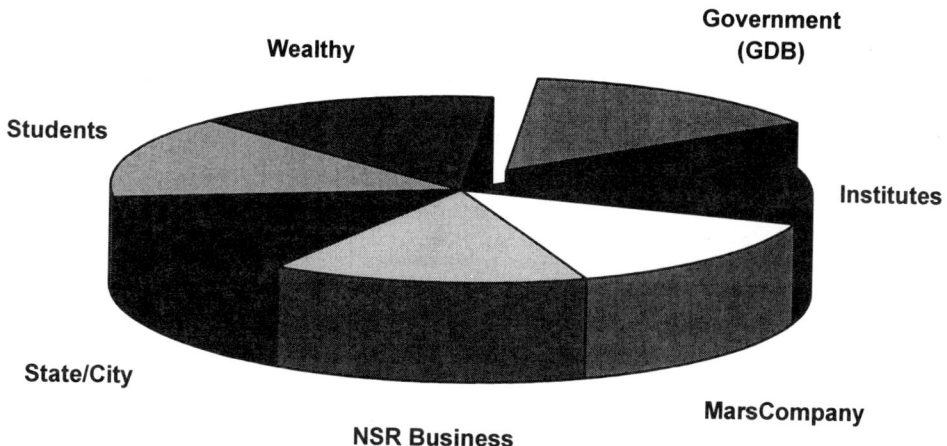

ONE-WAY TO STAY MINDSET

Another of the fundamental areas explored by the Planet Mars Home Page is called the "One-Way To Stay Mindset" which refers to the result of thinking about Mars as a place to permanently settle with humans, rather than a place for round trip visits.

Until recently, the realistic possibility of a permanent move to Mars has been ignored. Instead, our total focus or "mindset" has been one of tourism or sightseeing. Build a motorhome, take everything we need with us and go see what we can see, leave our waste and return home to Earth. However, during this headlong race to collect rocks from every heavenly body as space tourists, another mindset has emerged. The Planet Mars Home Page was created as home for the work and information required by those who operate under this other mindset, which we call "one-way to stay" or permanent human settlement of Mars.

A good example of the difference between our NASA type tourism missions and One-Way to stay permanent population of Mars is how we think about traveling to another place here on Earth. This is a good example because it illustrates the specific additional work required when you change your way of thinking about Mars missions:

If you want to go sightseeing or camping, you can just hop in the car or motorhome and go. This takes a minimum of planning. You stay in motels and campgrounds or the motorhome, tour all you want, and then return home.

If, on the otherhand, you intend to MOVE permanently to another location, there is MUCH more to consider BEFORE you leave. Sell the house? Are there jobs in the new place? Where will I live? How do I raise my family? Schools? Shopping? Environment? You can make your own list since most people and corporations have performed a move. In short, you need to acquire a place to live, reestablish utilities, medical facilities, communications, TV, registrations, deposits, certifications, find work, banks, legal services, and reach some consensus with other family members, relatives, friends, and business interests for this permanent move.

The information you need to know beforehand about where you are moving to must be far more detailed and certain than that needed for a camping trip. Furthermore, HOW you get there is secondary to all the other considerations you must explore and answer. What you take with you is also considerably more involved for a permanent move than for a sightseeing trip. The permanent move requires more extensive planning and action before AND after moving. The current and past space scientific tourism mindset requires only a small subset of the work required by a permanent move mindset.

Thinking and acting on the permanent one-way to stay mindset is difficult to keep in mind as we work on Mars mission details. This move verses visit emphasis requires a fundamental change in our thinking. Changing our thinking is fundamental to success. Furthermore, the move approach will also yield questions and answers useful to the government tourist groups.

Simply changing our mindset from camping to permanent move will force us to ask more comprehensive questions that each mission must answer in order to build upon experience, rather than just experiencing stand alone space camping trips requiring less preparation.

Going to visit Mars, establish an Antarctic type research facility where people can come and go, certainly has a purpose. It represents the government tourist mindset. Understandably governments will take this simpler, quick and costlier approach, since governments don't have to sell a house, worry about a place to move to, and create a new life as do people and corporations.

The corporate and personal point of view, ignored until now, is quite different. The historical government tourism approach to exploration is a small subset of the larger view in all aspects from planning to action to preparation of a "to do list." By actively including people and corporations in all aspects of permanent moves to Mars, we can tap a far larger resource, as well as provide better overall planning for all missions.

If you can take your motorhome and stay at established "camps", why then even consider moving there to stay, raise a family, and die? The government tourism approach obviously takes a great deal less planning, information gathering, and preparation. So why even consider the permanent move approach? From the above discussion, you already have some idea.

The Planet Mars Home Page was created to provide detailed answers to this global question and to provide a nexus for coordination of the information and activity concerning the one-way to stay mindset. It has been highly successful. Exploring the larger picture of human outreach has clearly shown the tip of an iceberg which is totally outside the government tourism mindset.

While governments and government-dependent big business continue to build their rocket motorhomes and supply their KOA camps from Earth using taxpayer funds, the other potentially larger group will be working on all areas of a permanent move to Mars. More difficult and comprehensive move work only corporations and individuals can do. by definition the profit motive is outside the comprehension of government. This does not mean that there cannot be cooperation and where needed, parallel development with shared information. A large number of people and groups must also be more involved in Mars settlement. Continuing as mere taxpayers or educational information users is no longer satisfactory.

Worldwide email has demonstrated that these people and groups understand they are part of a separate approach and mindset. They understand why the current tourism approach to space exploration cannot succeed to the satisfaction of the human race. Why must there be a parallel approach embodying the move mindset? What is the government approach incapable of providing? The answers are:

- **Economics**. Supplying people and camps from Earth is more expensive than self sufficient Martian towns with local populations.
- **Race Preservation**. Populating another planet is insurance against failure of the Earth ecosystem, culture or race, and
- **Substantive Information**. Information that can only be gathered on-site and information from questions that will never even be asked by tourists or answered robots.

All of these items embody major shortcomings if addressed only by the sightseeing government approach. Details are outside this paper, but it is clear that pursuit of these three goals will necessarily involve all parties in Mars development - not just those who would visit.

It is time to dare to think about a move to Mars. We as a society have not let ourselves even hold such a thought. If we did so, we would find that we can easily prioritize what information we need right now and what projects can be left for later by simply applying the 'one-way to stay' or 'move' mindset.

THE MARS ROCKET SURVEY

It must be stated up front that the survey has no official status, no academic procedures, no double-blind testing. It is simply the result of email responses to a web page that would be found only by those searching for information on Mars work. The site volume is small for statistical purposes but thousands of hits per month was far more than anticipated for a serious technical web site. The survey question was posted for two years from July 1996 through July 1998.

There are many issues addressed on the site. One example item on the "to do list" is that a heavy lift launch vehicle (HLLV) is needed to send the weight that crewed missions would require. Existing rocket power can send lower weight robotics missions but there is no current system on Earth to get a crew to Mars today. Several Mars capable rocket systems, liquid and solid, were in various stages of design and construction during Apollo. Only two Mars capable liquid systems were put into service by governments and none by the private sector. Those two were Energia (USSR) and Saturn V (USA).

To take the pulse of the world on this issue, the Planet Mars Home Page asked the following question:

There is NO heavy lift launch vehicle on Earth today to get us to Mars... "Should we pay to resurrect the ENERGIA or SATURN launch vehicles...or should we pay for a TOTALLY NEW heavy lift launch vehicle?"

The results were surprising. After 30 years and a whole new generation since the Moon landing, we expected 100% response to be a NASA government response as the ONLY space sector solution provider. For 30 years, the news media has nearly exclusively touted Shuttle and some "cheep access to space" (CATS) aerospace contractor events - reporting virtually no information on HLLV rockets of the size to go back to the Moon or on to Mars. It is clear, that without knowing why, the world public is more savvy than even space administration was in the 1950's and 1960's in highly technical areas.

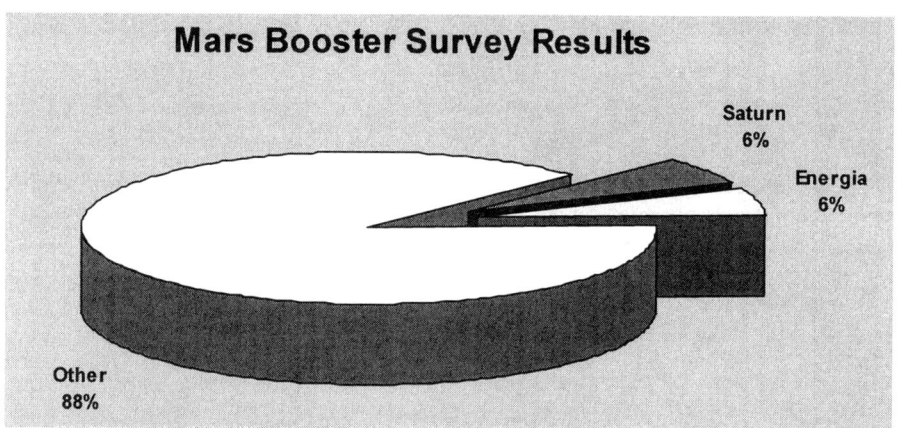

It is clear from this graphic, that some other Mars rocket system should be developed instead of an expensive upgrade of the Earth's only two systems that could deliver the necessary payloads to Mars for Human settlement. Of all the systems suggested, there were only two categories of system. The direct from Earth ground to Mars systems include the same single-shot scenario as the Saturn missions and were preferred by 62%. The other 38% reflect the emerging interest in multiple lifts to orbit with Mars rocket assembly in space as an alternative to the Saturn-Energia concept of one launch.

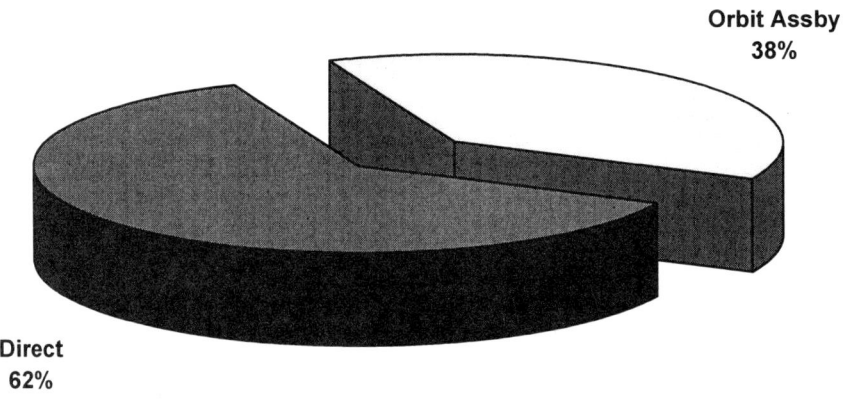

The survey recorded twelve different Mars human transfer rocket systems. These covered the range from old liquid technology through unused solid technology to what some would consider to be 'esoteric' such as the Searle Disk antigravity system. It is not known the extent that the respondents knew the costs, practicality, and timeframes that might be involved in their favorite system. Most did feel strongly about their choice, however, and some did provide some level of detail as to why they supported their choices. Esoteric or not, all categories currently have active groups working with their respective systems and have some level of worldwide support.

	Mars Rocket System
CA	CATS Orbital Assembly
EN	Energia Resurrection - Direct
ES	Electrostatic 1st Stage - Direct
MO	Moonbase Assembly & Launch
NA	NASA Shuttle Orbital Assembly
NS	All New Ground to Mars Solid - Direct
NU	Nuclear - direct
OL	Atlas & Titan alternate government liquids - direct
OS	1960'S PBAN Resurrection Solids - direct
RG	Rail Gun Orbital Assembly
RP	Rocketplane 1st Stage - Direct
SA	Saturn Resurrection - Direct
SE	Searle Physics - Direct
SS	Space Station Orbital Assembly
ST	Strap together Systems - Direct

Among the 62% supporting a single shot direct Earth ground to Mars system, eight different systems were proposed. The large response for a nuclear solution was a surprise, since most believed cost-effective power could be substantial enough to launch from the ground with this solution. Hybrid systems that used nuclear after orbital assembly were recorded in the non-direct group, so these are nuclear all the way.

The rocket plane group was a hybrid that launched the other chemical stages from a rocket plane platform that represented the first stage in the system. New solids represent the use of old and current technology in a new system similar to DOD solids. Strap-On systems generally were solids added to existing and upgraded liquid fuel systems although bundling of solids like the SRB to achieve the needed thrust and staging was also included in this category.

DIRECT to Mars
HLLV Types

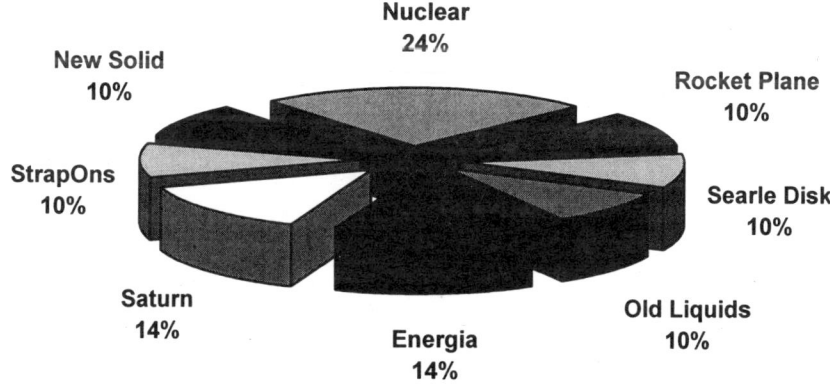

The 38% who support multiple shots with orbital Mars rocket assembly consisted of four different systems. The largest segment believed assembly on a space station was best without commenting on how the materials were to get to the station.

That the other three systems, rail gun, various cheap access to space (CATS), and Moon assembly representing 53%, were as large a segment as they were compared to space station assembly was not expected. The understanding that there are several ways to assemble in space *other* than a space station is significant.

ORBIT Assembly
Multiple Trips to Orbit

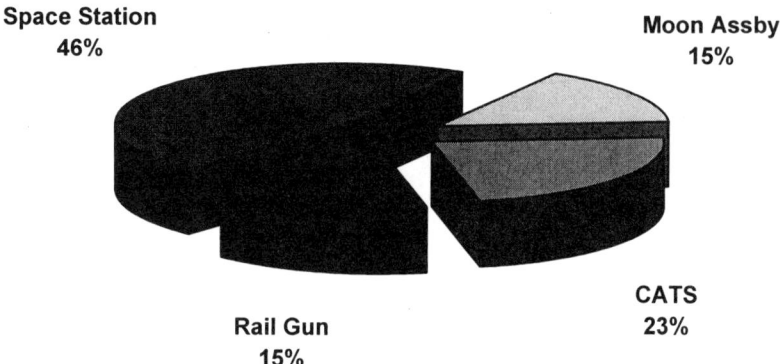

CONCLUSION

1. **Outreach**, cooperation and coordination must be greatly expanded to all Mars Interest Types.

2. There are six participating **Mars Resource Sectors** beyond just the government GDB sector.

3. It is essential to change our "**mindset**" from camping to a permanent human settlement move.

4. Two overall results of the **Mars Rocket Survey** are of prime interest:
   (a) The majority (88%) support is for use of current technology to build a new system rather than resurrect old systems.
   (b) The majority (62%) support is for a single shot system from Earth ground to Mars rather than orbital assembly from multiple shots.

The general belief of respondents was that action on all these targets will result in the lowest-cost, fastest solution to provide a permanent human settlement & transportation system.

To explore these results, the Planet Mars Home Page will post a new survey question for the next two years:

> *"Given that the least cost, shortest time to Mars launch of Human settlers is to use what we have leaned over the past 40 years to produce a simple, 4-stage hybrid Mars rocket(with 1st stage either rocketplane or chemical): Would you support such an effort? If YES: How much would you contribute per year and what do you expect in return. If NO: why not?"*

HOW (WHY AND UNDER WHICH CONDITIONS) COULD INTERNATIONAL COOPERATION REINFORCE THE CASE FOR MARS*

Richard Heidmann[†]

INTRODUCTION

In August 1998, the Mars Society held its founding convention in Boulder, Colorado. At the same time study groups were redefining the Mars Program at JPL, with the participation of colleagues from different nations, and the prospect for international engagement in future Mars exploration. This conjunction is quite meaningful in that it underscores the new deal for major space undertakings to come. And the start up of the Society was undoubtedly an excellent occasion to have some reflection about why and under which conditions international cooperation could reinforce the case for Mars.

BASIC AIM OF ACTION FOR A MARS SUPPORTIVE GROUP

As stated through its declaration, the Mars Society embraces the broad scope to push forward Mars exploration, beginning with robotic spacecraft but also, as soon as possible, with humans as well, and with the vision that it's the destiny of the human race to settle this habitable and live planet. Even if private initiatives and fund raising are relevant, both in the kick-off stage (by means of public excitement through brilliant initial results) and in the long term (when purely entrepreneurial initiatives will be made possible), it is evident that, basically, nothing really significant can be envisioned without political commitment and governmental budgets. So, to be practical, our action within the Society must be focused toward this target of public and political support; in other words, we must direct our efforts to raise general public interest and make Mars exploration politically desirable.

Such a statement may seem quite trivial. Not so much, if we recognize (and we have to!) that we are indeed up to now a rather limited community of people which, being over-enthusiastic, tend to forget that they have to convince other citizens not in the same state of mind... and to deal with politicians. Yet, to be effective, we must keep looking at the realities of this world. This is not to say, in any way, that we should deny our vision and content ourselves with short term and easily palatable achievements; on the contrary, our long-term "vista", our strong belief of Mars destiny, is what will give sense and strength to our proposals. But we have to remember that we don't speak to ourselves, but to people that could have views far from our own, and which have their own preoccupations. Simply said, we surely have to be aggressive, but in the same time must **keep looking pragmatically to political realities**. And,

---

* This paper represents the opinion of the individual author and should not, in any way, be interpreted as necessarily representative of the views or policy of SEP.

† Corporate Scientific, Technical and Industrial Evaluation, SEP, a division of Snecma (France). E-mail: RHeidmann@aol.com.

needless to say when speaking of the destiny of the human race relative to another planet, the global (international) nature of this topic is politically central.

POLITICAL REALITIES AT THE PUBLIC LEVEL

An Ambiguous Position for Space

The position of space in the public is quite ambiguous in that, on one hand, space activities have become routine business for them and have lost their nobleness, but, on the other hand, hints exist that prove interest and potential for excitement.

Without being overpessimistic, we have to recognize that a human Mars exploration effort can no longer be backed, at least at this moment, as were the first steps of man in space and the paroxysm of the Moon landing. This is not only true in the U.S., but also in Europe, and even in France, where space has, from the beginning, benefitted from a privileged position. There is no longer time to speculate on heroic tales of Mars conquest to make it happen, even if this is difficult to admit, and sounds sad for some! All we can hope is that this could change in the long run.

It is also true that space applications, as they spread in day-to-day life (mainly in the field of telecommunications), and fell in the realm of commercial business and fierce industrial competition, have lost most of their prestige.

Finally, we must confess that the most prominent and visible actors of space activity have not given the best image of the field, lending forceful arguments to opponents of expenditures on projects like the Space Station, to speak of the more important and meaningful in these days. Even if, as professionals or supporters, we legitimately can be proud of most of the achievements of the space age, we have to acknowledge some big mishaps and an age where soundness of choices, efficiency and effectiveness were not always the rule. So, we can understand why the opinion that space expenditures are not under control, wasteful, not to say senseless, is widespread. As a matter of fact, without facing this reality, we should not be able to take steps to reinstate our image.

But curiously enough, this bad situation for space affairs is not without signs for hope. Perhaps the more fundamental fact is that people are more and more aware of the acceleration of scientific and technological progress. Furthermore, a beneficial feed-back exists between stress of people, their need for hope, and mediatization of discoveries and breakthroughs. This is of course particularly true for all that goes on in the fields of health and biology, but also for environment control (perhaps more in Europe than in the U.S.) or for Earth resources depletion, climate control, and so forth... The net effect is a popularization of technical and scientific achievements, even in fields apparently more remote from people's immediate concerns, such as astronomy, on which we can truly rely to promote our projects. Closer to us, the astounding performance of Pathfinder on the Internet, if less profoundly significant, is even more directly encouraging. A large fringe of the public, worldwide, is yet a potential follower.

Major Public Concerns Should Not Be Ignored

This potential for interest and excitement is an opportunity that we should not miss by keeping a posture too much "Mars enthusiast-centered". On the contrary, we have to find how the case for Mars could best fit people's concerns, tapping on their need for good news, hope and, even, dreams. It happens that this can probably be achieved quite naturally, as most of these basic concerns are, in essence, concerns of humanity as a whole, and perceived as planet-wide problems. This is without question the case for environment and climate control, management of resources, development of telecommunications, and even for health, with the threat of global epidemics and the worldwide significance of major medicine breakthroughs. All of these topics share flavors: destiny of the human race, significance of Planet Earth and the value of science and technology, which we can rely on to promote the case for Mars.

If we recognize the constraints set forth by this situation, it becomes quite evident that the endeavor of human Mars exploration and the perspective of the settlement of a new world **cannot be sold to the public without the same kind of flavors and, hence, out of the frame of an international effort**. Note that this is also beneficial relative to another major wish of people, peace between nations.

Specific Points to Be Addressed

In advocating the case for Mars in front of the public we must address the critics against space activities which are specifically directed against it. At the same time, we must forward positive arguments, keeping in mind preferably those which bear significance for people's concerns and hopes (arguments purely related to planet Mars science, while of fundamental value, can be heard by only a small fringe of citizens!).

One of the most prominent critics is **the cost of space programs**. It is commonly stated that those projects are not worth the money, that they are pushed by the agencies and aerospace industries for their own profit, and that spending is not under control. Fortunately, this tendency has been somewhat reversed in recent years, especially thanks to the new NASA "faster, cheaper, better" paradigm; even if the real operational value and limits of this new way of doing space exploration business are probably not yet completely understood, and even if it can be felt as a restraining factor, its positive effect on the public helps us to rebut the critics. And the most recent successes obtained constitute forceful demonstrations that it can be done efficiently, and at an affordable and legitimate cost. But of course, when we speak of manned exploration, the reference in the public's mind is rather the Space Station, and there the problem is admittedly much more difficult to tackle... That's why it is probably wise to push at first an increased short term effort on robotic exploration, and to speculate on the thrilling results that will be obtained to sell the manned effort.

Another widespread opinion is that all those projects are indeed **useless for people**, and that money should better be spent on more utilitarian topics. That's where the demonstration of synergy between space science, more specifically planetology and exobiology, and major concerns about this planet (environment, climate, resources, health) has to be put forward. This could appear somewhat artificial and difficult to endorse to some. But it is in fact profoundly legitimate, as science and technology are more and more recognized as powerful means to deal with some of the more prominent problems of our civilization. It's a pity that this state of mind is shared by a part of the scientific community, even in our own ranks, which is against "big science". Yet, history of science demonstrates that "big science" projects work as powerful engines, pushing more science and budgets in their wake, whereas "big cancellations" are synonymous with loss of resources, not of redeployment of them!

Another frequently developed argument against man in space is that **it can all be done by robots**. The deployment of the current philosophy of the NASA Mars exploration program may well reinforce this belief, even more as it succeeds. It is not so easy to explain to lay men why astronauts are really needed to perform effective science on Mars and to convince them that ever smarter spacecraft cannot do much of it. We nevertheless surely have to forward firmly, and in the first place, this statement. But we must at the same time jump to more sensitive—if more eluding—arguments, such as "need to explore", "aim for mankind", "sense of achievement", and so forth. The trick is that it probably works best as a package, encompassing a large spectrum of human aspirations.

Place of International Cooperation in Advocating

From the preceding discussion, it is quite obvious that **globalizing** the prospect of Mars manned exploration and later settlement, that is putting it from the beginning in the frame of humanity accomplishments as a whole, eases considerably the situation of the case relative to public opinion.

As far as **costs** are concerned, even if bad feelings arise from some examples, people easily reason that sharing expenditures lowers the burden per taxpayer and eventually come to the conclusion that it is an unavoidable condition to make it affordable. Overcosts and possible loss of efficiency resulting from cooperation of several countries don't seem to be real concerns for them, compared to the advantages of sharing. So, provided that we are able to convince them that recent errors in international cooperation implementation can be avoided, this seems a quite basic and robust argument to rely on.

We have seen that the best way to manage the point of **usefulness** of space exploration for mankind is probably to exemplify the synergy between this effort and help for humanity global concerns through science and technology progress. Naturally, the consequence of this approach is that a Mars exploration program bears no sense if not perceived as a human endeavor, hence of international essence. **Internationalization legitimates the case for Mars by allowing to link it more closely to other more utilitarian scientific undertakings**.

Concerning the third point, namely **man versus robots**, we concluded that the most efficient way to promote manned exploration should be to wrap the purely scientific and rational arguments in a more sensitive humanitary package. The consequence is to put the aim of the effort at a higher ethical level, which, again, implies consideration of an undertaking of humanity as a whole. That's where international cooperation makes sense and develops its full strength as a vector for our ideas.

Of course, promoting international cooperation consolidates also considerably our case in that it proposes a new way to reinforce **peace and solidarity** between nations all over the world.

POLITICAL REALITIES AT GOVERNMENTS LEVEL

On the Side of the U.S.A.

The context in the U.S.A. presents some contradicting factors which have to be considered if we want to be able to appreciate the political situation relative to Mars exploration. The question of international cooperation and U.S. leadership will be discussed more specifically later. For the moment let us restrict the discussion to the factors which are likely to determine the willingness of a U.S. government to go, and to go alone or via international cooperation.

Space exploration is a matter where some specific characteristics of the American culture are very influential. The most prominent one is undoubtedly **the "new frontier spirit"**, which pervades the mind of most of the citizens of this country. Even if the "need to explore" can be recognized as a (genetic?) trait of human beings, the strength of this feeling in the U.S. still is amazing for people from other countries. It has certainly played a central role in American space program achievements. But, when lobbying politicians, it can backfire against us if we don't keep conscious that it is over-exaggerated within our community of space activists! In others words, U.S. citizens could probably be understood by their legislators when speaking of "the new frontier", but they must keep in mind that this is of little relevance relative to other political realities.

Another historical characteristic of the U.S.A. is their tradition of **isolationism**. This tendency can only be exacerbated in the field of space exploration, where American technological and programmatic predominance is unquestionable. This risk can be easily perceived in some talks in the community of space enthusiasts, which look somewhat arrogant or nationalistic. This of course is all bad if we are convinced that international cooperation is a prerequisite to the conquest of Mars. Fortunately this has been understood by NASA for a long time, and examples of bi-lateral or multi-lateral programs abound, which often have been quite successful (e.g. NASA-ESA programs such as Topex-Poseidon, Cassini-Huyghens, and the Space Telescope, where ESA has a 25% share).

It is unquestionable that the U.S. could go to Mars alone; they have the technologies in hand and they could tailor their budget for it, if political willingness was there. But **this is quite unlikely to happen**, for political reasons. A "nationalistic" remake of the Apollo demonstration is today politically inconceivable: if going alone, the U.S. would give a negative image of technological dominance, without any necessity as they are now recognized as the sole super power, at the moment. On the contrary, by making it possible (and leading it) as an international undertaking, they will reap many benefits, by reinforcing their influence on the world scene and their leadership within the advanced countries, and by giving an image of peace and wealth.

More prosaically, there are other reasons why a U.S. government and NASA will seek international participation in such an effort. The first one, of course, is the prospect of cost-sharing; despite the burdens of setting a cooperative program, the loss of efficiency which could result, the abdication of a part of nationalistic proudness, budget limitations lead to cooperation. This is perhaps one of the more deciding factors for legislators. The other one, to which agencies all over the world are quite sensitive, is the fact that international programs are politically much more robust, i.e. more difficult to cancel. Considering the fact that a manned Mars exploration program is, in essence, a long term program, it is of utmost importance to strengthen it as far as possible against political turmoil which inevitably will happen in a period on the order of a decade or more.

For Other Spacefaring Nations

The situation of other countries is quite different: they are less powerful, less technologically advanced, less engaged in space, and in a position of followers (even if signs of change appear, coming especially from Japan, France and the nations of ESA, which could well become the next "Martian" nations). Furthermore, the "need to explore" is not for them a cultural strength as it is for the U.S.A. This is why politicians of these countries have to be ever more attentive to the rationale of their decisions, in this field of long term (scientific) investment. **Their space activities should prove efficient** for the development of their nations. And that's where participation in an international effort such as the exploration of Mars is really politically appealing.

The first reason, of course, is that **they cannot afford** a comprehensive and durable space exploration program individually; cost as well as technological resources sharing through cooperation is unavoidable for them. That's why European countries have from the beginning merged most of their space efforts within the frame of the European Space Agency and sought cooperation with NASA.

Setting this mode of operations bears another profound incitement, in that cooperative space programs with advanced partners are perceived as an efficient way to **raise the technological level** of their industry, contributing to the development and the competitiveness of their economy. This perception is widely shared, and restriction of space efforts doesn't stem from doubts about return on investment, but only from budgetary constraints. It is in the interest of the U.S.A. to understand this motivation and admit that, through cooperation, they indeed take some risk to contribute to the advancement of potential competitors. This is the price to pay. And that's the one we paid in Europe when we decided to share technologies between all countries participating in the development of the Ariane rocket launcher; namely, beginning in the 70', when it all started, France was more advanced in this field than its partners, and had to spread a part of its know-how throughout Europe. But there was no other way to get independent access to space, and in the end, the commercial success of the enterprise has paid many times the investment. It would be the same story for Mars exploration!

Nations are also anxious to have **their scientists** in the relevant fields (geology, biology, robotics, and so on...) as much as possible involved in what is going on. Scientific communities and their governmental authorities really feel the threat to get overrun. Every affordable means to stay in touch with their American colleagues is considered as very valuable. That is the reason why it is so important for us to consider the scientific (and technological) stakes, because they are so significant when political decisions are to be made. And, again, there is no reason for the U.S. to get reluctant; first because those foreign teams are truly valuable partners for most of them, and second because competition in those matters are less critical than for technologies.

Finally, a major international space program has a real significance as **a means to ease tensions and stabilize the political situation**. This is even more important for the international community as it does not have the same capacity of intervention and influence as the U.S. has. And this is a point to which its public opinion is very sensitive.

SOME CONDITIONS TO FULFILL

So, a long term international space undertaking, such as Man to Mars, is in principle politically desirable and economically wise. Nevertheless, to be acceptable, practical and successful, such a program has to be implemented and managed taking into account the lessons learned from past bad experiences. Many pitfalls have to be monitored.

Reliable Political Commitments

We must first of all seek to set for the program sound and robust political foundations. Such a big and significant international undertaking could well be politically killed if one of its major participants happened to withdraw. The Space Station's long and chaotic history is not heartening on this matter... We have seen a program endangered by U.S. recurrent budgetary disputes, and by a major partner in a great difficulty to fulfill its commitments. We had to pass through a painful and frustrating multi-year negotiation process to set up the rules for cooperation and the management organization. On the other hand, ESA rules, either through so-called mandatory programs or optional ones, deserve attention as robust means to implement cooperative projects. In the case of mandatory programs, once the decision to proceed is made, each ESA country is obliged to contribute, so that there are no more budgetary contribution debates. For optional programs, it is the free decision of each nation to be a participant. But, in any case there is a real multi-year commitment to the program as a whole, within the frame of a budget and general rules.

A significant improvement of the political frame, when compared to the one reached for the International Space Station, is really needed if we want to protect a Man to Mars program from sudden termination. This can be achieved through:

-**A treaty** really committing participating countries; we have to realize that the ISS intergovernmental agreement (IGA) **is not** a treaty, but merely a "gentlemen's agreement". The U.S. pushed for that as to avoid the necessity of ratification by the senate... so good, but is that the best way to give a Mars exploration program the political robustness and status that it deserves? Not really.

-**A detailed intergovernmental agreement** setting the rules for the development and the operational phase of the program; the ISS IGA and related bi-lateral MOU seem OK from this point of view, even if we could dream of a somewhat simpler construct! But for sure sending humans together to Mars is not a trivial undertaking, and precise rules between partners is probably preferable to enforce sound cooperation.

-**Exclusive acceptance** of politically and economically reliable partners. When such a program will be implemented, the risk is to see would-be participants not really committed or able to fulfill their tasks, or to get political decisions made about their enrolment without any relation with their real capabilities!

-**A regulation** enforcing long term multi-year American commitment and clearing the recurrent threat, for their partners, to see the U.S. congress cancel the program every other year. Is it an excessive requirement? It doesn't seem so, if we take into consideration the significance for the partners of an unilateral and abrupt termination.

Efficient Project Implementation and Management

Even if we dream of it, a manned Mars program is not likely to be decided from scratch. Rather, the best and more robust way to it is to **push for a progressive—yet aggressive—approach**, with early, valuable and spectacular results. This is the realistic and efficient way to get the public more and more involved and supportive, and to direct governments toward a full commitment. In this respect, short term affordable robotics missions are fruitful. And early international involvement is of utmost importance in that it will demonstrate the value of cooperation and set the path for the great decisions. In the near term, a quick and constructive cooperative agreement on the 2005 sample return mission will pay immediately, even if science results are to be obtained only in ten years from now!

These preliminary missions, while raising public and political interest, will provide time, not only to refine the technical architecture of the project and to consolidate the technological assets of it, but also to set **sound and efficient management organization and rules**. In this respect we will benefit from the experience of the ISS (this contribution of that program seems to have been overseen);. By the way, there is no reason to estimate that a Man to Mars project is much more intricate and demanding than the ISS. The principles set forth in the ISS MOU between NASA ans ESA look sound; but a simpler, streamlined implementation could hopefully emerge when we have some years of real practice.

Fortunately, we probably can get rid of a "world space agency". On the other hand, it is recommended that the program, where different segments or subsystems responsibilities could be attributed to different industrial teams, stay under control of **a unique fully integrated project team**, empowered by a high-level executive committee. Nothing could be worse than having different parts of the project managed by separate organizations. This really needs substantial improvements from past cooperative practices, where distinct agencies were often contributing in parallel. This will no longer be practicable for an efficient management of such an undertaking.

We still have to **restore agencies's credibility** in planning, costing and controlling large space programs. Let's hope that lessons and improvements will be drawn from the ISS painful experience, and that we will able to greatly improve our way to settle and manage a big program such as Man to Mars. A lot of ingenuity and of political will is needed to get rid of poor negotiation and management performance.

Efficient Industrial Organization

Our focused attention to results, achievements, meaning and best ways to proceed must not lead us to underestimate the decisive role of industry in the making of it all! While the rationale and the decision process for this program are mainly of scientific and political nature, its achievement is basically **an industrial undertaking**. Drafting a sound, well-balanced, meaningful and politically appealing program is nothing if an efficient industrial organization is not set forth. Internationalization could, in this respect, be detrimental if sound basic principles were not respected, mainly if the industrial capacities of participating countries were weakened and companies' responsibilities diluted. These principles are nothing new in traditional business, and examples of successful international programs abound: Airbus airplanes, Ariane rockets, CFM jet engines... The question is for governments to be sufficiently far-sighted and strong to apply them. Here are some of the main points:

- Would-be participating firms should be selected by primes (and primes by the project team and executive committee) **on a competitive basis**. Easy to say, but not so easy to apply when a fair balance of involvement, relative to financial participation, is at stake! We Europeans have experienced this problem all along the Ariane program, and we have succeeded to manage it without endangering the technical and industrial soundness of the project. One of the tricks is to compensate, as much as possible, the discrepancies resulting from the major industrial choices, which have to be made on a rationale basis, by subcontracting among nations at inferior levels.

- Involvement of companies from participating nations should preferably be set forth through constitution of **competing international consortia**, tailored as necessary by adding balancing partners.

- These consortia should **reflect the technical structure** of the segment of the project they are supposed to deal with.

- The simpler overall organization should be sought.

- Minor participating countries could go "piggy-back" with these teams, or as suppliers.

- Relationship between industrial partners, as well as between primes and the project central office, should be **basically contractual**, and with no direct interference from the national space agencies.

- **Fixed price contracts** must be the rule. Sure, it is not feasible to get a total commitment from the industry for achievements ten years ahead, namely when not completely qualified technologies are to be used. But firm criteria of achievement can be defined for intermediate steps, in accordance with the level of risk involved. Incentives and penalties could be devised to reinforce the control of advancement.

INTERNATIONAL COOPERATION AND U.S. LEADERSHIP

Today, the U.S.A. is by far the leader in the field of space science, even if contributions from other countries (namely Europe, Japan) are really significant. At that time *they are* at Mars, and are engaged in a multi-year robotic exploration program which should culminate with the return of soil samples. America is also the country where the will to go there is the most widespread and where several well developed supportive groups for space (and more specifically for Mars exploration with the Mars Society) have emerged. Many of their citizens stand now in the front line to push forward. **Their wish to lead the effort**, or more precisely their feeling that they have to, is hence legitimate. It is quite certain that this position of leadership could not and will not be put into question in the frame of an international Mars enterprise; not only is the U.S. more advanced and more in front of the scene, but they also are supposed to be able to put many more resources than other partners in such a program.

Now, if we conclude that international cooperation is a must, partners not only **should be convinced to go**, but also to accept a form of leadership from the U.S.A. In this respect, the ISS could well yield the reference and experience we badly need! As yet stated, one of the more stringent conditions to get their adhesion is **that they could trust** this designated leader, mainly that Americans will not fool them, either by withdrawing abruptly or due to the incapacity to control costs and schedules. They also need to feel themselves to be in a fair and sincere **context of true cooperative spirit**, in which they are assured to reap the returns they are seeking: participation of their scientific teams, technology instillation and exchange, economic progress, political status on the international scene. Is the U.S. ready to play that game? It's sure that declaring, as President Bush did, that "it is America's *destiny* to lead" is not the best way to prepare for this scenario.

In conclusion, let's note that the case for Mars has to be made **a case for Mars within each potential participating country** (and not only for the country where it originated and where it is yet the most developed). This bears important implications for the form of our supportive action, which we have to think about. The point is that the profound differences in context between nations (cultural background, space history, economical and political situations) must be addressed, and the approach and messages tailored, both toward the public and the politics. For instance, it is clear that the strong "new frontier" cultural motivation of America is much less potent in "old" Europe. On the other hand, the desire to stay on the frontline of technological progress and to help economic development are first rank items.

CONCLUSION

Undoubtedly, the U.S. could do it alone. They have the required technologies in hand. Thanks to the clever improvements in mission architecture and philosophy devised during the last several years, the needed level of resources is no more than that of the Space Station or of the manned space program presently evolving. But, mainly for the political reasons that we have stressed, **it will not happen that way**. The conquest of Mars can only be understood, and hence supported, as an endeavor for the human race. International cooperation really appears as a must. Moreover it happens that it reinforces the prospects of the program. But the arcanes of implementation and the organizational pitfalls of such a scheme should not be overlooked! We should be reminded of the Lessons learned.

An aggressive but realistic and progressive approach should be embraced, with emphasis on early enlightening results (scientific... but also organizational and political ones). A private supportive group such as the Mars Society certainly has the capacity both to raise interest and to accelerate the process. The basic asset for this prospect is to demonstrate affordability and political desirability of the approach. As active members of this Society, we must **seek a sound balance between wild enthusiasm and attention to public and politicians concerns**. Not so easy.

WHY NASA MIGHT NEVER LAUNCH
A MANNED MISSION TO MARS
THE DEVIL'S ADVOCATE

Fred Kelly, MD

When we plan a manned mission to Mars we face formidable medical problems. However, I am confident that these problems can be solved with today's technology and knowledge.

A more troubling problem is that of funding. NASA must rely on political good will for its annual budget. In recent years, the political good will is vanishing and the budget is steadily shrinking. The glory days of unlimited budgets and unquestioning political backing are gone forever. We must find another way.

This paper suggests that a Band-Aid approach won't work. The National Aeronautics and Space Administration requires major surgery. The suggestion is to convert NASA into a non-profit international corporation called WORLD SPACE CORPORATION (WSC) and let it operate under a different set of laws. WSC could hold patents on discoveries and inventions, collect royalties and fees for service, accept government and private grants, and aggressively market products resulting from space research as well as products that can be manufactured in space.

NASA takes pride in its Technology Transfer Division that encourages private use of spin-offs from space exploration. These spin-offs from NASA research have spawned industries that have added billions to the national economy but have not added one red cent to NASA's operating budget. I call this THE GREAT NASA GIVE-AWAY.

WSC's charter would let it reclaim a portion of the financial benefits of space exploration and add the earned revenue to its operating budget. It would eventually have a legitimate chance of becoming financially sound and politically independent. Then, without NASA's political baggage, we might be able to proceed with a comprehensive program to explore and colonize the planet Mars.

Thank you for allowing an old NASA flight surgeon to take the podium for a few minutes today at this historic meeting. For most of my career, beginning almost forty years ago, I was a NASA flight surgeon, and we often played the role of "DEVIL'S ADVOCATE."

A few years back, I recorded an audiovisual history on Doctor Stanley C. White. He was the first flight surgeon assigned to NASA. It was still called NACA then, strictly an engineering organization. Stan's new boss, Doctor Bob Gilruth, didn't know what to do with a medical doctor so he gave him a copy of the RFP for the Mercury capsule to review. A few days later he asked for his comments. Stan said that he thought it was a great idea, but the astronaut should be able to see out of the spacecraft without getting out of his seat. And he probably should have some controls so he could at least get the thing out of orbit if the automatic system failed. The first design for Mercury was strictly a man in a can. Putting a man in the spacecraft just complicated the engineering problem. Flight surgeons have been complicating the engineer's life ever since.

We have come a long way since then and we've seen what man can do in space. But when we go to Mars; we still have some tough problems. Most of you are as familiar with these problems as I am. We've dealt with the same ones since the first Mercury flight: Environmental control systems, radiation protection, weightlessness, and isolation. The list goes on. I dare say the solutions to these problems touch every discipline in this audience. But the flight surgeon is usually the "DEVIL'S ADVOCATE."

Maybe that's changing. Let me read you a direct quote from a book by Michael Collins. Remember he parked the Apollo in a lunar orbit while Neil Armstrong and Buzz Aldrin walked on the Moon? He says: "I'm not quite prepared to wave the white flag of surrender at the medics, but I am willing to admit that their time has arrived. Finally, their concerns and precautions may be valid for the first time, and a Mars crew ignores them at its peril." He also says that: "Life support may be the long pole in the tent."

That's quite a concession from an early astronaut. If Mike were here, I'm sure he would deny it, but I see a lot of long poles in Michael's tent. I'll try to go through a few of them quickly because that's not what I really want to talk about.

ECS: What kind of atmosphere do we need in a Mars spacecraft, in a habitat, a Mars suit, a hopper or a Rover? Do we need the same atmosphere we have in New Orleans or is the atmosphere we have in Boulder good enough? Man can even survive on the top of Pike's Peak, and he can adapt to altitudes even higher if we tweak up the oxygen partial pressure a bit. Do we go with one gas or a mixture? If we plan to live off the land as Lewis and Clark did and as Bob Zubrin suggests, we'll find that inert gasses like nitrogen are scarce on Mars. However, early research tells us that a single gas environment increases the propagation of an accidental fire. I was on the surgeon's console during the Apollo fire and none of you want to repeat that night—especially on Mars. Gus and his crew might as well have been on Mars for all the help that we could give them.

The choice of an atmosphere for a Mars mission will be a compromise like so many other decisions in space flight. The only thing we cannot compromise is safety.

Radiation: How much radiation is too much? As the Devil's Advocate, I must say that all radiation is hazardous. The Mars mission will be leaving the protection of the Van Allen belt, so astronauts will be fair game for any radiation floating around. Do we really want to spend a lot of time in orbit around Mars? Do we want to increase our dosage by swinging around Venus on the way home? Can we build shelters in our transit vehicles and our base camp that will not interfere with operations? How much shelter is enough? Can we do more of the exploration and site construction with robotics controlled from the relative safety of our base camp?

We must do everything possible in mission planning, spacecraft design, and operational procedures to reduce radiation exposure to an absolute minimum.

Weightlessness: The gravitational force on Mars is one third that of Earth, but for the six months in transit the crew will be exposed to microgravity unless we use some method of producing artificial G. We know that microgravity effects every system in the human body. There is much that we still don't know, but man is a very adaptable creature. From our experience on the Mir space station we know that man can adapt to live in microgravity. When we return to Earth we have to pay the piper.

There are many countermeasures that will protect some of the body's systems from microgravity, but artificial G seems to be the most promising. If we can fool the body into thinking it's already on Mars before a Mars landing there should be no adaptive problems. The same goes for the return to Earth.

Isolation: I don't consider isolation as quite the problem that most psychologists do. Remember that we are dealing with an extraordinary group of people. They are self-starters. They have never had the time to do all they want to do. Transit time for them will be something like a Sabbatical leave. I'm sure they will have interpersonal relationships to work out, but these problems are solvable. Crew selection will be one of the long poles in Michael's tent.

I could go on, but I don't see any medical problems that represent a firm "NO GO" for a manned mission to Mars.

Then why might NASA never launch a manned mission to Mars? I believe the answer is spelled POLITICS. That's the real long pole in Michael's tent. NASA is, by definition, a political organization. NASA was born in politics. I don't think there is much doubt that military personnel and military rockets could have gotten into space sooner than Project Mercury did, but politics dictated that a civilian agency lead the way to space. When President Kennedy announced that we would land a man on the Moon and return him safely to Earth by the end of the decade, it was purely a political announcement. NASA succeeded because it had political backing and because a lot of good people laid their lives on the line to make it happen.

Today, the glory days of unlimited budgets and unquestioning political backing are gone forever. The NASA Administrator must go to Congress each year with hat in hand begging for enough funds to keep a diminishing number of programs alive for another year. Each year it's a harder sell. I'm amazed that Administrator Dan Golden has been able to keep the programs he has. I think he is the most dynamic administrator we've had since James Webb. Despite his efforts, NASA seems to be in danger of being reduced to an organization that must respond to the political whims of Congressmen who are more interested in the next election. We've all seen good scientific programs canceled for political reasons. We've seen government contracts awarded for political reasons rather than merit. It's nothing new. Politicians have always protected their turf. Do you think it was coincidence that the Johnson Space Center ended up in LBJ's home state?

There is no doubt that NASA under the heroic leadership of Administrator Dan Golden is on a scientific roll. Shuttle missions are adding daily to our knowledge of this planet and the near space environment. The Hubble Space Telescope is performing better than we had any reason to expect. It is looking far into deep space and giving us pictures of planets in other solar systems and of embryonic stars. Unmanned missions are giving us a better understanding of planets in our own solar system. We are finding evidence of water on our own moon and on a moon orbiting the planet Jupiter and now on Mars. Spin-offs from space exploration are bringing improvements to the quality of life in every nation on Earth.

No other agency of the U.S. Government has had such a profound impact on the Nation's past, present and future as the National Aeronautics and Space Administration, and certainly not for the dollar value. NASA has more than paid its own way for the last forty years. Thousands of entirely new industries have been spawned by spin-offs from discoveries and inventions related to space exploration. Billions have been added to the national economy and many small and large companies have grown prosperous using technology developed by NASA research projects. Unfortunately, this has not added one red cent to NASA's operating budget.

How do we solve this problem? For an answer I have to go to a little science fiction. Bear with me, this is not without precedence. After all, Jules Verne went to the Moon a long time before we did, and writers have been going to Mars since the last century. I believe it was Edgar Rice Burroughs who said that he wrote science fiction not to predict the future but to change it.

I don't know how many of you have tried your hand at writing science fiction, but I recommend it. It's therapeutic. Your systems always work and nobody ever argues with you. As far as I can tell there's not much money in it, so don't quit your day job unless you already have a good retirement. But it's fun.

I explored this problem in my new book, THE MARS JOURNAL. In the book, NASA is in a terminal spin. Russia defaulted on the international space station, and Congress did not have the votes to take up the slack, so the whole program went down the tube. The next year political opponents get the upper hand and, when it looks like NASA will not survive another congressional debate, the Senator from Texas comes up with a better idea.

The idea is to convert NASA into a non-profit international corporation called WORLD SPACE CORPORATION (WSC). Then he finds a dedicated Director and a Board of Directors with a proven record of fiscal management and innovative marketing skills who will run space exploration like a business. WSC operates under laws that allow it to hold patents on discoveries and inventions, collect royalties and fees for service, accept government and private grants, and market products resulting from space research as well as products that can be manufactured in space.

The Senator convinces each government that had supported the international space station to continue their support at the same level for a five-year trial period to see if the concept was workable.

Needless to say, at the end of five years, WSC is well on its way to fiscal independence and ready to launch a manned mission to Mars. You might want to read THE MARS JOURNAL when I finally get it published to see how they pulled it off and what they found on Mars.

Let's go back to NASA and a little bit of reality. I spent most of my life working for NASA in one capacity or another and I hope history will prove me wrong on this. I hope NASA will survive to plant its emblem on the surface of Mars, but I believe it has some serious flaws. I don't believe you can run a space program if you have to spend half of your time before Congress begging for money, or if you have to consider the political implications every time you propose a program or award a contract.

So how do we make NASA fiscally independent and free of political influence? I have one suggestion.

NASA and all other government agencies have a policy of sharing the results of any research they conduct. NASA calls it their Technology Transfer Division. I call it The GREAT NASA GIVE-AWAY worthy of Tom Brokaw's Fleecing of America segment.

I'll give you one small case in point. There are many others. I hold the patent for the floating EKG and EEG electrode. It resulted from research I did in the late 1950s while I was in the Navy. Before that time we were able to get a readable EKG only if the patient was in a shielded room and remained perfectly still. At Point Mugu we needed to monitor active pilots flying the F4-H/Sparrow III weapons system. The floating electrode worked. NASA refined a similar electrode and used it on Mercury and other space programs.

Now, every time you have an EKG, every time you have a stress test, every time you ride in an ambulance, or go into an emergency room, an OR, or an intensive care unit they slap electrodes all over your chest. And they charge you big bucks each time they do.

I didn't expect to benefit directly from this patent because I belonged to the Navy, and they've treated me well. But shouldn't there be a way for the Navy or NASA to benefit from this type of research?

Each year NASA publishes a book called SPINOFFS that documents the new technology that has been transferred to private industry. There are tens of thousands of examples from pacemakers to digital imaging, from CAT scans to MRI, from telecommunications to industrial robotics, to weather satellites. You get the idea.

If NASA could reclaim a portion of the financial benefits of space exploration rather than giving it all away, it would have a legitimate chance of eventually becoming financially sound and independent. Future space programs might be based on scientific rather than political merit.

I was able to come up with a solution in my book because I had some extremely smart and innovative people working the problem, and an exceptional politician backing the solution. Also, the writer was pulling all the strings. Now, I realize that the solution might not be so easy in the real world. But unless NASA is able to find a solution to its funding problem, I believe that the red, white and blue emblem of the National Aeronautics and Space Administration will never be planted on the Red Planet.

MARS NEEDS GUITARS*

M. R. Jardin†

A multi-city tour of rock concerts is proposed as a means to both intensify and focus support for space exploration and to raise money for space exploration research and development, particularly for Mars exploration missions. Many rock artists are genuinely interested in space exploration as is evident from their CDs, song titles, and performances. If several of these artists are brought together for a series of performances, the public relations benefit could be substantial as could be the fund-raising potential for specific Mars exploration projects or prize funds. The current popular interest in space exploration is sometimes paranoiacally misguided, but still provides a great opportunity to apply this approach to public relations for the space exploration community. This paper describes the type of shows that could be produced and discusses some quantitative aspects of the public relations and fund-raising benefits. Depending upon the calibre of rock artists involved, and the number of shows performed, the number of people reached through advertising and the news media would be in the range of one to ten million. The range of fund-raising potential is a strong function of many highly-variable parameters and could vary from a few thousand to a few million U.S. dollars.

INTRODUCTION

*It takes photon power/And eight minutes of an hour/To make it to our sun/And I know it sounds weird/But it'll take you four years/To make the next one/Expanding border/That's the sauce of chaos/And that's an order/.../So stomp your feet/And clap your hands/Get out of your seat/And do a little dance/Lift up your voice/And sing with glee/Now listen carefully to me/Desert your quarters/Behold the pie in the sky/And that's an order!/That's an order!/**That's an order!***

-- from the song *Pie in the Sky*[1], **Frank Black**

There is vast interest in space exploration in popular culture which, if tapped into, could be used to remind the world that we can and should be going to Mars. One way to achieve this is to produce a series of space-theme rock concert events to attract coverage from the popular

---

* This report is not related to Mr. Jardin's doctoral research at Stanford.

† Ph.D. Candidate, Department of Aeronautics and Astronautics, Stanford University, Stanford, California 94305-4035. E-mail: jardin@leland.stanford.edu.

media and to possibly earn enough money to fund a small hitch-hiker payload mission in the process.

As will be shown, producing such shows is entirely feasible. In the next section, several bands are listed that would be good candidates to assume headlining roles in the production of a space theme event. Since these bands have demonstrated personal interest in the exploration of space, it is reasonable to consider the possibility that one or more of them would be willing to help with the production and promotion of the event. Rock bands are notorious for helping to promote benefit concerts for different organizations, so why not for the Mars Society?

A rock concert is proposed instead of any other type of event because of the huge infrastructure of news media already in place to cover and promote rock music. With many different music magazines (e.g. Rolling Stone, Spin, B.A.M.), music cable channels (e.g. MTV, VH-1), and television programs in addition to more general news media, such an intriguing event as a popular rock band headlining a Mars exploration theme tour would certainly generate some positive press. With a multi-city tour of concerts, not only could the message be conveyed that *Human exploration of Mars is possible now*, but a substantial amount of funding might be raised.

In the next section, some of the connections between popular culture and space exploration will be illuminated. This is followed by a description of the type of inspiring space theme event that might be produced. In the third section, an analysis of what is required to produce such a show is presented.

SPACE IN POPULAR CULTURE

Take me somewhere cool/Take a long time/Take as long as it takes/Take me somewhere cool/I want to touch that face/I have seen the face/I have seen that vacant look/Oh, it's silly I know/Am I a fool?/.../I have waited my whole life/And I really want to go...

-- from the song *The Marsist*[2], Frank Black

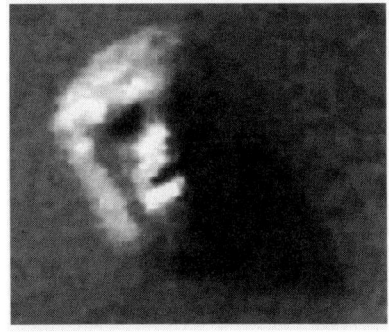

One can easily think of examples of popular culture's fascination with space. The first examples that come to mind might be the television shows: *Star Trek* (the original, *Next Generation, Voyager, Deep Space 9*), *Babylon 5, Third Rock from the Sun, The X-Files, Mystery Science Theater 3000, The Cape, Homeboys in Outer Space*, and others. Then there are the films (just a few of the more recent ones): *Contact, Independence Day, Mars Attacks, Apollo 13, Deep Impact, Armageddon, Species* (I & II), and many more.

It may not be as obvious as in the case of films and TV, but space exploration themes are also prevalent in music. To be convinced of this, just look around at any CD shop and take note of how many band titles, CD titles, song titles, and CD cover graphics are based on space themes. Or, try logging-on to your favorite on-line music service on the internet (e.g. http://www.cdnow.com) and searching for song or band titles with "space" or "Mars" in them.

A short list of current bands that the author of this paper listens to that are directly or peripherally associated with space themes appears in Table 1.

Table 1
ROCK ARTISTS IN SPACE

Artist	Degree of Space Connection	Artist	Degree of Space Connection
Frank Black (and Pixies)	**very strong**: Beginning with the last two albums from his previous band, the Pixies, all CDs include songs about the desire to travel in space, specifically to Mars! Song titles include: *Big Red*, *The Marsist*, and *Bird Dream of the Olympus Mons*. Third album, *The Cult of Ray*, named in reference to Ray Bradbury. One of his songs appears on the X-files show soundtrack.	Brian Eno	**moderate**: Wrote the ethereal and inspiring soundtrack to *For All Mankind*, a documentary of the Apollo program.
		Man or Astro-Man?	**amusing**: Album titles include *Live Transmissions from Uranus*, and *What Remains Inside a Black Hole?*
		Mars Needs Women	**amusing**: lots of retro-space graphics.
		Foo Fighters	**peripheral**: The band's name is derived from the WWII-era French term for UFOs. They contributed a song to the *X-Files* show soundtrack.
They Might Be Giants	**very strong**: TMBG were "Musical Ambassadors for International Space Year, 1992". Albums include *Apollo 18*, and *Why does the Sun Shine? (The Sun is a mass of incandescent gas)*.	Soul Coughing	**peripheral**: Contributed a song to the *X-Files* show soundtrack.
		Meat Puppets	**peripheral**: Contributed to the *X-Files* show soundtrack.
Love & Rockets (David J)	**strong**: Occasional references to space, including a song entitled *Holiday on the Moon*. Band member David J. has expressed great personal interest in space exploration.	Perry Ferrell (Jane's Addiction & Porno for Pyros)	**peripheral**: He created the ENIT festival (an idea taken from a book entitled *Cancer Planet Mission*) which was to "unite beings of various worlds in a festival of harmony, enlightenment and cultural exchange".
		Hoodoo Gurus	**peripheral**: The title of this paper happens to be the same as one of their CDs and songs.

WHY MARS NEEDS GUITARS

*...I got to somewhere renowned/For canals and the color red/Lot's of guys who shake their heads/Rhythmically to the sound/This ain't the planet of sound/.../I met a guy in a rover/He said it's one more over/It's just there where you're bound/****This ain't the planet of sound!***

-- from the song *Planet of Sound*[3], Pixies

Inspiration. Of course there's that pesky 99% perspiration required to actually get to Mars and back, but some inspiration among the masses is sorely needed right now or Mars will have to wait.

As illuminated in the previous section, popular culture is strongly influenced by visions of space exploration. The interest, excitement, technology, and even the capital—all the necessary ingredients—exist, but need to be put together. The trouble is that in spite of three decades of technological advancement beyond that required to take humans to the Moon and back, the public has been duped into believing that going to Mars is not possible right now. It has become a maddeningly vogue notion that there are more important demands on society's time and money than exploring space, and this is rarely ever questioned anymore.

Those with the technological background to know better don't subscribe to the pessimism and are ready to go, but there aren't enough scientists and engineers to pay for the exploration of Mars. It's everyone else who needs to be reminded that our destiny awaits us on the red planet. The time for unconventional approaches to public enlightenment has arrived. There are 5 billion screaming fans waiting for a good show. This is why Mars Needs Guitars.

THE CONCEPT

...But I dream of your red dress/Riding down these dry canals/Oh baby it's not Europe/But I'm sending you money/I'm sending you lots of money/So you can buy a ticket/You'll be my Martian honey all the days/They got one leaving today/And it's going away...

-- from the song *Lovely Day*[3], Pixies

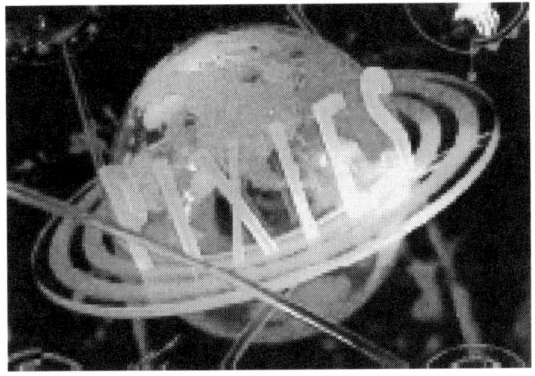

Since inspiration is the goal of Mars Needs Guitars, the event itself must be inspired much more so than a typical rock concert. The space theme should permeate every aspect of the event. In this section, one possible incarnation of a Mars theme concert is described.

Mars Needs Guitars is not a frivolous concept. In fact, a similar event to the one described below came very close to happening as a 25th anniversary of Apollo 11 show called *The Moon Rocks!*. The theater, financing, decorations, advertising arrangements, ticketing, and supporting artists were all lined up, but signing a headline band for the specific dates on or near the anniversary was not possible. Although that event did not happen, the experience gained in planning *The Moon Rocks!* demonstrated that producing this type of event is feasible.

The show concept outline is as follows:

- Theater decorated according to Mars theme, and filled with displays and demonstrations of how to get to Mars.
- Many video screens around the theater showing rapid clips of space exploration images (space cartoons, Apollo footage, Pathfinder images, space films, etc.).
- A space-art gallery will also be set up in the theater. [For *The Moon Rocks!*, NASA- commissioned artist Andreas Nottebohm was scheduled to display his work.]
- Space theme trinkets, t-shirts, posters, etc. are for sale throughout the theater.
- Space theme music playing before and after the show, and during intermission.
- The bar is serving such inspiring cocktails and drinks as: Rocket-fuel (vodka & Red-bull), Mars-tini, Sex-on-the-dust, the Red-eye, Marsen beer (Märzen-style beer), Hubble-Vision, Moonshine, Apollo Ale, etc.
- Comedian Master of Ceremonies performs 20-minute opening set. [The comedian that had been scheduled to do *The Moon Rocks!* Was known for doing Star Trek impersonations.]
- Modern dance performance of Tom Ruud's Mobile, a slow dance involving one male dancer and two ballerinas performed to music from the *2001: A Space Odyssey* soundtrack and evoking visions of weightlessness: One giant pirouette for mankind!
- Intermission
- An appeal by a spokesperson of the Mars Society for people to get involved in the drive to go to Mars.
- A short warm-up set by the Comedian M.C.
- Spectacular introduction of the headline band and 1.5 hour show [Visuals and effects made to look like sunrise from Mars orbit. Mock-up of Mars lander descends to stage]
- Post-show party.

Conveying the excitement envisioned for a *Mars Needs Guitars* show in a short conference paper may be an exercise in futility, so the reader is invited to allow his imagination to wander a little in what follows. Below is a stream-of-consciousness exposition of what one might be thinking and experiencing while at *Mars Needs Guitars*...

THE PRODUCTION

Producing a rock concert can, in some instances, be more difficult than rocket science because of the uncertainty involved. However, as in an engineering project, with careful planning, the show can go on.

A full cylinder of CO_2 dumped out to give the effect of a rocket launch!/Deafening applause/Red Madonna-smoke hanging over the stage like a Martian dust storm/Need another Mars-tini to wash down that astronaut ice cream -- tasty, but kind o' dry/The comedian M.C.'s Captain Kirk impression was insane!/Hit in the head with a Phobos beach ball/Did I really just see a slow-motion ballet?/Ballerinas are stunning!/Blue light, slow and ethereal motions.../The show storyline is incredible: 5...4...3...2...1...Initiate trans-Mars injection!/Image of the Sun rising over the surface of Mars, revealing Olympus Mons, Valles Marineris.../ Can't keep up with the images flashing around the stage... launches... dust-storms... little Sojourner stuck on Yogi... Shoemaker-Levy-9 crashing into Jupiter... Al Shepard shanking an iron shot on the Moon... The V2 rocket... Goddard... Marvin the Martian... O.J. Simpson in Capricorn One-- Ha-Ha!... Tickertape parade for the Apollo 11 crew/The band descends in a landing-vehicle mock-up! Cool!/The band rocks!

You can't get a suntan on the Moon/But I wouldn't mind a holiday there/They say, "You can't get a suntan on the Moon"/But I wouldn't mind a holiday there.

-- from the song *Holiday on the Moon*[4], Love and Rockets

From experience gained with the aforementioned *The Moon Rocks!* show, the steps involved in producing a 300 - 3000 seat show are as follows:

- Hire a concert promoter to contact agents, band managers, and for advice on all aspects of the show.
- Create a rough initial budget.
- Find a band that is interested in the show and available at the appropriate dates (be flexible!).
- Make arrangements to rent a theater.
- Solicit sponsors
- Finalize show dates and get contracts signed.
- Create a detailed revised budget.
- Make advertising plans
- Produce theater decorations, show posters, T-shirts, etc.
- Take care of miscellaneous production details (insurance, box office, security, back-stage catering, etc.)

When finished, a show budget might appear similar to the following example from a proposed 3000-seat theater show:

Table 2
EXAMPLE BUDGET

Example Budget

- Rent .... -$3500.00
- Rent (8% of ticket sales estimate) ..... -$3600.00
- Sound .... -$2000.00
- Lights .... -$1000.00
- Box Office (2 people) .... -$175.00
- Advertising (print media) .... -$2100.00
- Advertising (posters and art) .... -$1100.00
- Talent (Headline Band) .... -$5000.00
- Talent (M.C. and opening acts) .... -$1500.00
- Talent Backstage .... -$200.00
- Security (19 people) .... -$1400.00
- Production (stage mgr., stage hands) -$1850.00
- Insurance (0.40 per person) .... -$1400.00
- Petty Cash .... -$500.00
- Decorations (audio/visuals) .... -$2000.00
- Corporate Sponsorship .... +$4000.00
- Concert Gross ($18.00 per ticket).. +$36000.00

- net .... +$12675.00
- Promoter (25% of net) .... -$3168.75

- **Producer Net (75% of net) .... +$9506.25**

So, $9500.00 for one show... not bad! Well, not so fast. The $9500.00 estimate is for 2000 tickets sold. If the show sells out at 3000 tickets, then the net proceeds would be nearly

$30,000! However, if only 1000 tickets sell, then the producer is in-the-hole nearly $9000.00. While this risk is acceptable for commercial event producers, clearly some risk management would be necessary for an organization like the Mars Society.

Risk management may be accomplished in several ways. The ideal situation is that the headline act agrees to do the show as a benefit to the Mars Society, and therefore agrees to accept payment as a percent of net profit. Such arrangements may sometimes be made with theater owners too. Another way to minimize or remove financial risk altogether is to generate more sponsorship. Most rock concerts are sponsored by beer companies or other large consumer product companies. Since it is in the best interests of companies like Boeing and Lockheed-Martin to have strong public support for large space missions, they might also make ideal sponsors.

Yet another method for minimizing financial risk is through sales of peripheral, but lucrative items such as T-shirts and posters. The sale of just 500 T-shirts for a net profit of $10.00 per shirt ($5.00 wholesale cost, sold @ $15.00 each) generates another $5000.00. While total elimination of financial risk is difficult to achieve, with proper planning and use of the ideas outlined above, financial risk need not be a show-stopper.

THE MESSAGE

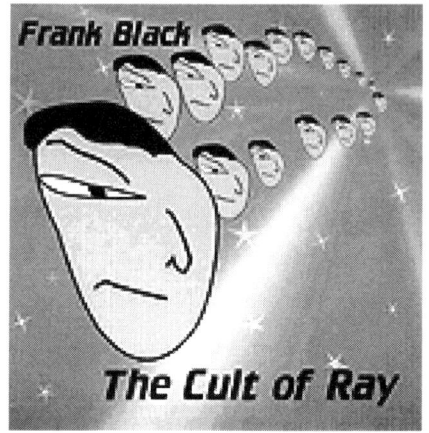

Have you heard about Big Red?/They even brought a beebread rig/To help the flowers in mean space/They're trying to make that place green/Hope the bees will take away the storm/Hope the trees will take away the storm...

--from the song *Big Red*[1], Frank Black

Produce an interesting and newsworthy show that combines popular culture with an appeal for Mars exploration, and the press will come. If some of the larger national and international media pick up the story, potentially millions of people can be reached. For example, the Nielsen ratings show that during a recent airing of a repeat cartoon show on MTV, nearly three million viewers tuned in. Typically, MTV will run news segments during show breaks several times a day for about a week. That's a lot of people to reach with a Martian message.

There is also the potential for receiving coverage from the national and international news media (e.g. CNN, MSNBC), local and national newspapers, magazines (e.g. Rolling Stone, Spin, Details) and radio programs. Large national newspapers such as The New York Times or USA Today have circulations over one million, and the larger national music magazines have circulations of over one hundred thousand.

Even without the benefit of media coverage, at the very least, each show with seating for between 300 and 3000 people in a major metropolitan area will reach tens of thousands of people through local radio and print advertisements, and possibly as many as one million. In the San Francisco Bay Area, for example, the local newspaper has a circulation of over half a million.

CONCLUSION

...Sun shines in the rusty morning/Skyline of the Olympus Mons/I think about it sometimes/Sun shines in the rusty morning/Once I had a good fly/Into the mountain/I will fall

-- from the song *Bird Dream of the Olympus Mons*[3], Pixies

The 30th anniversary of the first manned lunar landing is now less than a year away, yet the first Mars landing still seems to be decades away. Depressingly, the public is preoccupied with news about divorced silicone-filled actresses, congress-shootin' lunatics, and public officials with lipstick on their zippers. As the years progress, we will unfortunately have plenty of interesting tales of depravity, but there will only be one first-Mars-landing. That story has yet to be written.

It is time to try some unconventional approaches to raising awareness of the need to explore Mars. An idea has been proposed in this paper for producing a series of space rock concert events in an effort to reach a large number of people and to potentially raise some of the money to fund Mars research missions. It has been argued that by producing a show with enough interesting qualities, coverage by national media sources could take the Mars message to millions of people.

OTHER SPACE-INSPIRED RECORDS

Frank Black, *Frank Black*, 4AD/Elektra, 1993.

Pixies, *Bossanova*, 4AD/Elektra, 1990.

They Might Be Giants, *Why Does the Sun Shine? (The Sun is a Mass of Incandescent Gas)*, Elektra/A.D.A., 1993.

Love and Rockets, *Earth, Sun, Moon*, Big Time/RCA, 1987.

Space Hog, *Resident Alien*, Elektra Entertainment, 1995.

Man Or Astro-Man?, *Experiment Zero*, Touch &Go Records, 1996 (many other space recordings too).

Mars Needs Women, *Sparking Ray Gun*, WEA/Sire/Discovery/Antone's

Brian Eno, *Apollo — Atmospheres & Soundtracks*, EG Records, 1983.

Foo Fighters, *Foo Fighters*, Roswell Records, 1995.

Hoodoo Gurus, *Mars Needs Guitars!*, Big Time Records/EMI, 1985.

Various Artists, *Songs in the Key of X: Music from and Inspired by the X-Files*, Warner Bros., 1996.

REFERENCES

1. Frank Black, *Teenager of the Year*, 4AD/Elektra/Virgin, 1994.
2. Frank Black, *The Cult of Ray*, American Recordings, 1996.
3. Pixies, *Trompe Le Monde*, 4AD/Elektra, 1991.
4. Love and Rockets, *Express*, Beggars Banquet, 1985.

MAR 98-017

YES, BUT WILL THE PEOPLE SUPPORT US? ENGAGING OUR CUSTOMERS IN THE MARS EXPLORATION ADVENTURE

Humboldt C. Mandell, Jr., Ph.D.

Sending people to Mars will require a high level support from the American public. Although NASA has now identified affordable ways to send humans to Mars in the near future, the extent of public support is largely unknown. What would the American people want NASA to do *if* they were fully informed of what can be done? The NASA Strategic Plan demands a better understanding of what our customers, the citizens of the nation, expect us to do. The Administration has called for a new "Dialog with America," to involve the people more fully in the determination of what their government does in all of its agencies.

To address these issues, NASA is taking some unprecedented steps, and has formed a "Customer Engagement" group, which is actively seeking ways to better communicate with the people, to inform them of what is possible, and to engage them as customers and stakeholders in an organized campaign to promote human Mars exploration. A first step is to find out what an enlightened electorate would want NASA to do.

There have always been polls to measure public attitudes, but, for many reasons, they have had little influence in the casting of public policy in space. A primary reason is that they draw information from an uninformed population.

But there is now new technology for measuring public attitudes, in this case, public support for space exploration. These methods involve the sampled population actively in the deliberation process, and will tell us what an enlightened population would want NASA to do. Deliberative Polls are expensive, and new funding mechanisms have been developed, including a new Foundation for Space Exploration.

Many things remain to be done, including the development of sound business plans. The skills required to address these issues, and to involve our customers more closely, do not all exist within NASA. To that end, new partnerships are already being developed with traditional and non-traditional sources. The academic community is playing a major role, and dialog is already being held with the international community.

INTRODUCTION

Occasionally overlooked in our zeal to send humans to Mars is the question of who will pay for it. By now, it is community folklore, or at least conventional wisdom, that there can never be another "Apollo" program, wherein the President of the United States issues a mandate, and that mandate is supported by both houses of Congress, to send humans on a new all-American space adventure. Without debating that issue, it is clear that a return to Apollo budgetary levels is unlikely, and that any realistic plan for sending humans to Mars must not rely on those levels of funding from government sources.

NASA now has a major thrust to identify "affordable" Mars missions. Affordability is in the eye of the beholder, but a realistic assessment of affordability would be for a program to cost no more than allowed by current NASA budgetary levels, plus whatever the international community and the private sector are willing to contribute.

Missions which meet these criteria have been identified, but to date, the costs have not been verified. Neither has NASA fully come to grips with the fact that, to achieve affordable costs, the cultural changes initiated by Administrator Goldin must continue and even accelerate.

Some critics say that NASA is indeed incapable of mounting a human mission to Mars, for a number of reasons, one of them being the inability to make a major change to the management paradigm.

There are others who say that the human Mars mission should be done privately, and that when humans eventually go to Mars, the motives will be purely commercial.

Others say that the human Mars mission must be international, that the United States alone can not mount such an effort.

Although I personally disagree with these points of view, it is not the purpose of this paper to debate them. But it must be conceded that it will be extremely difficult for NASA to make the necessary management changes, harder, perhaps, than the considerable engineering challenges involved.

Joe Rothenberg, the Associate Administrator for Space Flight in NASA, has stated four conditions which will have to be met before humans can explore Mars. These are: a compelling scientific or exploration rationale; a strong potential for high commercial return; strong public support; and credible (affordable) cost estimates. Credible costs, according to Mr. Rothenberg, plus two of the remaining three conditions, will be mandatory. He is undoubtedly correct. This paper will deal primarily with seeking and cementing the necessary public support, and assuring program affordability, a necessary ingredient in public support.

THE NASA CUSTOMER ENGAGEMENT TEAM

Currently, no one really knows what the customers want. NASA is not alone among Federal agencies in not understanding its customers well. The Administration has recently called for a Dialog with America to be accomplished as a follow on to the successful National Performance Review.

In NASA's case, do the people support human Mars exploration or not, to the extent that they are willing to pay for it? No one knows. There is abundant anecdotal evidence to support the fact that the people are highly interested. The billion "hits" on the Mars Pathfinder web site are often cited as evidence. There has been enormous press coverage of Pathfinder, Mars Global Surveyor, and Lunar Prospector missions. Television polls show strong support for human space exploration. But these are only interesting trivia, and certainly not the evidence required to convince those who must appropriate the necessary funding to do so.

There are large uncertainties which must be answered. Detailed demographics of public support must be measured. We must know what the people would want us to do *if they knew what we are able to do and what the true costs to them would be*. We do know that there is a large public misperception of how much money NASA spends relative to other Federal agencies.

Not only must the customers and other stakeholders in the mission be understood; they must be educated in what is possible and what it would cost them. Only then can they make the enlightened decisions required to say what they want the Congress to do with their money. The customers must become involved in the process. Toward that end, NASA has appointed a Customer Engagement Group to find out what an enlightened electorate would want NASA to do in planetary exploration, to enhance the ongoing constituency building process, and to involve the customer in planning what we do. Only if that is done will the necessary funding be made available.

IMPORTANT QUESTIONS: THE NEED FOR A BUSINESS PLAN

A number of important questions must be answered for our customers before humans will travel to Mars. When should humans go? (It is considered a certainty that they will go some day.) What should they do while they are there? What is the role of the Moon, and lunar exploration, if any? What is the role of the private sector? What are the reasons for private sector involvement? How can money be made? What is the proper role of the international community, both private and public sectors? What are the proper roles of governments? What technologies should governments fund? How much will the venture cost, and how can the money be raised? How should it all be managed? How should the ideas be marketed? How should the public be engaged? Is new legislation required? What is the role of NASA in all of this, and how can the necessary cultural changes be made within the NASA institution to enable all of this to happen?

A business plan is needed to begin to answer these questions. NASA is pursuing such a plan with in-house and academic assistance.

The plan for going to Mars will probably resemble more closely the startup plan for a multi-national company than for a traditional space program. NASA has very little expertise in the writing of business plans, particularly ones involving public/private partnerships and the cooperation of several nations and their industrial sectors. The complexities presented by the many cultures, languages, ways of doing business, national and international laws, and governmental differences make the development of the business plan one of the major challenges of the entire program.

Couple this with NASA's new, lower-cost ways of doing business, where the agency does not itself fully understand the implications of imposing new management practices on its forty-year-old bureaucracy, and the problem becomes even more complex.

So NASA has recognized that it genuinely needs outside skills and expertise to complement its own. Many of those skills are available in the academic community.

Skills required include business planning (domestic and international), public affairs, public administration, marketing, advertising, and communications. It may only be within the academic community where *all* of the needed skills reside for a plan requiring so many disciplines.

The "NASA Means Business" project will enable business schools to participate in a "hands-on" manner in the actual planning of the NASA Mars Exploration enterprise. Several prestigious universities and management experts associated with these universities are expected to participate. The results will be a fully integral part of the NASA program management planning process. This activity, entitled "NASA Means Business," will have the additional effect of building a larger constituency and advocacy for the Mars exploration mission.

NASA intends to participate fully in the development of the business plan, and will provide access to experts in all aspects of mission planning and vehicle design. And because of the high levels of interest by NASA management, it is expected that key NASA managers will be directly involved in this project.

EARNING PUBLIC SUPPORT

The Customer Engagement Group is implementing a systematic approach to answering the key questions, and assessing and developing public support for human space exploration, in support of strategic planning activities currently underway in the NASA Office of Space Flight and at the Agency level.

A private consulting firm has been hired to provide some of the expertise which is not resident in the Agency, in particular the skills needed to assess the needs and desires of NASA's customers and stakeholders, and to assist in developing the campaigns to engage these groups.

And a large advertising firm is becoming closely involved with NASA, providing us with badly needed expertise in how to better get our messages out to our customers.

DETERMINING WHAT THE CUSTOMERS NEED AND DESIRE

It is presumed that once NASA knows what its customers need and desire from the agency, it will deliver high value products, and will be rewarded by resources. But previous efforts to determine customer needs have not been very successful, for a number of reasons. First, respondents have often been self-selected, resulting in very biased results. Even when randomly chosen, respondents are often essentially ignorant of the factual background of the questions which they are asked, and, with knowledge, would answer quite differently.

But there is now new science to allow better results from polls. The Deliberative Poll™, developed by Dr. James Fishkin of The University of Texas, has been successfully utilized in Europe and the United States to measure the true attitudes of the electorates when they are familiar with the issues. A large sample is chosen completely at random, representing all ages, ethnic groups, and economic strata of the nations. The sample is physically assembled at a common point, administered scientifically-developed instruments, exposed to experts pro and con on all issues, and allowed to debate and deliberate each issue in depth. Then the sample is re-polled to determine shifts in attitudes. What has been shown by this process is that an enlightened group will have substantially different attitudes from the same group prior to their enlightenment.

But Deliberative Polls are expensive, primarily because of the costs of bringing all of the people together at a common point and feeding and housing them.

Therefore, NASA and others are in the process of developing funding mechanisms. Funding may come entirely from the private sector, but at this point, some public funding has not been ruled out. A private foundation, the Foundation for Space Exploration, is actively involved in the process of fund raising.

LEARNING WHAT THE STAKEHOLDERS NEED AND WANT

Stakeholders such as the Congress of the United States are equally important. Their needs and wants must also be addressed. Therefore, NASA must focus on the issues which are important to the key stakeholders, and demonstrate how the NASA programs address those issues. If other programs are needed, they must be initiated. If some are irrelevant to the needs and wants of both customers and stakeholders, they must be eliminated and the resources utilized on things which are important to the customers and stakeholders.

This will also require expertise outside of the usual NASA in-house skill set.

INVOLVING OLD FRIENDS AND NEW ONES IN UNCONVENTIONAL WAYS

NASA recognized early in the planning process for a Mars initiative that the skills required to bring the activity to fruition do not all reside within its own organizational walls. Already, substantial dialog is being held with technology developers in government laboratories and the private sector. The Department of Energy has a resident engineer at the Johnson Space Center to assist in identifying technologies useful to Mars exploration.

NASA Space Act Agreements (SAA's) are being written with the private sector to bring innovative (and proprietary) technologies to bear, and to thereby reduce the risks and costs of Mars exploration.

Very meaningful dialog is also being held with the international community and a number of promising partnerships have been identified, which should significantly decrease the financial burden on the United States taxpayer.

Because of its current hiring freeze, and a downsizing of the work force, there are some critical skills which NASA does not have. SAA's are also being pursued to create partnerships to supplement the NASA skills. One such agreement involves a Mars Precision Landing institute created at The University of Texas at Austin to provide experts who can supplement the NASA work force and solve the critical problems associated with close proximity landing of two or more spacecraft on the Mars surface.

The academic community is also involved in in-line NASA mission design activities, further involving that important sector of our customer base in the process of developing the human Mars mission. For example, designs of Mars drilling devices, developed by Texas A&M University in conjunction with consultants from the petrochemical industry, have shown significant promise.

And, as stated above, the academic community will be offered the opportunity, beginning in the Fall of 1998, to participate in the development of a business plan for the NASA Mars mission.

NASA has traditionally had a number of university partners, and intends to not only maintain but to expand and enrich these relationships. To assist in aligning university activities more closely with those of the human space exploration program, a new HEDS-UP (Human Exploration and Development of Space University Partners) project has been initiated, managed by the Texas Space Grant Consortium and the Lunar and Planetary Institute.

AFFORDABILITY

If Mars missions are to happen at all, they must be affordable. Affordability is somewhat in the eye of the beholder, but the general assumption is being made that whatever NASA funding is required will have to fit within its current budget or less.

This will be a major challenge. Earlier cost estimates, made with cost models based on previous NASA experience, produce cost levels which can never be funded. Approximately an order of magnitude reduction in costs will be required from those produced by the models. NASA research has established that costs are much a function of the management style of the organization performing the job. Products are being sold to the private sector for prices far below those which NASA has had to pay. The difference is in how the items are procured, or how the development programs are managed. NASA knows the parameters of low cost management. But a major challenge will be to implement the very substantial cultural changes required to make low cost Mars programs possible.

This challenge must be met, however, if NASA's customers are to perceive that Mars exploration is indeed affordable. Partnerships are being pursued with experts in cost analysis and organizational cultural change to assure that when the time comes to send humans to Mars, the people of the nation will be convinced that the venture will be affordable, and will have confidence that NASA can manage it well.

LEADERSHIP AND PARTICIPANTS

The Customer Engagement Activity is led by Mark Craig, Deputy Director of NASA Stennis Space Center, Donna Shirley, Manager of the Mars Program at JPL, and Humboldt Mandell of the Exploration Office at the NASA Johnson Space Center. Other members have been drawn from several other NASA centers. This group reports to the Human/Robotic Exploration Team steering group, jointly chaired by Norman Haines of JPL and Doug Cooke of JSC, and is working very closely with the NASA HEDS Strategic Planning Group to insure that all activities are fully coordinated.

Objectives are to have detailed plans in place, and programs such as the Deliberative Poll™ underway in time to be completed prior to the year 2000 elections.

CONCLUSIONS

Although many of the technologies required to explore Mars are currently available, or well on the way to fruition, there is still a lot NASA needs to know, particularly about the attitudes of its customers and stakeholders, and a lot of outside help that will be required to successfully pursue a leadership role in human Mars exploration. Although there are program architectures now in place which satisfy the affordability criteria, it will be a very hard sell to get permission from our customers and stakeholders to do the job. Unprecedented and unconventional approaches will be mandatory if we are to succeed. We have already begun the process of involving new people in new and exciting ways.

A large part of the problem is public perception. The general population, which is our primary customer, has little of the knowledge which will be required for it to be willing to provide the required resources. So NASA and all who want to see humans go to Mars in our lifetimes have first to engage the customers in the adventure.

The engagement process begins with knowing what the customers and stakeholders need and want from us, and what they would want if they knew what we could do for them, and at what cost. A Deliberative Poll ™ is therefore planned to gain some of that knowledge. But the process does not begin and end with polling. NASA and its partners must do research into the expressed needs, priorities, and desires of key stakeholders, and assess all of its programs in that light.

To accomplish all of this, a number of unprecedented and unconventional activities have been initiated to better involve our customers and stakeholders in the grand venture of human exploration of Mars. NASA is already teaming with many outside of the aerospace industry, including the academic community, other nations, technologists, and other interested parties, to acquire the critical skills which it does not have within its own institution. A Customer Engagement Group has been formed to lead this effort.

The hardest problems in getting to Mars may not be the engineering challenges. NASA knows full well that it needs to continue the successful management reforms initiated by Administrator Goldin in order to be able to produce affordable products. It knows well that it must have the permission and support of its customers. And it knows well that it needs help.

HUMAN MISSION FROM PLANET EARTH: TECHNOLOGY ASSESSMENT AND SOCIAL FORECASTING OF MOON/MARS SYNERGIES

Eligar Sadeh* and Evan Vlachos†

This paper advances scenarios for an evolutionary approach to the establishment of a Human Mission from Planet Earth (H-MFPE) involving exploration and utilization of the Moon and Mars. Of critical importance, in this regard, are the concepts of robotic/human and Moon/Mars synergies. The technological, scientific and sociopolitical impacts and consequences related to this mission are presented and discussed.

INTRODUCTION

Space exploration is (and will be) the grandiose search for and interaction with our destiny. Explorers, discoverers and seekers perpetuate the space odyssey; explorers venture into the unknown of the cosmos; discoverers pursue cumulative knowledge about the Earth and space; and seekers search for underlying models and causal factors to explain Earth system and cosmic physical processes.

The goal of this paper is to outline a systematic process of technology assessment and social forecasting to advance a framework for specifying Human Mission from Planet Earth (H-MFPE) scenarios involving exploration and utilization of the Moon and Mars. Technology assessment and social forecasting are accomplished through the following tasks: (1) identifying historical trends and developments that have shaped sociopolitical rationales of human space exploration; (2) articulating technological, scientific and sociopolitical impacts and consequences of H-MFPE by modeling critical variables; (3) forecasting scenarios relevant to the evolution of this mission; and (4) assessing policy implications.

1.0 TRENDS and DEVELOPMENTS

1.1 Coming of the Space Age

The origins of a space exploration mission are rooted in a synthesis of public imagination, the individual will of scientists and engineers in fostering technological innovation and the international imperatives of the Cold War and the ensuing "space race" between the U.S. and Soviet/Russia.[1] Science fiction, such as Jules Verne's *From the Earth to the Moon* (1865), Edward E. Hale's The Brick Moon (1869) and H. G. Wells' First Men on the Moon (1901), sparked public imagination about space exploration. These writers provided detailed accounts on how humans could venture into space and land on the Moon. The visions of these writers evolved from science fiction to science fact with the pioneering development in rocketry that

---

* Eligar Sadeh is a NASA Fellow and University Instructor in the Center for Engineering Infrastructure and Sciences in Space (CEISS) at Colorado State University, Fort Collins, Colorado. The author can be contacted at esadeh@lamar.colostate.edu.

† Evan Vlachos is a Professor of Sociology and Civil Engineering at Colorado State University, Fort Collins, Colorado.

was led through the efforts of Robert H. Goddard in the U.S., Herman Oberth in Germany and K. E. Tsiolkovskiy in Soviet/Russia during the first half of the 20th century.[2]

The beginnings of the space age are earmarked by the Soviet/Russian launching of Sputnik in 1957 and the orbiting of Yuri Gargarin in 1961, and the U.S. response to these events pertaining to the establishment of NASA in 1958 and the Apollo missions to the Moon from 1969 to 1972. These state-government supported endeavors implied that the dreams of scientists and engineers, fueled by public imagination, were in synergy with the policy imperatives of the massive power state. More pragmatically, scientists and engineers had at their disposal the mobilization of state resources to implement their ideas. The visions of the former and the political needs of the latter combined, and once the political decisions were made, the spread of a space age infrastructure was automatic.[3]

Within this context, NASA adopted the "Wernher von Braun paradigm" of space exploration. Wernher von Braun, largely acknowledged as one of the foremost rocket scientists[4] and public advocates of the space age in the 1950s and 1960s, first wrote about the possibilities of spaceflight in a series of articles published, from 1952 to 1954, in *Colliers*.[5] These articles described the scientific and technological developments necessary for space exploration involving orbital space stations and human-tended lunar and Martian bases. This vision became part of and continues to underpin NASA's planning for a human space exploration mission. In large part, the vision of a human space exploration mission persists in U.S. public policy because it is wedded to an American ideology and mythology of pioneerism and frontierism.[6]

1.2 Military Competition

The coming of the space age spawned U.S. desires for global preeminence and the preservation of national security in space. This was a by-product of the missile age and served as a platform for furthering political and military competition with Soviet/Russia. The priorities for the Eisenhower Presidential administration were to maintain intercontinental ballistic missile (ICBM) research and development (R&D) to provide for effective nuclear deterrence and space reconnaissance satellite R&D as a means to penetrate and defeat the Soviet bloc. There was a low estimate of the political, military and symbolic aspects of human space exploration in the years before Sputnik; civil space exploration was valued primarily for its scientific merit.

Two steps were taken by the Eisenhower Presidential administration that reflected this view. First, the U.S. space exploration mission was delinked from the military space mission. The decision to launch a civilian scientific satellite in cooperation with the 1957-58 international geophysical year (IGY) was indicative of this philosophy; the overriding concern was to maintain civilian dominance over space exploration.

The Soviet launching of Sputnik in 1957 challenged this conception. Sputnik represented Soviet technical and military parity with the U.S. and challenged the assumptions of American military and fiscal policy raising doubts about U.S. security and economic prosperity.[7] President Eisenhower's concern was how to outcompete the Soviets in technology and science without sacrificing the essentials of U.S. freedom and prosperity rooted in limited government involvement and private initiative. Sputnik elevated national prestige as an important stake in world politics linking civil space achievement with society's ability for organization and technological revolution as necessary to maintain freedom and stability.[8]

By 1958, due in large part to public and Congressional pressures, NASA was established. Space technology was not only important, but also urgent and inevitable given the nature of the Cold War. The creation of NASA was a break from Eisenhower's laissez-faire approach to governance and a step toward state-directed mobilization of science and technology. NASA's raison d'être was competition with the Soviets and not science. It brought to fruition the idea that state managed R&D produced economic and military vitality.

Concomitantly, Eisenhower retained the delinkage between the civilian space efforts and the military ones. In effect, two parallel space programs came into existence: one that is scientific and symbolic of American values of democracy and openness and the other that is closed and concerned with military applications (e.g., reconnaissance and missile development) to maintain strategic superiority in the Cold War confrontation with Soviet/Russia. In fact, Eisenhower utilized the former program as a political "smoke screen" for the development of the latter.

1.3 Putting a Man on the Moon

The coming of the space age, Sputnik and military competition brought about the development of two points-of-view within the space and political communities. One group, comprised primarily of scientists and engineers, viewed space for its scientific and technological merits. A second group, consisting of Congress and the Presidency, linked space exploration to the "fluid front of the cold war." The merging of these two outlooks was instrumental in President Kennedy's challenge in 1963 to land a man on the Moon. Politically, this implied competition in human space exploration with Soviet/Russia and a strategic outlook that viewed civil space accomplishments as important in world politics.

From 1958 to 1959, the bases of human spaceflight shifted from the military to NASA. Given Eisenhower's ambivalence about human space exploration, NASA operated in a political and policy vacuum. In such an environment, NASA planners chose the idea of a Moon mission in 1961, on technological grounds, as a logical successor to project Mercury; NASA went ahead with technological and scientific plans for a lunar landing a full two years before Kennedy's political commitment in 1963.[9]

A "Moon mission" offered the best focus for developing space technology for humans to operate in space. By the mid-1960s, NASA had the technological and scientific infrastructure in-place to realize the science fiction dreams of Verne, Hale and Wells. This indicated the response of political institutions to an emerging set of technological possibilities and allowed, in specific, for project Apollo to proceed not only on the merits of national prestige, leadership and security, but also on technological and scientific criteria.

1.4 Space Age America

The "Moon-Ghetto" metaphor epitomizes "Space-Age America." This metaphor is based on the notion that if we can go to the Moon (and apply the necessary organizational and technological tools to accomplish that end), then we can apply those same skills and tools solve the problems of the Ghetto (i.e., crime, poverty, civil rights, social welfare, environment, etc...). A "Space-Age America" is one where science and technology, emanating from the civil space program, can be harnessed for peaceful purposes to solve social and environmental problems and to foster education in the sciences and engineering. James E. Webb, NASA's administrator from 1961 to 1968, sought to create such an America based on the "large-scale endeavor" that safeguarded the democratic process.[10]

More specifically, "Space-Age America" pertained to the NASA-industry-university nexus forged by Webb. Public-private partnerships were established to enable the large-scale technological development needed for project Apollo. These partnerships provided jobs and skills to diverse areas of the country through the location of NASA field centers and their contractors and subcontractors. Educationally, Webb instituted the Sustaining University Program (SUP) as a vehicle for relating NASA to democratic purposes. The university was seen as a repository of knowledge that could be harnessed for public goals and general societal problem solving.[11]

1.5 International Cooperation

The space age came into being in a context of international cooperation directed at sharing knowledge among scientists. This is exemplified through U.S. and Soviet/Russian satellite and scientific programs that were part of the 1957-58 IGY.[12] As a result of the IGY, the U.S. belief in "openly" sharing scientific knowledge[13] and U.S. Cold War foreign policy interests, the *National Aeronautics and Space Act* of 1958, which led to the establishment of NASA, incorporated international cooperation in its declaration of policy and purpose: "cooperation by the United States with other nations and groups of nations in work done pursuant to this Act and in the peaceful application of the results thereof should be conducted..."[14]

International space cooperation is linked to a state's symbolic and functional interests.[15] Symbolic interests are political in character encompassing prestige, propaganda, policy legitimization, enhanced policy influence over other actors, desires for international accountability, world leadership and national security. Functional interests are economic and technological. Economic interests deal with promoting industrial autonomy and economic competitiveness and bringing about economic savings or cost burden sharing (i.e., maximizing national and economic benefits). Technological interests are valued in connection to technological and scientific capabilities that are realized either through the augmentation of capabilities vis-à-vis contributions in noncritical path components, coordination of separate nationally approved projects with independent capabilities that are complementary, interdependence involving interstate cooperation in critical path and infrastructure components through technology transfer or integration denoting joint and shared technological R&D.[16]

Throughout the Cold War and its aftermath, human spaceflight cooperation has symbolized stable and friendly overall international political relations between the U.S. and Soviet/Russia.[17] The period of political détente between the U.S. and Soviet/Russia, for instance, was symbolized by intergovernmental agreements (IGAs) on space cooperation. IGAs on space cooperation, concluded in 1972 and 1977, led to cooperation on the Apollo-Soyuz Test Project (ASTP). In similar ways, the emergence of interdependent interests between the U.S. and Russian space station programs, formalized by the 1998 IGA on the International Space Station (ISS),[18] are symbolic of U.S. foreign policy interests in the post Cold War era.[19]

Examples of human spaceflight cooperation based on functional interests, that enhance technological capabilities, include both the space transportation system (STS) and ISS. Canadian development of the robotic arm remote manipulator servicer on STS plays a critical role in the ability of STS to retrieve and deploy payloads in low Earth orbit (LEO). The functional cargo and energy blocks and service module provided by the Russian Space Agency (RSA) and the mini-pressurized logistics modules developed by the Italian Space Agency (ASI) are both necessary to the proper functioning (i.e., in the critical technological path) of ISS.

2.0 MODELING the SOCIOPOLITICAL SYSTEM

2.1 Critical Variables

There are a number of critical variables that impact the prospects for Moon/Mars exploration. These variables include: (1) national prestige and leadership; (2) exploration and scientific knowledge; (3) economic growth and development (e.g., space commercialization and privatization); (4) enabling technologies and technological innovation; and (5) international cooperation. The five sets of variables are examined in the context of their significance from the Apollo to the post-Apollo and present eras. In turn, this examination serves as the basis for assessing the range of impacts (short-term effects) and consequences (long-term effects) for the present and future prospects of H-MFPE.

During the Apollo era (1961-1969), the U.S. space program was governed by a paradigm that consisted of three central tenets: (1) political ethos; (2) exploration ethos; and (3) technological ethos. The political ethos was premised on national prestige and national leadership; the exploration ethos linked exploration, discovery and seeking; and the technological ethos, based on the idea of a "Space Age America," provided for economic, technological and educational prosperity.

NASA planning assumed that political support would continue unabated and with little controversy, that the exploration ethos would continue to spark the American imagination and that science and technology would be viewed as the basis for the building of a "Space Age America" that could be applied to solve social ills. During the 1960s, the ambient conditions of the political system, within the U.S. and abroad, supported these political, exploration and technological worldviews. These views, to an extent, were reinforced with the successful Apollo moon landings. By the end of the 1960s, however, these assumptions were no longer valid in radically changed domestic and international political circumstances.

The post-Apollo era (1970-1985) is earmarked by a decline in support for human space exploration.[20] Fulfilling the challenge of placing a man on the Moon and the foreign policy of détente, which ended the space race and eased Cold War tensions respectively, resulted in an emphasis on science, economics and enabling technologies for the development of a human spaceflight infrastructure. Human spaceflight was wedded to space utilization and a "mission to infrastructure" in LEO; ASTP, Skylab, STS and ISS exemplify this course of action.

This implied that the exploration ethos became a utilitarian ethos were other social and political concerns predominated public policy. Science and technology became increasingly viewed as "autonomous" forces that could be not controlled or guided to the benefit of humankind. This was compounded by the fact that the application of technology could not solve social ills (e.g., shortcomings in President Johnson's "Great Society" programs), was very often found to be destructive to the environment (see Rachel Carson, Silent Spring, 1964) and was used for military purposes (e.g., Vietnam War).

Space, from a utilitarian outlook, offered a new perspective on Earth and as such was seen as a platform for dealing with Earthly priorities. The Nixon and Carter Presidential administrations marked a watershed, in this regard, in that rather than pursue prestige and leadership through human space exploration achievements, the U.S. should lead in practical scientific and technological capabilities that have an economic return.[21] Even though the rhetoric and metaphors in support of the Apollo political ethos and exploration ethos reemerged during the Reagan and Bush Presidencies, concrete political action, in terms of increased funding and policy initiatives, was absent. Moreover, Reagan and Bush intensified the theme of space utilization by supporting commercialization and privatization of space activities.[22]

As a result, NASA's proposed policies and plans were out-of-synch with the political system and budgetary resources. NASA as an organization and government bureaucracy also underwent changes in its cultural make-up that led to planning problems and errors of judgment (e.g., the decision to launch the space shuttle Challenger in January of 1986). By way of illustration, NASA went from a largely R&D culture during Apollo to an operational one afterwards; from a frontier mentality and the propensity to assume risk and failure to a utilitarian (applications and operations) outlook and the propensity to avoid risk; and from a "cult and culture" of the engineer to a bureaucratic based culture.[23]

The present era (1986-1998), denoted as the post-Challenger period, is characterized by both space utilization and space exploration themes. NASA, for example, remained committed to a utilitarian "mission to infrastructure" typified by the on-going STS and ISS programs. At the same time, the Space Exploration Initiative (SEI), put forward by President Bush in 1989, sought a return to the exploration ethos of the Apollo era.[24] SEI called for human missions to the Moon and Mars and for the establishment of a human-tended base on the Moon.

Albeit, SEI remained elusive—that is, in search for a rationale that would provide the needed political support—and thus, died a political stillbirth, robotic exploration activities were renewed. NASA's unmanned Mars exploration program, initiated in 1996 with the Mars Pathfinder and Mars Global Surveyor, for example, is developing missions over the next decade that will testbed technologies (and enhance our knowledge of Mars and the search for the origins of life) important for human missions to Mars. This utilization-exploration duality is best expressed through NASA' s 1998 vision statement:

> NASA is an investment [utilization] in America's future [exploration]. As explorers, pioneers and innovators [exploration], we boldly expand frontiers in air and space [exploration] to inspire and serve America [utilization] and to benefit the quality of life here on Earth [utilization].[25]

With the ending of the Cold War in the early 1990s, human spaceflight and exploration, in the view of many observers, can only proceed with international cooperation.[26] ISS clearly illustrates the saliency of international cooperation for human-based space endeavors.[27] Today, mission concepts for human-tended lunar bases and human missions to Mars are put forward on the assumption that they will be implemented through international space cooperation.[28]

Since the end of the Apollo era, a fundamental concern of the space community has been the search for critical variables to support a human space exploration mission. This has spawned a number of studies and reports. NASA's post-Apollo plans, for example, called for resources to implement the development of a space shuttle, orbital space station, nuclear space tug, human-tended lunar base and human expeditions to Mars.[29] In the 1980s and 1990s, a series of reports and initiatives to advance human space exploration missions were proposed as well.[30]

As a whole, these reports justify future space program scenarios in nationalistic terms (prestige, leadership, technological and economic benefits). For example, SEI was justified on a number of such rationale factors: national prestige; advancing science education; developing technologies; commercializing space; and strengthening the U.S. economy.[31] *The Ride Report* (1987) provides a systematic analysis of the U.S. civilian space program to show how the U.S. has lost its leadership position in space especially as it relates to maintaining a human presence there.[32] To this end, a space strategic development plan for the 21st century is developed based on restoring U.S. leadership status. This requires that the U.S. have capabilities, which enable it to act independently and impressively when and where it chooses.

In the *NASA Strategic Plan* (1998), four space strategic enterprises (Human Exploration and Development of Space, Space Sciences, Earth Sciences, and Aeronautics and Space Transportation Technology) are established. The strategic plan focuses on developing these enterprises to meet the goals of various governmental (President and Congress) and domestic public constituencies with the ultimate benefactors being policy makers, science communities, aeronautics industry, other governmental agencies, public sector and academic communities within the U.S..

Others have argued, among them Carl Sagan and John Logsdon, that tangible pragmatic type variables are ultimately inadequate to support human space exploration. In other words, it is argued that what is needed is a renewal of the intangible exploration ethos.

> But, we are the kind of species that needs a frontier—for fundamental biological reasons. Every time humanity stretches itself and turns a new corner, it receives a jolt of productive vitality that can carry it for centuries. There is a new world next door. An we know how to get there.[33]

> The primary justification of space exploration lies in the imperatives of human nature. Preparing to leave Earth is an appropriate rationale for human activities in space if the broad sweep of history rather than the fiscal and political constraints of a particular time are taken into account.[34]

2.2 Range of Impacts and Consequences

The range of impacts and consequences related to the critical variables in support of Moon/Mars exploration can be represented by a cross-impact matrix as shown in Table 1 in the Appendix. This matrix considers each set of variables in cause-effect type relationships. These relationships include:

(1) if the primary cause for Moon/Mars exploration is national prestige and leadership then the effect on exploration and science is the exploration ethos; on economic growth and development is the expansion of the aerospace industry; on enabling technologies and technological innovations is commercially viable technological spin-offs; and on international cooperation is to minimize any possible scientific and technological interdependencies;

(2) if the primary cause is exploration and science then the effect on national prestige is the promotion of education and leadership in the sciences and technology; on economic development is increased government and industrial investment in basic R&D; on enabling technologies is space–based technological innovations; and on international cooperation is to maximize the sharing of scientific knowledge;

(3) if the primary cause is economic growth then the effect on national prestige is national (vs. international) oriented space missions; on exploration is R&D directed at robotic followed by possible human space exploration; on enabling technologies is technological development that enables human missions to the Moon and Mars; and on international cooperation is the coordination of nationally approved missions and the augmentation of capabilities in robotic and human space exploration;

(4) if the primary cause is enabling technologies then the effect on national prestige is the promotion of global leadership in science and technology; on exploration is robotic space exploration directed at scientific discovery and technology demonstration; on economic growth is the

expansion of the aerospace industrial sector; and on international cooperation is the restriction of technology transfer and intellectual property rights; and

(5) if the primary cause is international cooperation then the effect on national prestige is to promote foreign policy goals; on exploration human exploration of the Moon/Mars; on economic growth economic interdependencies; and on enabling technologies the coordination of nationally approved missions directed at R&D space-based technological development.

This cross-impact matrix can be applied in a number of ways to investigate the primary and secondary factors of importance for H-MFPE. For instance, if the goal is to advance a scenario of human space exploration of the Moon and Mars then a number of critical variables of significance can be identified. In this case national prestige, economic growth and international cooperation, as the primary causal drivers for Moon/Mars exploration, would realize (i.e., cause effects) the renewal of an exploration ethos, and the utilization of robotic exploration and technology R&D as precursors to enable human space exploration. These effects are highlighted in Table 1.

3.0 SCENARIOS

3.1 Enabling Technologies

The need for enabling technologies implies that limitations in technical capabilities for establishing a lunar base and undertaking a human mission to Mars exist. Given this assumption, there are a number of enabling technologies that can be identified. These include more economical earth-to-orbit transportation systems; cost-saving (in terms of mass) technologies such as aerobraking and autonomous landing; the development of in-space engineering assembly and launch operations; improvements in surface operations in terms of planetary habitats, energy use and in-situ resource processing; life support systems technologies; improved understanding of physiological changes due to exposure to micro- and hypo- gravity environments and to cosmic radiation; and nuclear or other non-chemical based forms of propulsion.[35]

A "synergisms" scenario for the development of enabling technologies is considered herein. Synergisms refer to the use of automation and robotics (A&R) to enable H-MFPE and the establishment of a mature lunar base as a precursor for human missions to Mars. A&R technologies can support H-MFPE by accomplishing scientific objectives, such as regolith assessment and remote sensing of environmental data, and testing technologies in aerobraking, autonomous landing and in-situ use of resources; and a lunar base can serve as a platform to testbed (for reliability and sustainability) critical technologies in planetary habitats, life-support systems and in-situ use of resources important for sending humans to Mars. A human-tended lunar base would also provide important operational, managerial and international cooperative experiences to carry out human exploration of Mars.

3.2 Resources

It is anticipated that the foreseeable trend in resources for human space exploration, as measured by past NASA appropriations and funding projections from 1997 to 2020, will most likely follow minimal linear growth patterns (e.g., $2 to $4B every five years). These resource trend projections are shown as scenarios #3 and #4 in Figure 1 in the Appendix. The trends would allow, at the very least, the continuation of an aggressive robotics exploration program of the Moon and Mars.

Currently, NASA has plans over the next decade to launch every twenty-six months two robotic probes to Mars. These missions are planned with one objective (among others) of demonstrating technologies important for human space missions to Mars. Assuming that this focus on Mars continues and that NASA funding trends follow scenario #3 (linear growth of $4B every five years), resources for a human mission to the Moon or Mars become available by 2020. This also assumes that there is a political decision to undertake such a mission, that it will remain part of NASA's plans and that NASA will have developed the critical enabling technologies. It is only in the case of upper limit resource trend projections (which are not likely unless a crisis event or the discovery of life on Mars or elsewhere takes place)—that is, exponential growth in resources (trend #1) or liner growth at $6B every five years (trend #2) depicted in Figure 1— that NASA could immediately undertake missions to the Moon and Mars.[36]

3.3 H-MFPE Evolution

A scenario for an evolutionary approach to H-MFPE is based on continuous expansion in mission requirements directed at the development of human-tended lunar and Martian bases. This scenario encompasses four strategic development stages spanning from a simple encampment to mature bases. These four development stages include: (1) exploratory; (2) pioneering; (3) outpost; and (4) mature bases.[37] The exploratory stage involves discovery about the space, lunar and Martian environments in order to gain greater knowledge about how humans can safely live and work in space. This stage consists of unmanned robotic surveys and short-term human missions. Examples of such an effort include the robotic probes that NASA has sent to the Moon (e.g., Surveyor, Lunar Prospector) and Mars (e.g., Vikings, Mars Pathfinder, Mars Global Surveyor), the Apollo Moon landings, and STS and ISS in terms of understanding human adaptation to the space environment (e.g., physiological effects of microgravity, exposure to cosmic radiation, sustainability of closed life support systems).

Objectives of the exploratory stage include the use of robotics for site selection for exploration and the establishment of a base, sampling and testing of regolith and planetary subsurface materials, geochemical assessment, gathering of site topographic data, measurement of radiation and micrometeoroid impact, seismic and gravity data, testing of remote sensing and communication systems and testing of lunar rovers. Realization of these objectives will require short-term (fourteen-day lunar day cycle) Apollo-like human missions to the Moon.

The pioneering stage deals with building upon the knowledge gained in the exploratory stage to establish a base encampment. Two fundamental goals are essential to this stage: (1) operational testing of human habitat deployment; and (2) testing of in-situ resource utilization. The deployment and testing of infrastructure enabling technologies, such as inflatable habitats shown in Figure 2 in the Appendix, must be undertaken as a precursor for the eventual deployment of human habitats.[38] The objectives to realize these goals include further expansion of the tasks form the exploratory stage. This includes regolith handling and moving operations, robotic construction and mining equipment testing, use of solar energy and batteries for energy storage, introduction of nuclear reactor for power supply, testing of heat rejection systems, initial tests of hydroponics plant growth, intensive geological assessment, testing of robotic extraction of lunar resources and initial exploration trips with lunar rovers. Realization of these objectives will require medium-term (forty-two days, two lunar day cycles) human missions to the Moon.

The outpost stage marks the transition from an encampment to the establishment of a permanent infrastructure geared toward achieving a self-sufficient base. Crews of ten to twenty astronauts with duty tours to the Moon of up to six months is envisioned. The major task for this stage is the deployment and construction of an operational lunar biosphere. This entails life support outfitting and shielding of the human habitats deployed in the pioneering stage. Life support outfitting is accomplished by integrating engineered closed controlled ecosystems (ECCES) type technologies into the habitats, such as physical/chemical air/water recycling systems, expansion of hydroponic plant growth and use of regolith as a plant growth medium.[39] Radiation shielding and micrometeoroid protection is provided by either covering the habitat with a layer of regolith of about three meters thick or by inserting the habitat within lava tubes or craters. The construction of an operational biosphere also requires the development and use of a high-level of A&R. Additional objectives to further develop the technology for a flexible, expandable and permanent lunar base infrastructure include mining and extraction of in-situ resources as permanent ongoing activities, the initiation of manufacturing techniques in lunar hypogravity, construction of photovoltaic solar farms for energy production and extended rover exploration trips to evaluate the construction of an astronomic observatory on the Moon's far side.

The mature development of a base is the most advanced stage before settlement and colonization. A mature base with a permanent population of up to fifty astronauts with duty tours of up to one year is envisioned. The primary goal of this stage is to establish a permanent lunar infrastructure capable of reaching an acceptable level of self-sufficiency. This implies that the infrastructure is capable of sustaining an ECCES system as shown in Figure 3 in the Appendix. Objectives of the base are to bring the tasks outlined for the outpost stage to the level of continuous activity. This could involve the manufacturing of products using lunar indigenous resources for export to Earth and expansion of the photovoltaic solar farm.

Additional tasks include the construction of an astronomical observatory on the far side of the Moon and a lunar space port for launching of robotic/human missions to Mars and for the exploration of the outer solar system and beyond. An important aspect of this stage is to apply the knowledge and technologies developed in the base stage to enable human missions to Mars and to eventually establish a Martian base. The fundamental scientific objective for such a base is to search for the origins of life (i.e., possible microbial life forms (or fossilized remnants thereof) on Mars).[40] Ultimately, the transition from a mature base stage to lunar/Martian colonies takes place when the base is economically viable and self-sustaining (i.e., independent and autonomous from Earth).[41] According to Willy Z. Sadeh, colonies implies the establishment of a "humanity extended home in space."[42]

4.0 SOCIOPOLITICAL IMPLICATIONS

4.1 Political and Policy Evolution

The evolution of space politics and policy, since the coming of the space age, reflects two themes: (1) space policy is evolutionary in that it has responded to dramatic political events, such as the launching of *Sputnik*, and has undergone dynamic policy changes (from confrontation to cooperation) over the course of the past fifty years of the space age; and (2) space policy is part of and interacts with broader public policy processes.

The evolution of space politics and policy can be represented by a framework as shown in Figure 4 in the Appendix.[43] Space politics involves the process by which historic conditions, rationales, public opinion and space advocacy coalitions interact with and impact agenda-setting; actors and institutions (Presidency, Congress and Bureaucracy) interact with and impact public policy formulation and implementation; and policy outcomes bring about policy change (e.g., the emergence of commercialization and privatization of space activities).

Space policy concerns the courses of action taken to achieve politically and technologically determined outcomes. These outcomes include science policy, environment, economics and commerce, law, international cooperation, and military and intelligence.

The evolution of public policy over time takes place through policy change. On this basis, space public policy processes the past fifty years of the space age, represented by T_1 in Figure 4, has involved the mobilization of governmental resources, actors and institutions (Presidency, Congress and Bureaucracy) based upon historic-conditions rooted to the geopolitical Cold War confrontation with Soviet/Russia and an exploration ethos based upon the "Von Braun paradigm". Currently, space policy processes are shifting to one where nongovernmental actors, such as private corporations and commercial enterprises and ventures, will increasingly play a role based more so on historic-conditions that view cooperation and collaboration as important to advance a utilitarian ethos of economic exploitation of space. This "paradigm-shift" is denoted by T_2 in Figure 4.

The implication of this shift for H-MFPE lies in the realization that national space agencies will have to partner with commercial and private entities.44 One viable policy option would be to hand over NASA's "mission to infrastructure" (i.e., STS and ISS) to private industry. This could free up 30% to 40% of NASA resources for space exploration and could allow for H-MFPE even in the lower resource trend projections discussed in section 3.2.45

CONCLUSIONS: RECONCILING IMAGINATION and REALITY

As public policies are implemented, they invariably require some sort of reconciliation between vision and reality...Modern governments require the reconciliation of imagination, popular culture, and real events.46

Human space exploration is rooted in imagination and popular culture that is perpetuated by the role of exploration, discovery and seeking in the evolution of human history. Space represents the search for a utopia and an escape from the "dystopian" tendencies of Earth and its Earthly problems. This makes public policy for H-MFPE especially difficult; as Howard McCurdy has stated above, public policies require a reconciliation of our utopian dreams with the present reality here on Earth.

Public policies for the realization of H-MFPE need to overcome the inherent problems of this reconciliation. Effective formulation and implementation requires that our political and social systems become more proactive (as opposed to reactive). Instead of reacting to a *Sputnik* type event, these systems need to envision what future they desire in space and then take the steps to realize that future. Such steps would necessitate the building of a new spacefaring culture that views human expansion into space as necessary for the development and evolution of the human species. More pragmatically, how public policies plan and manage uncertainty, in terms of institutional capacity building, estimating risk, and political impacts, consequences and trade-offs, will ultimately determine the course of events as to whether or not H-MFPE endeavors are realized.

ACKNOWLEDGMENTS

The support of the NASA Space Grant College Program in preparing this paper is gratefully acknowledged. Eligar Sadeh and Evan Vlachos would also like to thank Willy Z. Sadeh for his invaluable support, professional guidance, friendship and inspiring vision.

APPENDIX OF TABLES and FIGURES

Table 1
H-MFPE CROSS-IMPACT MATRIX OF IMPACTS AND CONSEQUENCES
(Source: Eligar Sadeh)

Effect: Cause:	National Prestige & Leadership	Exploration & Scientific Discovery	Economic Growth & Development	Enabling Technology & Innovation	International Cooperation
National Prestige & Leadership		Exploration ethos promoting human spaceflight	National economic expansion in technological sector-- aerospace industries (applied R&D)	Space technology directed at applicable commercial spin-offs	Scientific and technological interdependence
Exploration & Scientific Discovery	Promotion of education & leadership in the sciences & technology		Increased investment in basic R&D	Space-based technological innovations	Maximize scientific sharing of space-based knowledge & information
Economic Growth & Development	National-based space exploration missions	R&D directed at robotic and human space exploration		Technological development that enables human space exploration of Moon/ Mars	Coordination of nationally approved missions & augmentation of capabilities in robotics & human space exploration
Enabling Technology & Innovation	Promotion of global leadership in science & technological innovation	Robotic space exploration directed at science & technological demonstration	National economic expansion in technological sector-- aerospace industries (applied R&D)		Restrictions on technology transfer & intellectual property rights
International Cooperation	Promotion of foreign policy goals	Human exploration of Moon/ Mars	Economic interdependence in science & technology	Coordination of nationally approved R&D space-based technology programs	

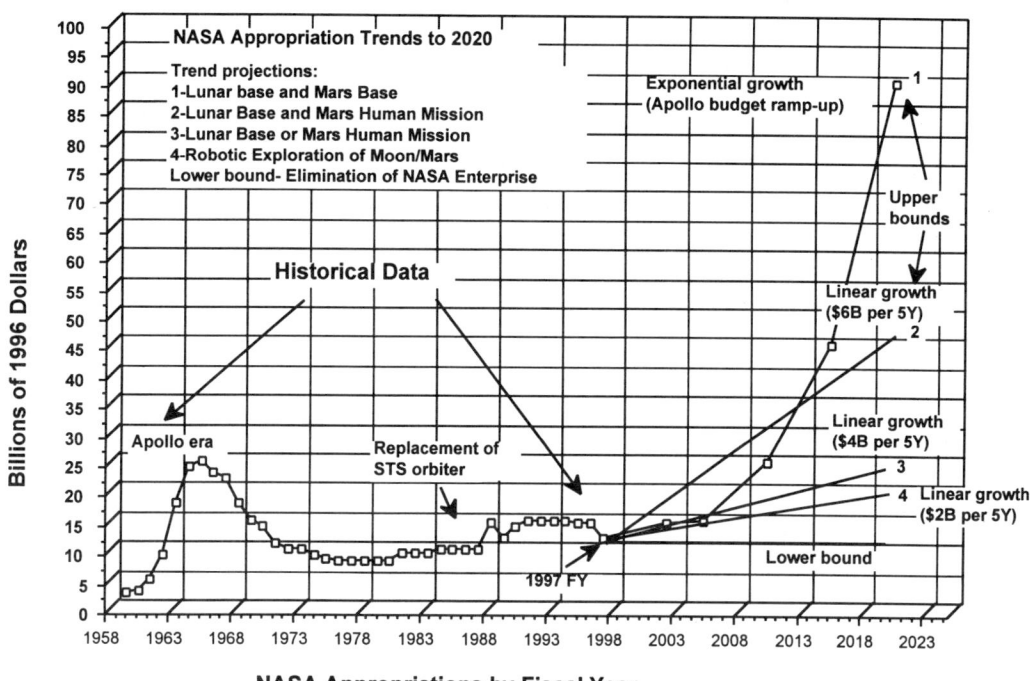

Figure 1 H-MFPE Resource Trend Projections. (Source: Eligar Sadeh)

Kevlar Membrane Material: 0.3 to 0.5 mm thickness
Pressurized Vessel: 10 PSI
Compact and Deployable: 6 x 6 x 3 m
Light Weight: 200 Kg per module

Figure 2 Inflatable Habitats for Planetary Surfaces. (Source: Willy Z. Sadeh)

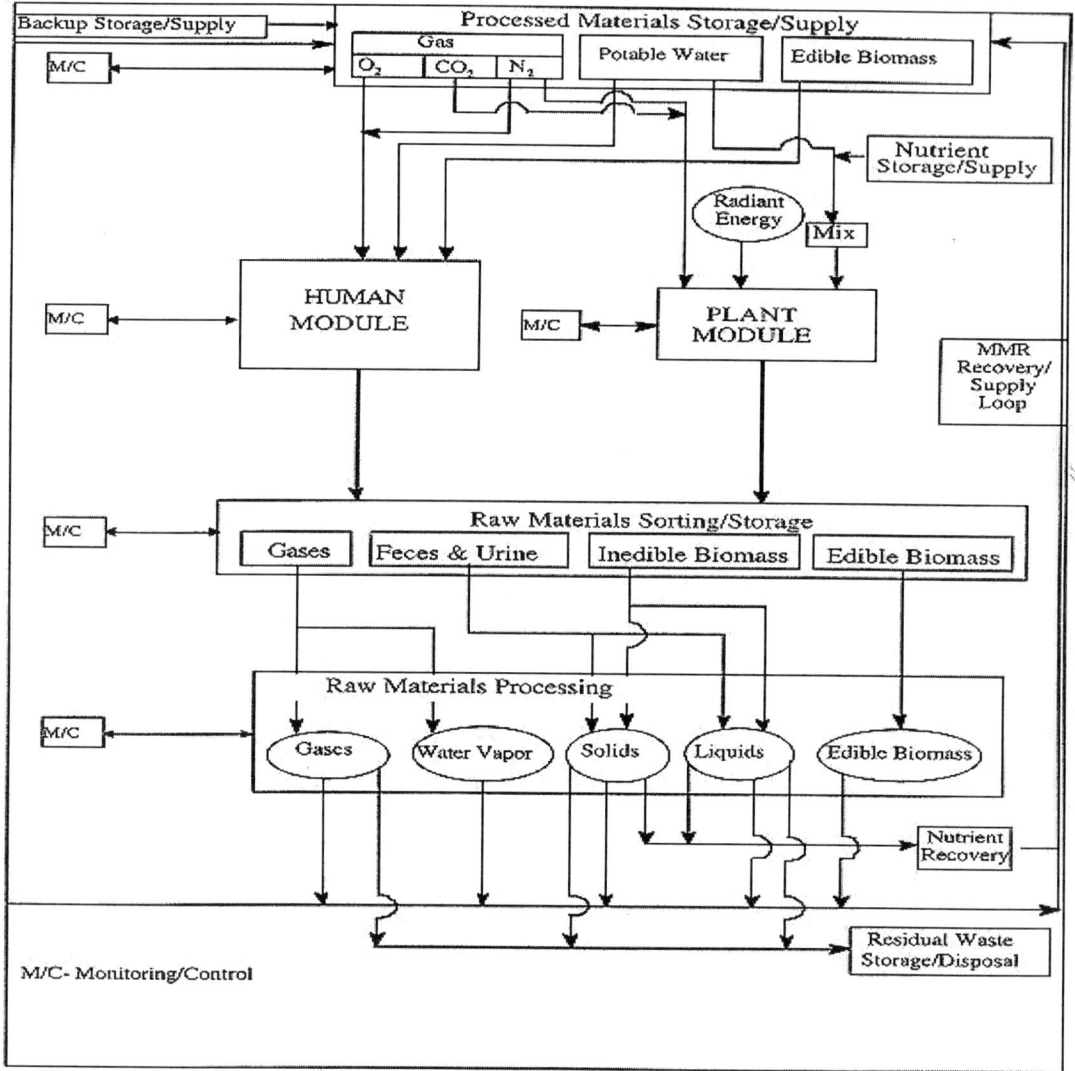

Figure 3 Engineered Closed Controlled Ecosystem Model.

(Source: "Bioregenerative Life Support Systems for Long Term Space Habitation: A Conceptual Approach," *Life Support and Biosphere Science* 2, no. 3/4 (1996): 161-168. Co-author with Willy Z. Sadeh.)

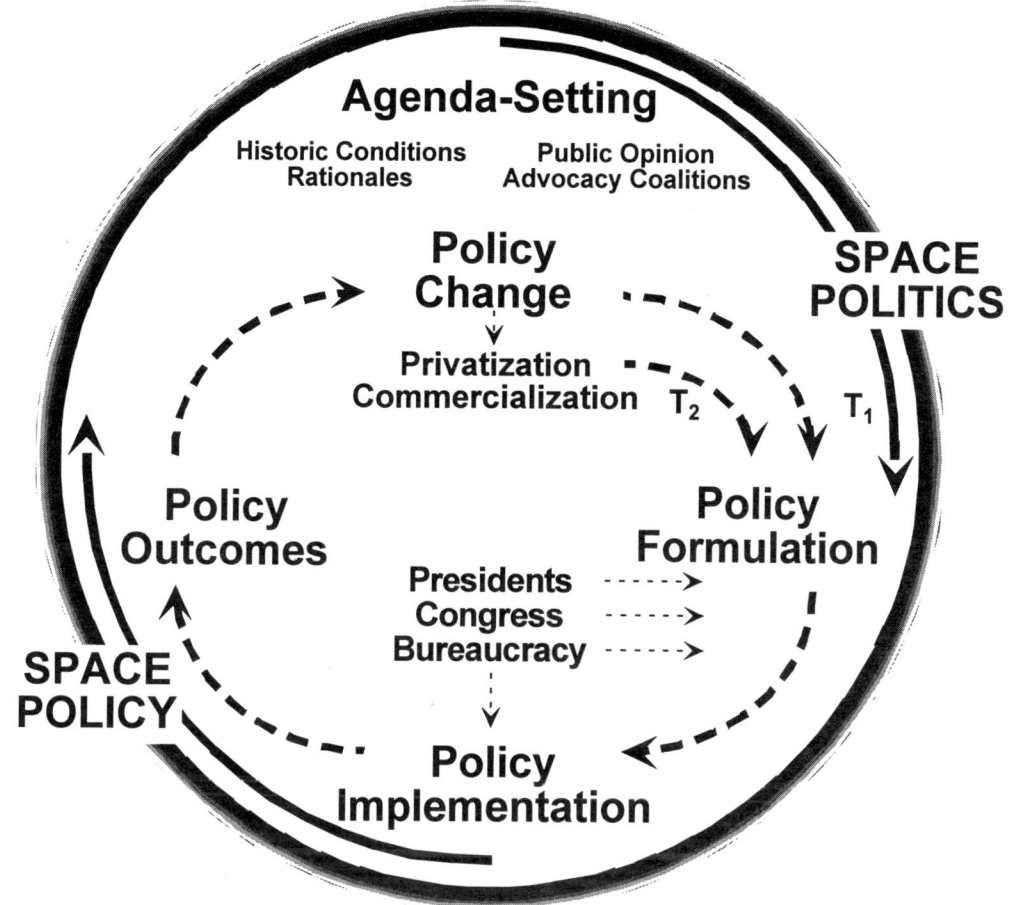

Figure 4   Evolution of Space Politics and Policy. (Source: Eligar Sadeh)

REFERENCE NOTES

1. See Walter A. McDougall, ...*The Heavens and the Earth: A Political History of the Space Age* (Baltimore, Maryland: John Hopkins University Press, paperback edition, 1997), 3-13; Howard E. McCurdy, *Space and the American Imagination* (Washington, DC: Smithsonian Institution Press, 1997), 1-52; John M. Logsdon, ed., *Exploring the Unknown, Selected Documents in the History of the US Civil Space Program, Volume I: Organizing for Exploration*, NASA History Series (Washington, DC: Government Printing Office, 1995), 1-207; and Roger D. Launius, *NASA: History of the Civil Space Program* (Malabar, Florida: Krieger Publishing Company, 1994), 3-16.
2. See works by Goddard, Oberth and Tsiolkovskiy in *Exploring the Unknown*, 59-139.
3. See McDougall, ...*The Heavens and the Earth*, 12.
4. In addition to Wernher von Braun, Sergei Korolev of Soviet/Russia was also considered a foremost rocket scientist of his generation. See James Hartford, *Korolev, How One Man Masterminded the Soviet Drive to Beat America to the Moon* (New York, New York: John Wiley and Sons, 1997).
5. See Wernher von Braun, "Can We Get to Mars," *Colliers*, April 30, 1954, 22-29; "Man on the Moon the Journey," *Colliers*, October 18, 1952, 52-59; and "Crossing the Last Frontier," *Colliers*, March 22, 1952, 23-29, 72-73. *Colliers* was a popular newsmagazine and these articles were directed at the layperson.

6. For this historical interpretation see McCurdy, *Space and the American Imagination*; Robert Zubrin, *The Case for Mars* (New York, New York: Free Press, 1996); and James L. Kauffman, *Selling Outer Space* (Tuscaloosa, Alabama: University of Alabama Press, 1994). Also, see Dale M. Gray, "Current Space Development as a Manifestation of Historic Frontier Processes" (paper presented at the 49th International Astronautical Congress, Melbourne, Australia, September 28-October 2, 1998). There are others who have argued that the American expansion into the Western frontier has been misapplied to the American civil space program. For this particular view, see Patricia N. Limerick, "Imagined Frontiers: Westward Expansion and the Future of the Space Program," in Radford Byerly, Jr., ed., *Space Policy Alternatives* (Boulder, Colorado: Westview Press, 1992), 249-262.

7. See McDougall, *The Heavens and the Earth*, 132-140.

8. See McDougall, *The Heavens and the Earth*, 195-209.

9. See John M. Logsdon, *The Decision to go the Moon: Project Apollo and the National Interest* (Cambridge, Massachusetts: The MIT Press, 1970), 39-62.

10. See James E. Webb, *Space-Age Management, The Large-Scale Approach* (New York, New York, McGraw Hill Book Company, 1969), 89-108.

11. See W. Henry Lambright and Laurin L. Henry, "Using Universities: The NASA Experience," *Public Policy* (1972): 61-82.

12. For a discussion of the International Geophysical Year (IGY) and its importance to scientific collaboration in space, see Arnold Frutkin, *International Cooperation in Space* (Englewood Cliffs, New Jersey: Prentice-Hall, 1965), 17-23; and Dodd L. Harvey, and Linda C. Ciccoritti, *US-Soviet Cooperation in Space* (Miami, Florida: Monographs in International Affairs, Center for Advanced International Studies, University of Miami, 1974), 1-22.

13. Moved by an inherent "idealism", U.S. scientists thought of cooperation in space science as a good means for reducing Cold War tensions. See Homer E. Nevell, *Beyond the Atmosphere: Early Years of Space Science* (Washington, DC: U.S. Government Printing Office, 1980), 299-318.

14. Work done pursuant to the *Space Act* includes the expansion of human knowledge related to the atmosphere and space, improvements in the safety and efficiency of aeronautical and space vehicles, the development of space vehicles capable of carrying supplies and living organisms into space and long-range scientific studies about the cosmos and Earth. See "National Aeronautics and Space Act of 1958," Public-Law 85-568, in *Exploring the Unknown*, 335, 339.

15. See Stephen M. Shaffer and Lisa R. Shaffer, "The Politics of International Cooperation: A Comparison of US Experience in Space and in Security," *Monograph Series in World Affairs* 17, no. 4 (Denver, Colorado: Graduate School of International Studies, University of Denver, 1980), 11-13.

16. See Lynn F. H. Cline and Jeffrey D. Rosendahl, "An Assessment of Prospects for International Cooperation on the Space Exploration Initiative," *Acta Astronautica* 28 (1992): 391-399. As of 1998, there are no applicable examples involving NASA where joint and shared technological R&D programs exist.

17. See Matthew J. Von Bencke, *The Politics of Space: A History of US-Soviet/Russian Competition and Cooperation* (Boulder, Colorado: Westview Press, 1997), 1-7.

18. See "Agreement among the Government of Canada, Governments of Members States of the European Space Agency, the Government of Japan, the Government of the Russian Federation, and the Government of the United States of America Concerning Cooperation on the Civil International Space Station," January 29, 1998. Also, see Lynn F. H. Cline, "International Negotiations: The International Space Station Agreements" (paper presented at the 47th International Astronautical Congress, Beijing, China, 7-11 October, 1996); and Lynn F. H. Cline and Graham Gibbs, "Re-Negotiation of the International Space Station Agreements- 1993 to 1997" (paper presented at the 48th International Astronautical Congress, Turin, Italy, 6-10 October 1997).

19. In 1993, U.S. President Clinton linked ISS to U.S. foreign policy interests. These interests were directed at integrating Russia into a partnership among traditional allies in the U.S., Europe, Japan and Canada and safeguarding Russian capabilities in human spaceflight. Cooperation with the Russians is also linked to facilitating Russia's economic transition to capitalism, strengthening Russia's emerging democratic political system, maintaining Russian aerospace industrial capabilities and providing incentives for the Russian scientists and engineers to stay at home through participation in space station hardware development. In exchange for agreeing to participate in the space station program, President Clinton secured Russian adherence to the Missile Technology and Control Regime (MTCR) aimed at preventing the transfer of cryogenic launch vehicle technology to third world states.

20. Public and political support dwindled resulting in a deterioration of NASA's budget from a high of 0.8% of GNP in 1966 to approximately 0.2% of GNP from 1978 to 1986.

21. See John M. Logsdon, "Presidential Leadership, Congress, and the US Space Program," (Keynote Talk, NASA-American University Symposium, March 25, 1993), photocopy, 14-15.
22. Commercialization of space activities includes expendable launch vehicles, telecommunications and Earth remote sensing satellites.
23. See Howard E. McCurdy, *Inside NASA: High Technology and Organizational Change in the US Space Program* (Baltimore, Maryland: John Hopkins University Press, 1993).
24. See *America at the Threshold: America's Space Exploration Initiative, Report of the Synthesis Group on America's Space Exploration Initiative*, by Thomas P. Stafford, chairman (Washington, DC: Government Printing Office, 1991).
25. See *NASA Strategic Plan* (Washington, DC: NASA Headquarters, Government Printing Office, 1998), NASA Policy Directive (NPD)-1000.1, 8.
26. See, for example, Bruce Murray, "Can Space Exploration Survive the end of the Cold War?" *Space Policy* 11, no. 1 (1991): 23-34; Kenneth S. Pederson, "Thoughts on International Cooperation and Interests in the Post-Cold War World," Space Policy 12, no. 3 (1992): 205-220; Eligar Sadeh, James P. Lester and Willy Z. Sadeh, "Models of International Cooperation in Human Space Exploration for the 21st Century," *Acta Astronautica* 34, (forthcoming 1998); and Carl Sagan, *Pale Blue Dot, A Vision of the Human Future in Space* (New York, New York: Random House, 1994); US-Russian Cooperation in Space, Office of Technology Assessment, Congress of the United States (Washington, DC: Government Printing Office, April 1995), OTA-ISS-618.
27. ISS is an international cooperative venture involving U.S., Russia, Europe (France, United Kingdom, Belgium, Netherlands, Denmark, Norway, Spain, Germany, Italy, Sweden and Switzerland), Japan, Canada and Brazil. The space station is planned for completion in 2004 with a fifteen-year operational capability. See NASA, "International Space Station Assembly Sequence (5/31/98: Revision D)," *NASA Facts* (Houston, Texas: NASA Lyndon B. Johnson Space Center, June 1998).
28. See, for example, Sagan, *Pale Blue Dot*; George W. Morgenthaler and Gerald J. Smith, eds., *International Exploration of Mars* (Paris, France: International Academy of Astronautics, 1996), compiled by the International Academy of Astronautics.; *The Case for an International Lunar Base* (Paris, France: International Academy of Astronautics, 1990).
29. See "Space Task Group, *The Post-Apollo Space Program: Directions for the Future*, September 1969," in *Exploring the Unknown*, 522-543.
30. See *Pioneering the Space Frontier, Report of the National Commission on Space* (New York, New York: Bantam Books, 1986); *Leadership and America's Future in Space, A Report to the NASA Administrator*, by Sally K. Ride (Washington, DC: Government Printing Office, August 1987); *Report of the 90-Day Study on Human Exploration of the Moon and Mars, NASA Report Prepared for the NASA Administrator* (Washington, DC: NASA, 20 November 1989); *Report of the Advisory Committee on the Future of the US Space Program*, by Norman R. Augustine, chairman (Washington, DC: Government Printing Office, 1990); and *America at the Threshold*.
31. See *America at the Threshold*.
32. See *Leadership and America's Future in Space*.
33. See Sagan, *Pale Blue Dot*, 285.
34. See John M. Logsdon, "A Sustainable Rationale for Human Spaceflight," *National Forum* (Summer 1992): 31.
35. See *Exploring the Moon and Mars: Choices for the Nation*, Office of Technology Assessment, Congress of the United States (U.S. Government Printing Office: Washington, DC, July 1991), 29-48.
36. One concern that must be considered if NASA was to receive immediate funding for a H-MFPE is that of enabling technologies. In other words, if enabling technologies are not first developed, then the prospects that huge masses (a "death star" or a "battle star galactica") would have to be launched into space with all the life support, fuel and protection on-board for the duration of the planned mission(s) is enhanced. This was one of the fundamental problems with the Space Exploration Initiative (SEI).
37. See Willy Z. Sadeh and Marvin E. Criswell, "Infrastructure for a Lunar Base," *Advances in Space Research* 18, no. 11 (1996): 139-148.
38. See Sadeh and Criswell, "Infrastructure for a Lunar Base," 139-148; and Willy Z. Sadeh and Marvin E. Criswell, "Inflatable Structures for a Lunar Base," *Journal of British Interplanetary Society* 48, no. 1 (1995): 33-38.

39. See Willy Z. Sadeh and Eligar Sadeh, "An Integrated Engineered Closed Controlled Ecosystem for a Lunar Base," *Advances in Space Research* 20, no. 10 (1997): 2001-2008; and Eligar Sadeh and Willy Z. Sadeh, "Bioregenerative Life Support Systems for Long-Term Space Habitation: A Conceptual Approach," *Life Support and Biosphere Science* 2, no. 3/4 (1996): 161-168.

40. The search for the origins of life is one of six fundamental questions of science and technology that provide a philosophical underpinning for why NASA exists and a foundation for NASA's goals. See *NASA Strategic Plan*, 3-5.

41. The mature base stage represents the emergence of a spacefaring civilization. Such a civilization is best exemplified by Krafft Ehricke's "extraterrestrial imperative." This imperative is based on the idea that humans expand into space for the same reasons that life moved through evolutionary stages—to grow by developing new capabilities. See Marsha Freeman, "Krafft Ehricke's Moon: A Lush Oasis of Life," *21st Century Science and Technology* 11, no. 2 (1998): 19-33.

42. Attributed to the late Dr. Willy Z. Sadeh, Professor of Space Engineering, Center for Engineering Infrastructure and Sciences in Space, Colorado State University.

43. For a comprehensive overview of space politics and policy see Eligar Sadeh, James P. Lester and Peter L. Hays, eds., *Space Politics and Policy: An Evolutionary Perspective* (Westport, Connecticut: Praeger, forthcoming).

44. See Bencke, *The Politics of Space*, 159-184.

45. An example of this shift is represented by NASA plans for the X-33/Venture Star reusable launch vehicle (RLV). Plans call for the use of NASA funds for R&D to enable the development of a working prototype with the intent that the private aerospace industry world then commercialize and develop an operational RLV.

46. See McCurdy, *Space and the American Imagination*, 2-3.

MAKING IT HAPPEN

Jonathan Stabb

Humans are, by nature explorers, adventurers, and conquerors of frontiers. It's in our hearts and minds and gives us purpose. In our drive to explore outside our Earthly domain, we seek who we are and where we stand in the universe. This exploration imperative seems all too familiar to this audience, but is often lost, changed or diffused outside this group. The challenge before us is to awaken the exploration wanderlust in others, specifically those that will take action. The action taken must be structured, specific, and sustained if we are to succeed.

Successfully sending humans to Mars and returning them safely will require great institutional support and true leaders. History has shown that if both of these are present, successful exploration is almost guaranteed. The Portuguese in the fifteenth century were tremendously successful in their explorations due to a collaborative national support and unique leaders. The Portuguese chose to explore in a progressive, systematic program, learning from the unknown and adapting as they went. The geography of Portugal further ingrained the exploration imperative into the minds and actions of its people. Looking away from civilized Europe, and out onto the Atlantic, the Portuguese people and their monarchs embraced exploration as an enriching endeavor. The leader of the initial explorations never made any grand expeditions himself but drew people together, organized resources and pointed the way.

This leader, Prince Henry the Navigator, son of King John I, was according to Daniel J. Boorstin,

> "...a curious combination of a bold heroic mind and an outreaching imagination, with an ascetic stay-at-home temperament. Frigid to individuals, he was passionate for grand ideals. His talents of obstinacy and his power to organize proved essential for the first great enterprise of modern discovery"[1]

Henry excelled in three areas all having analogies to Mars exploration development. Henry realized that before the actual expedition, exploring the unknown had to first be done in the mind. Although we know a great deal more about our destination as compared to the Portuguese, we need leaders to envision the specifics of the voyage. Secondly, Henry understood that the ability to come home was essential if people were to be enriched. This drove him to redesign the sailing vessels for the close continent exploration, rather than open seafaring. His 'caravel' ships had shallow drafts enabling greater access for trade, pointed better into the wind aiding in the return voyage, and were repairable. Operability and maintainability are quintessential on a voyage into the unknown as well as returning explorers home. Thirdly, Henry understood the power of accurate charts in exploration. The unknown, in those times, could be discovered only be clearing marking the boundaries of the known. This effort is currently underway through robotic exploration of Mars, establishing boundaries and limits for Mars mission designs. The Portuguese commitment to exploration did not end with Prince Henry's death in 1460, but continued throughout the fifteenth century under the reigns of King Alfonso V, and King John II.

In opposition to Portugal's continued institutional support, the Chinese imperial institutions could not sustain an exploration imperative. It was not for lack of seafaring resources or leaders, as the Chinese navigators (Cheng Ho in particular) had already sailed beyond the China Sea, around the Indian Ocean, down the east coast of Africa and beyond by the early 1400's, but rather a fickle national commitment to exploration. As Daniel J. Boorstin states,

> "Fully equipped with technology, the intelligence, and national resources to become discoverers, the Chinese doomed themselves to be the discovered"[2]

The underlying purpose of the massive Chinese expeditions, some consisting of hundreds of ships and tens of thousands of men, was to show the world that China *did not* need anything from the West. The voyages were to demonstrate the power and prestige of the new Ming dynasty. The Chinese, in opposition to European explorations, did not seek to return new ideas or goods to be incorporated into their society. Anything new or different was absorbed into the context of existing imperial folklore. Without the infusion of ideas and eventually capital, the Chinese emperors did not promote a second generation of expedition leaders nor exploration. In fact, they went in the exact opposite direction;

> "Cheng Ho's grand seventh voyage was China's last. His return home in 1433 marked an end to his country's organized seafaring adventures. An imperial edict in that year, and others that followed (1449, 1452) imposed increasingly savage punishments on Chinese who ventured abroad"[3]

These two examples and the context in which they occurred, illustrate the dependence of successful exploration on enabling leaders and institutional support. In defining our context and in arguing our case for exploration, these terms need to be clarified. For this paper, institutional support will be defined as having an empowered subset of the general public, the sponsoring organization and funding source(s) aligned towards the common goal of sending humans to Mars. Though this paper is presented with the government in mind, analogies may easily be made to the private sector. The empowered subset of the general public could be thought of as members of the Mars Society or Stockholders in a private corporation. The sponsoring organization, NASA in the public sector, could be replaced again with the Mars Society or a private corporation. The funding source, tax dollars approved by Congress, could be equated with the allocation of funds by a CEO or the head of a private organization.

Leaders in either model must be able to orchestrate actions within and between these organizations. Early leaders must focus on engendering support, while other leaders must focus on project develop and execution. The model, shown in Figure 1, demonstrates this two-phase approach.

We have, for the most part, achieved institutional support and implementation leadership for limited unmanned exploration of Mars with NASA as the sponsoring body. However, we need look no further than NASA's Strategic Plan to see erosion of institutional support for specific human exploration. NASA's current Strategic Plan describes a multitude of priorities. The Plan is both a response to Congressional and Executive Office's priorities and a framework for Agency action. We see a multiple path approach in the Agency's vision statement:

> "NASA is an investment in American's future. As explorers, pioneers, innovators, we boldly expand frontiers in air and space to inspire and serve America and to benefit the Quality of life on Earth"[4]

There is an emphasis both on low orbit and atmospheric activities, and on exploration. The Plan then defines the Agency's mission:[5]

> "To advance and communicate scientific knowledge and understanding of the Earth, the solar system, and the universe and use the environment of space for research"

"To explore, use, and enable the development of space for human enterprise"

"To research, develop, verify, and transfer advanced aeronautics, space, and related technologies"

Again we see the Agency being very careful in its purpose. "Enabling the development of space for human enterprise" can mean any number of space related activities. These words were carefully chosen to allow flexibility and not force a commitment to a specific course of action. However, it is just this flexibility which would doom any Mars imperative. The words must be changed, if this sponsoring organization is to conduct human exploration. Whether the sponsoring organization is governmental or private, the mission statement must be clear and directed. A statement such as, "Determine through visitation which near Earth bodies and planets can support human civilizations" would suffice.

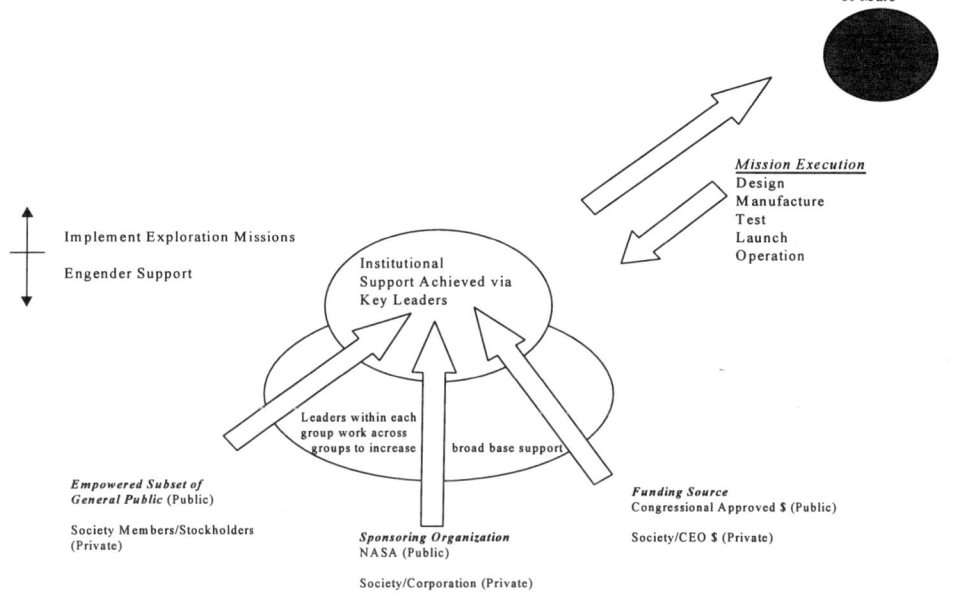

Figure 1

The Plan is the basis for establishing NASA priorities. It contains sections on Domestic Policy, Foreign Policy, Political and Public Support, and Key Factors. It provides for the re-organizing of priorities based upon changes to any one of these areas. Since we are the Public, then we need to awaken the inherent spirit of exploration that lies within each of us. The Agency will change its Plan, and activities only under duress from external forces. If the Agency allows inclusion, then let's be included. Since the Plan states that the public is the ultimate beneficiary, should not they be included in the development of the Plan? If the Agency is serious about International cooperation, should they not also be given an opportunity to impact the Plan?

Without NASA, Congress, or Executive Office Leadership courageous enough to set the Mars priority, the people must organize and make their voice heard. The people of the United States, particularly, can have significant impact on the direction Congress and NASA follows. Although our citizens have always had the power to influence, direct and even create policy, they rarely exercise that power. The current feeling among many Americans is one of, "I have no voice" or "I don't have the time", or any number of other excuses. However, the framework is set for the people's wishes to be heard, and, more importantly, acted upon. The current interaction between the general public, NASA, and Congressional or Executive Office decision-makers is shown in Figure 2.

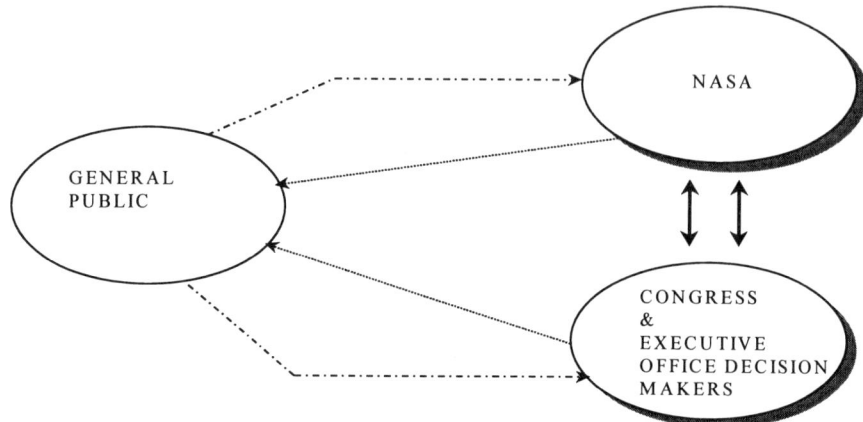

Figure 2   Current Interaction.

The short double arrows on the right represent the current strong relationship NASA has on Congress and the Administration. NASA promised to deliver current Programs and Projects on time and within budget to the Congress. This is the agency's primary focus, and has committed little to the people. When a success occur, the dashed line represents NASA's communication of that success to the general public. This is done in a passive mode for the most part, where those in the general population having an interest can visit websites, or purchase Space related publications. If the Agency believes that some success is truly noteworthy, then a Press conference or other press event may be conducted. More often than not, these Press conferences serve the political master more than the public. The dashed line going from the Congress and Administration back to the general public represents the passive means used by our elected officials to report back to their constituency. Once again, it is up to the motivated citizen to verify if those elected officials are performing according to their promises and governing goals.

The dotted-dashed lines going from the general public back to NASA and the Congress represent the general public's current influence and communication of their dissatisfaction with those groups. NASA and Congress see this communication flow as votes and survey results, and certainly not a driver of policy or financial decisions. The interested public, to this point, has usually had very little say in the direction which these organizations execute space activities. One notable exception is the recent victory by the Planetary Society and Mars Society in getting Congress to restore funding for the Mars 2001 mission. This direct and targeted action by the people is exactly what needs to happen with increasing frequency.

It is easy to simply change the model, transferring the dotted-dashed lines into solid ones, but extremely difficult in the real world. Paradigms shift, power is transferred, master becomes slave. But let us change the model and see what it means (Figure 3)

In our new model, the empowered and growing subset of the general public now flows information and opinions to specific members of NASA and the Congress/Administration. This information includes the empowered public's *priorities for the Agency*, specific strategies for implementing those priorities and in broad terms what amount of resource should be dedicated to those priorities. The empowered public now demands accountability from NASA and Congress/Executive Office officials to act upon these strategies. Members of both organizations must still 'champion' the public's wishes, but now they have a viable force behind them. As other fractions of the general public increase (or maintain) their influence upon NASA and Congress/Executive Office, the champions must demonstrate, convince, explain, the Mars priority on a daily basis and bring the Mars community into the fray. When the empowered public is sufficiently large, leaders within the sponsoring organization no longer have to try to determine the 'compelling reason' for human Mars exploration, *the people demand it.*

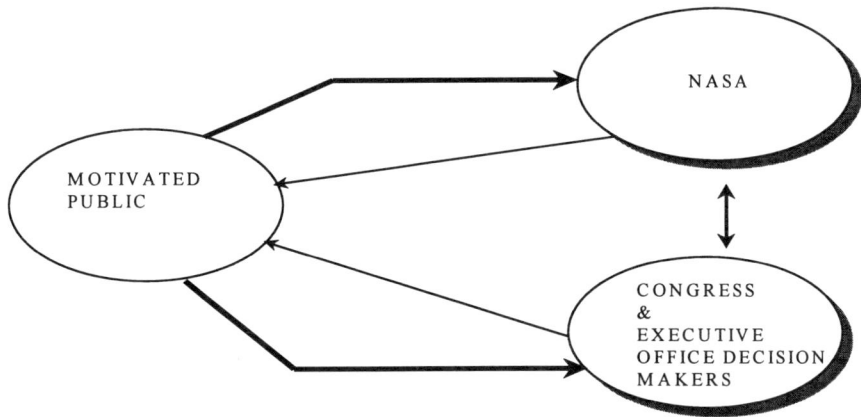

Figure 3   Desired Interaction.

The transfer of power from within the sponsoring organization to the empowered public will not occur until leaders within each group have a span of influence in both groups. These leaders must be clever in their approach, working within each group's culture to engender support. Clear target audiences must be contacted, made to care, and spurred to action. How are we to do that? Let's examine some possible routes.

There already exist a large number of individuals who wish to send humanity beyond Earth. They visit the Air & Space museum by the ten of thousands and pay movie-makers millions of dollars to bring space adventure into their lives. They go to work everyday for aerospace companies or pursue knowledge and scientific research at our universities. Increasing the number of people who will take action is the primary goal. An easy way to do this is to set up a touch screen survey, much like the ones in place at many businesses (Einstein Bros. Bagel stores, for example). Museums, theme parks, or movie theaters immediately come to mind as potential venues. As one walks out of these venues, the screen can query the interested person with four or five quick questions. I suggest something like these:

1. Do you feel Human Exploration of Mars is important? (Y/N)

2. Do you feel NASA should be trying to achieve this goal? (Y/N)

3. Would you like more information of current and future plans for Mars exploration? (Y/N)

4. Would you like to send these answers to the head of NASA and members of Congress which control NASA funding? (Y/N).

The final query would ask for additional personal information (name, address, phone, and e-mail) and if the respondent wishes to be part of the planning for the Mars endeavor. The key is to make the respondents immediately feel part of the endeavor, and show them that their participation matters.

Many of us also witness the dramatic change from conference ebullience to workplace drudgery. It is painful for me to see a group of skilled, motivated, intelligent people toiling through uninspiring work. The answer is again, not easy. Current work must be completed and is relevant to our organization's current goals and may be supporting Mars exploration projects. When I discuss human Mars missions with co-workers at the Kennedy Space Center, the response is generally, "When do I have time to change the direction of the Agency?" or "That's too far in the future". Champions within the sponsoring organization must understand this and make it easier for workers to have meaningful input and impact in Agency policy. Not only must we become known as "Those Mars people" at work, but also outside of work. We should share the excitement we feel for human Mars exploration with family, friends, neighbors. We should take the time to educate those that want to know more. This can be done by forming local chapters (of the Mars Society) or by starting a 'Mars Club' in our communities. These clubs would provide the needed structure for focusing member action and for recruiting new members.

If we are to be successful in engendering support, our leaders must first get people's attention. Charisma is essential. No matter how good one's argument is, no matter what the data says, a detached bureaucrat will inspire nothing. Secondly, these leaders must succeed in doing. Nothing loses potential supporters quicker than promising action and not delivering. If the leader of the agency exploration office states we are going to fly a human related experiment on a planned robotic mission, it better happen, or credibility is lost, regardless of the reason. Leaders must articulate the Mars imperative adroitly, tailor the argument to the audience, and sustain the vision. Most people will follow real leaders, not because of their organizational standing, but because of what they do and the vision they have.

Once we have succeeded in defining the Mars exploration mission(s) and have funding commitment to conduct it, a different type of leader is required. The leader of this team, must keep the Program focused and *implement* lessons learned from previous programs. The Program Management can not be large and encumbering, the design and test requirements can not be constantly changing, and the workforce must be Program dedicated. The leaders must be blunt and to the point keeping external factors way from the workforce. The leaders must heed the advice provided from the project workforce including field and manufacturing personnel. These leadership challenges are numerous and varied, but should never be underestimated. Engendering support is the battle now, but the technical development and operational deployment of Human Mars explorations are themselves, heroic tasks.

I believe history's patterns of exploration provide clear models for today's actions. We have the necessary leaders gathered here to engender support, align our institutions, and eventually conduct successful human Mars expeditions. People, ultimately, drive all institutions and can create the fertile environment for enriching exploration. We must start today, continue tomorrow, and make the future one in which Humanity grows beyond its Earthly home.

REFERENCE NOTES

1. Daniel J. Boorstin, The Discoverers: A History Of Man's Search To Know His World And Himself (New York: Vintage Books, 1985) 159.
2. Boorstin, 201.
3. Boorstin, 199.
4. National Aeronautics and Space Administration, "NASA Strategic Plan: 1998 NASA Policy Directive (NPD)-1000.1," 8.
5. National Aeronautics and Space Administration, 8.

A SOCIALLY SUPPORTABLE MARS COLONIZATION PROGRAM: EARTH MARS AMBASSADORS

Philip A Turek[*]

Earth is a shrinking world. Thirty years ago the Earth was seemingly a vast world of nearly limitless resources; now it is but a small planet of acutely limited resources as perceived by the general public. Talk of exploring and colonizing Mars increases steadily as new data streams in from probes at the red planet. During the Apollo missions the public was merely a passive spectator. This paper advances a new approach to Mars's exploration, one that provides for the direct, sustained, and increasing involvement of individuals and groups with the success of the program. The project's mission statement is to establish a new biosphere on the surface of Mars capable of supporting a new civilization. The method of the project (named Earth Mars Ambassadors, or EMA) is to start preparing future colonists and ambassadors NOW as the first phase. The second phase is to launch a manned crew to Mars for a long duration combined exploration/colonization mission. The critical third phase splits the first manned crew on Mars into one segment that remains on Mars as colonists while one segment of the crew returns to Earth and one segment of the crew comes to Earth as ambassadors from the new colony on Mars. Innovative techniques sustain public involvement throughout each phase. Tasked with applying Martian experiences, perceptions, and innovations toward the environmental restoration and management of Earth, the first Martian ambassador to Earth could arrive as early as 2014. Students will determine whether this program succeeds or fails.

INTRODUCTION

The public's general perception is that our nation's space program lacks a strong vision and has no clear focus. This perception partly results from the contrast of current space activities with the Apollo program. The advantage of the Apollo program was that its clear vision focused all activities efficiently on landing a man on the moon and returning him safely to the Earth by the end of the 1960s.

While Apollo engaged the public's imagination, few could directly *participate* in the project. (Hundreds of thousands of people worked on the Apollo program, but not as members of the public; rather, they were under contract.) American taxpayers were spectators during Apollo, passively watching the moon landings on their television sets. A graph of public involvement during Apollo would look like the pulse of a square wave: zero public involvement before launch followed by saturation media coverage from launch to splashdown, abruptly ending at splashdown with a dry spell of months with no further coverage. As there was no real direct public involvement with Apollo, public interest in the program faded quickly after the first moon landing. This implies Mars exploration programs that don't involve the public won't be publicly funded.

At the same time that the American public perceives our space program to be lacking in vision and focus, they exhibit a broad and occasionally intense interest in space exploration.

---

[*] Physics Teacher, Whitney High School, 16800 Shoemaker Avenue, Cerritos, CA 90703

For evidence of this, look at the amount of coverage given to various space topics over the past year. Movies such as *Contact* and *Deep Impact* were both scientifically plausible and generally popular. The discovery of probable water ice on the moon stirred the public's interest. The Mars Pathfinder website received 700 million hits during the first four months of its mission [1]. More recently, the Mars Global Surveyor image of *The Face* on Mars made front-page news last April. Mars Global Surveyor captured *The Face* in *response* to public interest in the matter. In the 1960s Americans only thought of space in terms of our competition with the Soviets. Americans today have broader interests in space than in 1961.

Figure 1   Public interest in Apollo ended when Apollo 11 safely splashed down (NASA Photo).

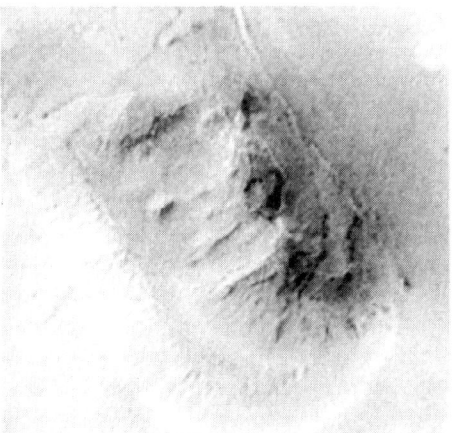

Figure 2   Public interest affected the utilization of spacecraft resources around Mars (NASA Mars Global Surveyor image of *The Face*).

PROGRAM LAYOUT

"The future belongs to those who believe in the beauty of their dreams." – Eleanor Roosevelt

The strategy of project Earth Mars Ambassadors (EMA) is to involve key elements of society—particularly students, environmentalists, and congressmen—in the planning and implementation of a manned mission to Mars. Each one of these three particular elements has the capability to block the realization of a human mission to Mars. Working together, the three elements could ensure the success of a manned mission to Mars.

Students	Environmentalists	Congress
Teachers	General Public	International Participants

Figure 3   EMA needs environmentalists and other elements of society for success.

To meet the requisite that these elements become directly involved with a manned Mars mission, it's necessary to invent a unique mission. Most people discussing Mars view the arrival of humans at Mars as the goal of any mission. EMA views the arrival at Earth of humans from Mars as the essential goal of its mission. This distinction makes Mars exploration socially supportable.

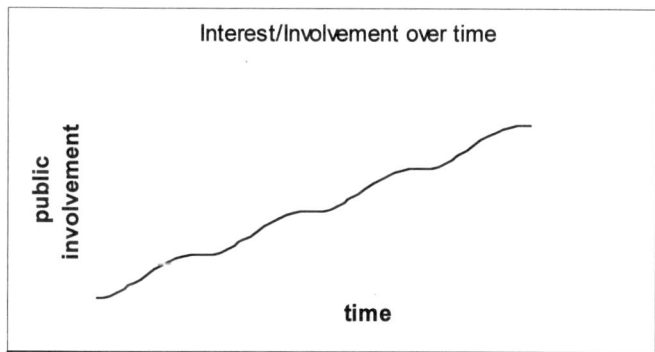

Figure 4   EMA increases public interest in space exploration by increasing public involvement.

MISSION STATEMENT

Earth is filled with Life. Her teeming masses, her restless swarms of Humanity crowd each other in the eternal competition for room to grow. If left confined to this Earth, then this planet's growth will be stunted, the competition for Earth's dwindling resources will become increasingly fierce, and the quality of Life for all but the most elite will steadily decline.

Therefore, the network of Life on Earth needs a new planet. Such a planet should provide the raw materials needed to build a new ecosystem capable of supporting a new civilization. The world's best and brightest scientists, engineers, and farmers are invited to build a new ecosystem capable of supporting a new civilization on Mars. Ideally, as many of Earth's diverse life forms as possible are to be adapted to living on this new planet. The experience gained from the endeavor is to be applied to the restoration of Earth's environment, maximizing the benefits of this undertaking to the quality of life on both worlds.

EMA is divided into three phases to facilitate the inclusion of disparate segments of society. A summary of EMA's three phases is presented in the table below. (A more detailed version of Table 1 is located at http://www-sci.lib.uci.edu/SEP/STSHTML/PTurek/emahome.html.)

Phase 1 EARTH		Phase 2 MARS	Phase 3 EARTH
Be sent out in teams to explore Earth.		*Explore and begin to colonize Mars.*	*Use your wisdom to help solve problems on Earth.*
Grade	*Territory*	Unmanned rovers and a base will precede humans to Mars. Technical details and logistics are congruent with Dr. Robert Zubrin's *Mars Direct* plan. Earth directly participates with the astronauts on Mars. At least one astronaut-turned-colonist will remain on Mars permanently as the first Martian (colonist).	Astronauts return to Earth. At least one astronaut-turned-Martian-colonist comes to Earth as the first Martian ambassador to Earth. The US Congress selects the train route(s) used by the Martian ambassador. The Martians are tasked with helping to restore Earth's environment by sharing perceptions and innovations.
K-3	Classroom, school grounds		
4-6	Neighborhood		
7-8	County		
9-10	State		
11-12	Country		
13-16	World		
17-18	World in-depth		
Astronaut	Near Earth space, Moon		
Ambassador	Mars colonist		

Table 1 EMA program layout breaks the first manned exploration of Mars into three phases. Students explore Earth during phase one, to become astronauts who explore Mars during phase two. At least one astronaut-turned-colonist comes to Earth as the first Martian ambassador during phase three.

EMA STUDENT EXPERIENCES TO DATE

During all three phases of EMA the students/astronauts/ambassadors keep journals of their experiences and perceptions. In 1993 a team of EMA students was dispatched from Santa Ana High School one Saturday. Their mission: collect and chemically analyze water at a natural pond located approximately 10 miles from campus. It began well, but an unexpected problem came up. Heavy rains two days earlier had rendered the trail through the woods to the pond impassable. Faced with this challenge the team worked out a way to reach the pond without getting their feet wet. They successfully completed their mission before lunch.

Figure 5 A flooded wooded trail was an unexpected challenge on this EMA field trip.

This simple example highlights the importance of the phase one field trips. It's one thing to sit in a room and read about what's outdoors; it's quite another thing to actually go out there and experience it directly. Unexpected challenges occur routinely in any outing. By going through these experiences and facing these challenges the participating students are that much better prepared for facing the unexpected challenges they'll eventually encounter on Mars.

Figure 6 Students from California and Maryland merged to form one team, and in the process, explored the Earth and themselves.

In 1996 a team of students at Whitney High School was tasked with merging with a team from a high school on the other side of the country to compete in a space colony design contest. Student Robert Lopez kept a journal of the experience, detailing his trip from a Los Angeles high school to a Maryland high school to the competition site at Kennedy Space Center in Florida, and (most importantly) back to his high school in Los Angeles. Robert became adopted by the other high school during the days he spent with them. His perceptions of life in southern California changed appreciably as a result of the brief time he spent living in a different environment. You can read Robert's journal yourself, as he's posted it at http://home.earthlink.net/~ral1120/ema/ema.htm. Robert's growth away from home enabled him to perceive his home environment more clearly. This illustrates why the restoration of Earth can't be accomplished by simply doing EMA phases one and three.

Some might argue that involving students in EMA wastes the students' time. Fortunately, EMA phase one is of great benefit to its participants even if phase two never occurs. EMA is about personal growth. Students explore their environment and themselves. They experience the interactions of being part of a team, and they develop the self-confidence to act independently when the need arises. They also see how they fit into the big picture of life on Earth, and become aware of the social and physical challenges facing the inhabitants of this planet. In short, EMA is a means for students to grow and reach high levels of maturity and wisdom, and that is something that every teacher (who still cares about their students) longs for. This fundamental objective of EMA phase one and its universal appeal to educators accounts for the support this project receives from the Science Education office of the University of California Irvine.

Figure 7   Dr. King Huber lectures on geologic processes to a team visiting Devil's Postpile, California. Note the journal writer at the stump.

PUBLIC INVOLVEMENT TARGET

Someone once claimed that Christopher Columbus set sail for the New World with only 3% of Europe's population supportive of undertaking voyages of exploration [2]. Although this value is almost certainly somewhat arbitrary, it is instructive to use it as a metric for measuring the progress of project EMA. Not knowing better, the current target for EMA involvement is 3% of the U.S. population. The U.S. population will be roughly 265 million in 2010. Eight million citizens are three percent of 265 million. Eight million of the nation's 70 million students (K-16) amount to 11.4% of the student population [3].

Suppose each school participating in EMA contained 100 students actively involved with the project in 1998, and that this number of students increased by three students per year at a given school for the next 10 years. Further, suppose that each school recruits the participation of two to three new schools each year (to a maximum of 80% of all U.S. schools). Assume, also, that in any given year no more than 80% of the schools in the program renew their participation the following year. Beginning with two EMA schools in 1998, this growth rate results in 8 million students involved with the project in 2009 at 62000 schools throughout the nation. The chart is shown in Figure 8; the data is listed in Table 2.

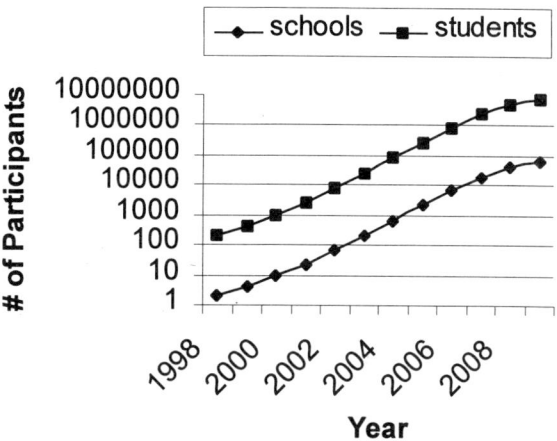

Figure 8   Student involvement in EMA could exceed 8 million U.S. Students by 2009.

Year	1998	1999	2000	2001	2002	2003
schools	2	4	9	23	67	209
students	200	412	954	2507	7504	24035
growth rate	2.0	2.25	2.59	2.92	3.12	3.23

Year	2004	2005	2006	2007	2008	2009
schools	675	2180	6779	19523	41193	62201
students	79650	263780	840596	2479421	5313897	8086130
growth rate	3.23	3.11	2.88	2.11	1.51	

Table 2 Estimated student involvement in EMA from 1998 through 2009. Growth rate peaks around 2003-2004.

Given that the youngest NASA astronaut is 28 years old, most of these EMA students will still be too young to enter into astronaut selection and then emerge from training in time to meet the target launch date of 2009. It may be possible to accelerate the EMA student involvement rate in the early years to qualify more candidates for the target launch date, or slip the target launch date. There will be an abundance of EMA students in the pipeline for successive missions.

WHITNEY PILOT PROGRAM

Whitney High School (in Cerritos, CA) has agreed to implement EMA as a pilot program during the 1998-99 academic year. All eighth grade science students (Whitney encompasses grades 7-12) will receive an EMA overview briefing during the Fall semester. All Advanced Placement (AP) U.S. History students will take part in an in-depth instruction and discussion period on EMA in roughly the middle of the school year. (The event is structured as an extension to the study of the 1890s leading to an exploration of the political aspects of EMA.) Monthly meetings open to faculty and students interested in becoming involved with EMA will be enacted. EMA students will take fields trips generally within the greater Los Angeles area. Some of the students may undergo cultural exchanges with students at other high schools, though this is difficult to arrange. (The goal is to have students attend other schools long enough for them to appreciate and understand the different environment.) Whitney's library staff has agreed to store journals generated by EMA students who've undertaken field trips. In a related activity, Whitney students will most likely continue to compete in a contest called SpaceSet, which entails broad exposure to the challenges of completing a major advanced aerospace design project.

Figure 9 A joint Russian/NASA Ames rover attracts two EMA students. 1997 International Planetary Rover Conference, Santa Monica, California.

INVOLVING ENVIRONMENTALISTS

Some initial work was done years ago in identifying the potential compatibility of specific environmental organizations with EMA. The most promising group was The Nature Conservancy. They are pragmatic, apolitical, and admirably successful in their efforts to protect endangered species from extinction. Aside from some brief first contacts with local managers (years ago, at that) no interactions have occurred. The potential for mutual benefit is high; the task of forging links between environmental groups and EMA remains dormant.

U.S. CONGRESS

Applying the 3% rule of thumb to the U.S. congress implies the active support of a minimum of 16 congressmen (3% of 535 members of congress) is required before EMA can even begin to stand a chance of being seriously considered by congress. EMA phase three calls for a tour of the United States by the Martian ambassador.

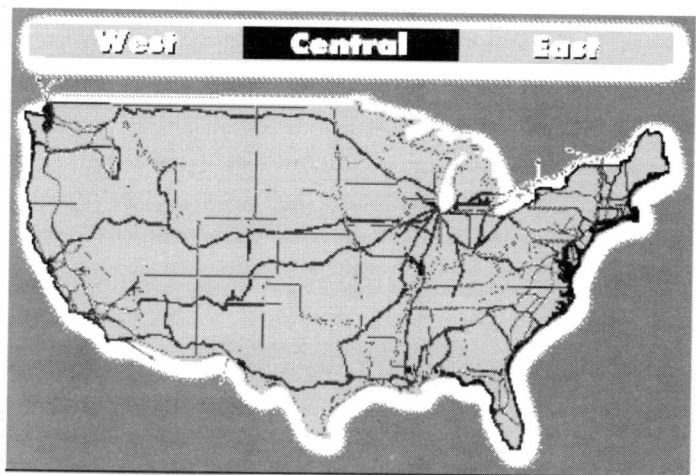

Figure 10   Congress selects the Martian Ambassador's Train Route across the U.S.
Lines indicate Amtrak's main routes.

Practically speaking, the ambassador is likely going to be initially quarantined (just as the Apollo 11 astronauts were quarantined when they arrived on Earth). In addition, the ambassador will likely be initially poorly adapted to functioning in Earth's gravity. To meet these three requirements – tour the U.S., while quarantined, all while still somewhat physically impaired – travel by train was selected. The technical details are unimportant here; the role of congress centers on the train route itself. The ambassador's landing point and train travel route are to be decided by congress. The train route will unavoidably become convoluted by competing political pressures. Thus, members of congress have the opportunity to become fully engaged with this project.

INTERNATIONAL PARTICIPANTS

No significant progress has been made in this area at this time.

THE 80/20 RULE AND GENERAL PUBLIC INVOLVEMENT

NASA focuses on science. Science is good, but humanity's interests range far beyond scientific curiosity. What was it *like* to walk on the Moon? How did the trip change your perspective on life? These questions fall outside NASA's program, yet they are the commonest questions asked by a curious public. A manned mission to Mars ought not to be restricted to purely scientific observations; the public's needs exceed that.

EMA phase two addresses this need with the 80/20 rule. Eighty/twenty – the percentage allocation of mission resources. Eighty percent of the time, mission resources (rovers, astronauts, satellites, communication bandwidth, ...) are dedicated to scientific objectives – the standard NASA mission profile. Twenty percent of the time these resources are allocated to art, culture, philosophy, social relations, and even politics. This opens up a myriad of possibilities for direct public involvement with the first manned mission to Mars.

ADDITIONAL INVOLVING MECHANISMS

There are more mechanisms than just the 80/20 rule for directly involving the general public with the exploration and colonization of Mars. A lot of environmental restoration work needs to be done – on a local scale, and more than environmental groups alone can handle – by the beginning of EMA phase three. Baseline data regarding the conditions of life on Earth (particularly cultural characterizations) need to be observed, recorded, and preserved during EMA phase one. Families on vacation could tour the proposed Martian ambassador train routes, contributing their preferences to the congressional deliberations. A government scholarship program for EMA students is desirable. The scholarship should specifically encourage EMA students who've earned Master's degrees to go on to earn two additional Master's degrees in two disparate disciplines, thereby ensuring broad and in-depth educational backgrounds for the EMA phase two colonists.

Figure 11 Families on vacation by train: Susie records her impressions in her journal, Johnny plays with a scientific gadget, while their parents study an ambassador route map (Amtrak Photo).

CURRENT EMA STUDENT DISTRIBUTION CHART

Over the past five years perhaps 500 students have been exposed to the overall concept of project EMA. Most of them left it at that. Some, however, seemed to retain more than just a fleeting awareness of the concept. The estimated numbers and distribution of those students now in college are listed in Table 3.

College/University	# of EMA Students	College/University	# of EMA Students
UC Berkeley	18	USC	3
UCLA	5	Pomona	2
Stanford	4	UCI	2
Harvard	4	Cornell	1
UCSD	3	Duke	1

Table 3 Estimated College EMA Student Distribution in 1998.

STEPS TO IMPLEMENTING EMA

Can we do this? The answer is, it depends on the students. They have the potential to make this project happen. The strategy is outlined below.

1. Build up a trained core of EMA students.
   - Start with science teachers and students.
   - Branch to involve math, history, and English teachers and students.
   - Link teachers and students between schools.

2. When EMA students reach college, they link up with EMA students from other high schools.
   - College EMA student chapters form and then involve their faculty.
   - College EMA students revisit their high schools to inspire the new student body to be more involved with EMA.
   - College EMA students vigorously promote EMA as a real, international program.

3. The energy and enthusiasm of the world's students determines whether EMA flourishes or dies.

FOR MORE INFORMATION

The author can be reached at: pturek@laedu.lalc.k12.ca.us

The EMA website is located at:
http://www-sci.lib.uci.edu/SEP/STSHTML/PTurek/emahome.html

REFERENCES

1. JPL press release of Oct. 1997. See NBC News article at [http://mars.jpl.nasa.gov/press/msnbc/index.html] (Aug 6, 1998).
2. John L Mathews, Fullerton, CA. 1990. Private communication.
3. U.S. Bureau of the Census, *Statistical Abstract of the United States: 1992* (112th edition.) Washington, DC. p.8, pp.135-149 (1992).

THE GEN-X RALLYING CRY? TO MARS!
Gen-X Needs a Cause, and That Cause Should Be Space

George T. Whitesides*

Gen-X is the first generation in history with the chance to send one of its own to Mars. Here is why we must go.

We are a curious generation, both lucky and adrift. Blessed with prosperity, peace, and education, we are on the verge of inheriting the strongest country in the world, at one its greatest moments. In just a few years we will hold an unparalleled opportunity to do something great for mankind, to make a deep and lasting mark on history.

There is only one problem: as a generation, we have no grand aspirations for our future. Yes, some of us care deeply about issues like the environment or service, but is there a goal out there that we seek together, that unifies and inspires us? Corner one of us, look us in the eye, and you'll see that the answer is no.

This much-publicized ennui has its roots in the happy condition of our land. Living in a world without advancing evil, we have had the luxury of turning inward. Raised in a country in which the great injustices have been at least legally remedied, we have been left to work out the details of battles fought before our time.

With some hesitation, we admit that we wish we did not live in such uninteresting times. Privately, we yearn for a cause of our own, a grand public effort that would take us beyond the video-game society in which we have been raised.

What we have not realized is that there is a cause out there for us, one that is everything that we secretly hope for.

It is space exploration in general, and Mars in particular.

Making space our cause, and Mars our goal, would bring us together under a grand, peaceful endeavor. It would teach us about our universe in ways that we can barely imagine, and it would spur technology to greater heights. It would be a great gift to our children, both for their education and their future. To go to Mars would transform us from a generation that history will forget into a generation that history will revere.

Space is the next frontier of humanity, and we are the next generation. Mars, an uninhabited planet whose environment is closest to our own, waits, empty, for our arrival. If ever there was a match made in heaven, this is it.

So why haven't we realized it? It is not hard to understand. To date our relationship with space has been defined by two things: our parents and the Challenger disaster.

---

* The author works for a space business company and serves as a Trustee for Princeton University. He is 24.

Gen-X has been told since birth that the glorious moments in space history came before our time. We learned about the Apollo moon landings from endless TV documentaries, heard again and again from our parents how much pride and wonder they felt while watching Armstrong's steps.

Our defining moment with space, in contrast, was not stunning success but tragic failure. Challenger is our generation's tragedy, our moral equivalent of JFK's assassination. Where were you? I was in my sixth grade classroom, and I remember looking at Kevin Frasier and feeling loss for the first time.

With this background it is clear why we feel that space is not our generation's issue. Space was our parents' triumph, and our tragedy.

But space is our future too. And if we choose, Mars can one day be our present. We have much to gain.

It will give us a goal and a destination, at a time when too many of us feel neither.

It will yield wide-ranging scientific and technological benefits, at a time when Earth's environment seems precariously delicate. By learning about Mars' environment, we will learn valuable lessons about our own.

It will give us spiritual benefits, teaching us about our role and place in the universe, at a time when many of us feel lost. Maybe we learn for sure whether there was life over there. Maybe we realize that we can live beyond our Earth. By going to Mars we will learn more about ourselves.

It will reinvigorate our nation's education system, at a time when it could use some new life. Skeptical? Apollo has already done it once. By pushing the boundaries of science we will spark interest in its fundamentals.

Finally, in a time when the World Wide Web and genetic manipulation are drawing us towards a future none of us can be sure of, making Mars our goal will put us on a path in which we can all believe.

The funny thing is that we are interested in space; our media habits prove it. By now we've all heard that the Mars Pathfinder mission web-site was the most visited site in history. Well, who uses the Web? We do! Gen-X's film and TV entertainment—from Apollo 13 to Contact to Star Trek to Tom Hank's new series—is rife with space. It may be that at some level we already realize how space answers our needs.

The technology and cost for a Mars mission are within our grasp. We start by building on the achievements of our parents and grandparents, from Apollo to the Shuttle to Mir. To that fine foundation we add recent technological innovations—from computers to advanced materials—that the aerospace industry has been too slow to adopt. When we add it up we will see that the cost is far less than we expect, and far less than NASA has predicted in the past.

To focus our minds we should set a date for our departure. I know the perfect day: January 28th, 2011, the twenty-fifth anniversary of the Challenger explosion. What better testament to the astronauts who died—and to the character of our generation—than to answer that tragedy with this resounding triumph?

We have the technology, the interest, and the motivation. We have the reasons: spiritual, educational, and technological. Now we must take our blessings and do something truly great with them. In this historic moment of peace and prosperity, we have the chance to marshal our resources and take the next step in our exploration of the universe. The greatest achievement in human history is within our grasp. We have only to decide to go. So, Gen-X, what are we waiting for? To Mars!

Chapter 5
VOICES OF YOUTH

TV crews were everywhere.

OUR FUTURE ON MARS FROM MY PERSPECTIVE AS A TWELVE YEAR OLD

Kathleen B. Bohné

My name is Kathleen Bohné and I am twelve years old. I am here today because I think the possibility of going to Mars is the most exciting thing that has ever happened to the human race. I really want to see this happen and I hope to be a part of it.

I have been interested in astronomy for what seems to me a long time and a few months ago my Dad and I went to hear Dr. Zubrin speak. It was exhilarating. The possibilities seemed boundless and I felt my excitement grow with everything he said—a lot like that Christmas Eve feeling!

Last year *Life* Magazine listed the 100 most important events and people of this millennium. My sisters and I spent hours pouring over this magazine. The invention of the printing press was number one of course. Newton's discovery of gravity, Columbus's journey and Einstein's Theory of Relativity also made the list.

I can easily imagine the editors of *Life* sitting around a conference table in 2998 arguing about what numbers two and three will be but surely the colonization of Mars will be number one. Even considering the pace of technological miracles that we seem to be living with, what could compare to the impact this will have on the world as we know it?

Think of the questions this raises. The "how?" "when?" "who?" and oddly enough, to some people, the "why?" This seems to me to be the easiest question to answer. Because it's there!

There are certainly similarities between the age of exploration during the 15th and 16th centuries and our ambitions today. They were driven by a pioneering spirit, a quest for knowledge, a desire to spread their religious beliefs, and in great measure, greed. Gold, silk, spices and the wealth they brought were indeed great motivators! I don't imagine anyone is anticipating bringing Martian riches back in the name of the queen! But we do still have that pioneering spirit and drive for knowledge. We want to know what's out there and find a way to make it our own—another kind of greed I suppose.

Mars has always been a place of wonder and mystery to earthlings, and the current idea that there could be humans on Mars in less than a decade is, needless to say, a mind-boggling thought. It has been one of mankind's wildest and most dazzling dreams for so long, and now it finally seems to be coming true.

As I am only 12, I could be one of the first people to set foot on the rusty Martian surface. I might climb the Twin Peaks or walk along the dried up channels that split the soil. My children might taste the first crops grown on Mars or be among the first to stand in the bottom of that enormous canyon system, Valles Marineris. My grandchildren could see a permanent human settlement on Mars and might even live in it. The possibilities are limitless with technology progressing at its current rate.

I often wonder what everyday life would be like on Mars. Would things be so alien that everyday activities such as showering would become strange adventures? Or would it be relatively similar to life on Earth, despite the obvious differences. Of course a lot of the answers to questions such as these depend on our approach to colonization.

Will we attempt to terraform Mars or will we live enclosed in a large "bubble" where the atmosphere and temperature have been adjusted for humans. This would make Mars seem like a vast, red desert. Exploring could only take place in specialized suits and vehicles. Daily life would boil down to a fight for survival. And how strange not to have life all around us. We are used to living things and movement—plants, animals, crashing waves. On Mars there is no life that we know of, certainly none we could see or feel. We would be the only living souls.

If terraforming became a possibility things would change dramatically. Eventually by warming the poles to thicken the atmosphere and by oxygenating the planet, everyday life could almost mirror life on Earth. We could go from mere survival to truly living. Of course, we would end up exporting all of our problems as well. It would be natural for us to try to make Mars a utopia. But whose idea of utopia?

I wonder what exporting "human nature" will mean and I guess only time will tell. Will we change in the process? Human nature seems to have remained basically the same at least throughout recorded history. I'd like to think we're a little more civilized and humane, but perhaps not.

I do think that having a new frontier is vital. It seems to bring out the best in us as individuals and as a nation. It keeps us always looking outward. Having this in our future is like having the promise of a great surprise yet to come. One that we know is special, vital and completely wonderful but still ultimately mysterious.

These are major questions which will only be answered in the years to come, but we can at least try to lift the veil of mystery about living on another planet by sending humans to Mars. This will be the first step to reaching the pinnacle of the human imagination, making another planet our own.

I hope you will all keep fighting for this and I can't wait to join you. Thank you very much.

THE HAKLUYT PRIZE LETTER

Adrian Hon[*]

Dear Sir,

I am writing to you as a part of the Mars Society Hakluyt contest, in which students have to make the case for initiating a humans to Mars program. Sending humans to Mars could prove to be one of the most important events of the next century; however, it is not a decision made easily. I hope to outline some of the arguments for and against a Mars program in this letter, perhaps in some way helping you to make an informed judgement about the merits of such an endeavour.

Firstly, why should we send humans to Mars? Why not send machines? The recent Pathfinder mission was remarkably successful, quick and cheap. However, as with all machines, Pathfinder and those which will follow it can only perform certain tasks and have a limited operating lifetime. Humans can carry out a much wider range of experiments and the presence of humans and their problem solving skills would overcome any barriers, as would their ability for instant decision making. Pathfinder, on the other hand, suffered a long communications delay due to the distances involved between Mars and Earth. On a machine, there can be only so many backups and redundancies. The benefits of humans over machines are justifiably great enough to warrant the extra cost, risk and time. Besides, these are only reasons of practicality. Only humans will be able to accomplish the true utilisation of the resources of Mars.

The sight of a man walking on the Moon brought dignity and hope for mankind, a defining moment in our history. In the same way, men walking on Mars would give a tangible symbol of international co-operation and of how far the human race has come, but also prove to be the starting point of regular manned missions to Mars, leading to the ultimate goal of colonisation. The advantages of the colonisation of Mars would be many-fold—a new frontier would be created for mankind.

When America was colonised the entire world reaped the benefits from the 'New World'. The cost of the colonisation was great, but the new ideas and technology the Americans created far outweighed the expense. The famed 'Yankee ingenuity' was born when the colonists had to change their ways and were under pressure from the untouched land into thinking up new ideas for living and working which changed the world. Colonists of Mars will be faced with new challenges and in true human spirit and ingenuity create new solutions, solutions which could again revolutionise the way we live.

Scientists have remarked that Mars appears to be so suitable for human colonisation that it seems to have been put there on purpose. Situated close enough to the sun to receive sufficient light and heat for comparatively trouble free colonisation, Mars also possesses huge resources of water in the north polar ice cap and the permafrost. In relative terms, this far outclasses the stores of water recently discovered on the Moon. It would certainly be easier to establish a colony on the Moon due to its proximity, but in the long term, the costs would soar

---

[*] West Kirby, Wirral, United Kingdom.

above those of colonising Mars. A colony, in order to cut down on transport costs, needs to produce its own food—this is a necessary prerequisite. The Moon would not be able to produce its own food, as the elements required by plants for growth are simply not present on the Moon. On Mars, however, colonists would be able to find every essential element and compound they would require present in the regolith (the Martian 'soil') in abundance.

Vast leaps in aerospace technology are being made every year, driving prices down for components. We can already see cheaper and faster next-generation space planes on the horizon. The same technology could be adapted to suit the purposes of the humans to Mars mission, technology that exists now, not ten years or twenty years in the future. Using the experience gained during the construction of the International Space Station, and building on the international co-operation that made it possible, the financial burden on each participant could be lessened, allowing countries to take part in what would be perhaps the most important human event for centuries.

With each successive mission to Mars, the costs would grow lower and lower as manufacturing techniques are perfected and components are standardised then mass produced. Indeed, it is even possible that humans on Mars could mine fuel from the atmosphere and regolith to act as propellant for the journey back to Earth. Using increasingly efficient technology, the costs of such a program to Mars would undoubtedly be large, but certainly not on such a scale comparable to the Apollo missions. A manned mission to Mars is not some sort of utopian dream, neither for that matter is colonisation. Both are entirely possible now.

There will always be significant opposition to any type of space exploration. People argue that we have enough problems on Earth as it is, hardly warranting expenditure on 'useless' trips into space which provide no 'practical' benefits, an argument which was proved true with the 'flag and footprints' vastly expense Apollo program. Many have not, however, considered the situation if we do not explore and expand. The pressure on us to change and innovate has disappeared—we are becoming bound in our ways while there is no frontier, nowhere to begin anew. We are living in a closed society, and the signs of stagnation are beginning to tell. In the next century, Earth will be running out of resources and running out of living space. Over 5 billion humans are alive at this moment, and the number is increasing by a quarter of a million daily, even against the best of our efforts. Mars, with its untold quantities of resources and potential to become habitable for humans could go some way to helping the problems of Earth.

The first step to this would be to initiate a humans to Mars program, and the decision can be made now; we could see the first man on Mars less than fifty years after the first man on the Moon. For this decision to be made, leaders of the political and scientific worlds will have to work together towards a common goal, the furthering of mankind. I hope you will appreciate how much this program can mean.

This letter was sent to:
President of the United States Bill Clinton
Prime Minister of Great Britain Tony Blair
The Leader of the Opposition William Hague (Great Britain)
The NASA administrator Dan Goldin
ESA Director General Antonio Rodota
British National Space Centre
Merseyside West European MP Richard Corbett
Minister for Science, Industry and Energy John Battle
Wirral West MP Stephen Hesford

Chapter 6
EDUCATION AND THE ARTS

Mars artist Michael Carrol (left) exchanges thoughts with astrophysicist and science fiction author Greg Benford.

Air Force Captain Lynnanne George brought a group of Cadets from the Academy

MAR 98-024

SECONDARY MARS EDUCATION: FASTER, BETTER, CHEAPER

Thomas W. Becker[*]

A combination of timely factors has made public education about Mars possible on a far greater scale than ever before. The (1) Information Superhighway via the Internet and the swift movement of private information allows fingertip data at the push of a computer key. Together with a broad range of (2) up-to-the-minute relays of continued findings by university and NASA scientists, and the (3) availability of online educational lessons and easily purchased CD-ROM disks slide sets and a full range of other related materials all allow daily instruction of space-related technology education almost at a moment's notice. Space education can assist considerably to help young people understand the next stage in human evolution when humans build the first colony on the Martian landscape.

"The woods are lovely, dark, and deep, But I have promises to keep, And miles to go before I sleep, And miles to go before I sleep."
Robert Frost

The traditional classroom of the 20th century with its pin-up posters, wall maps and written tests, is gone forever. Current and ongoing advances in more realistic subject matter and methods of bringing about learning offer remarkable opportunities for education to take place faster, better, and cheaper for both student and school district. These advances tend to make textbooks, paper and pencil desk exercises, and chalkboards laborious and cost ineffective barriers in the view of enlightened young people who hunger for information about the next civilization in space. Young people know their future is in space and they want to know why they aren't being told more about it (Figure 1).

DILEMMAS OF THE INFORMATION SUPERHIGHWAY

Traditional urban classrooms are being replaced at a rapid rate in the face of computers operating at internet speeds which can disseminate the projects and findings of scientists and engineers while they are taking place. Students and teachers have access to a truly enormous fund of online, frequently updated information from which they can extract data for individual study, student deskwork and classroom use. In some areas of the nation, on the other hand, the pace of change to computer-driven knowledge is slow. We do not all travel the Information Superhighway at the same speed.

People who work in the space community, especially in large cities, seldom can appreciate the dilemmas faced by young people in small towns, especially throughout farmlands in the Midwest and plains regions. Operating capital for small school districts is limited, students are spread out across large geographic areas, community libraries are small and distanced, and student access to and use of major computer systems are highly difficult at best.

---

[*] Space Technology Educator and Consultant, University of Missouri-Columbia.

Figure 1 The future emphasis for everyone is education. Sketch by the author.

Some rural school districts are not yet on the Internet. For people in these communities, knowledge about where to get teaching materials about space topics is difficult to acquire. Daily computer work is severely limited and students often have no opportunity to take data from the Internet home with them. One 10th grader in rural Missouri, for example, said she was allowed to print out only one page a week from the school's computer—under supervision. Another student said he was allowed to use the school computer only when teachers, administrators and clerical staff weren't using it; another said his mother had to drive him 20 miles to the nearest library to use a computer. Some students have difficulty gaining access to a decent computer and/or a printer of good quality. Images from the Internet could be worthless if printed out on an inferior printer.

Although situations such as these are changing, the rate of change is slow and uncertain. Still, the Internet is the only chance young people have to keep abreast of what is happening in the world of science and technology that is fast overtaking the American culture. As adults and educators, we must broaden that chance to give young people more opportunity not only to use computers but to learn how to create right on the keyboard (Figure 2).

European students (Figure 3), like American students, are faced with similar problems as I learned teaching five seasons at the European Space School for Sixth Form students headquartered at Brunel University in Uxbridge, England. European industry and politics are very supportive of technology education and allow space-related curriculum to become part of the national education structure. Curricula in "Earth and space sciences" are mandated across the United Kingdom and in countries on the mainland so that pure space topics can be taught at will.

Serving as Principal Lecturer at the British Space School from 1989 through 1994, I delivered illustrated lectures about Hypersonics And Spaceplanes, Hubble Space Telescope, Shuttle Orbiter Technologies, the Voyager missions, and also taught two extended courses: Remote Sensing, and Comparing Earth And Mars.

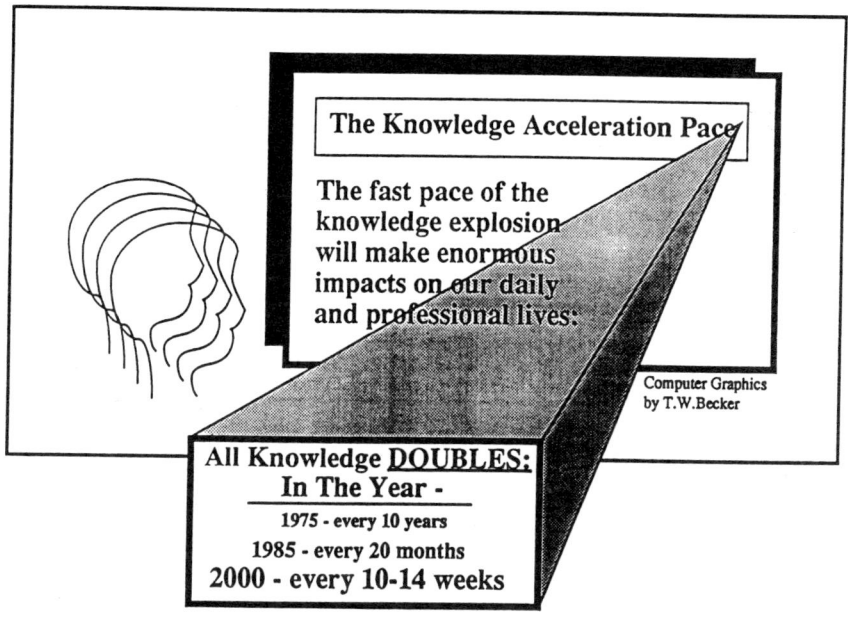

Figure 2   The pace of the knowledge explosion. Sketch by the author.

Figure 3   European Sixth Form students study a course in Earth/Mars comparative planetology in Space School at Brunel University. Photo by the author.

National space schools for high school-age students are now operating in Australia, Brazil, Britain, Canada, France, India, Mexico, Pakistan, Poland, and the former Soviet Union. The schools are strongly supported by industry, education and government and are meeting on regular schedules throughout the year. A number of these countries have asked me for teaching materials and copies of my published articles over the past decade; some have asked me to teach in their schools.

INTERDISCIPLINARY AND NON-TRADITIONAL APPROACHES TO MARS SPACE EDUCATION

Space technology education is not difficult; it's just different. Simply stated, space education is the teaching of the fund of knowledge that allows humans to live and work in the space environment (Reference 1). Certainly Mars is a different world but it still is a sphere and it has geological processes and geography counterparts similar to Earth, although the similarity ends there. Everything on Mars is much bigger.

Teaching about Mars and other space topics requires the acceptance of two non-traditional approaches to the education process (Reference 2). The first is a *Technology Preparation System* (Figure 4) driven by three basic objectives:

- Preparation of a knowledgeable labor force
- Continual acquisition of new knowledge and new technology
- Constant exploration, exploitation, and commercialization of the space frontier.

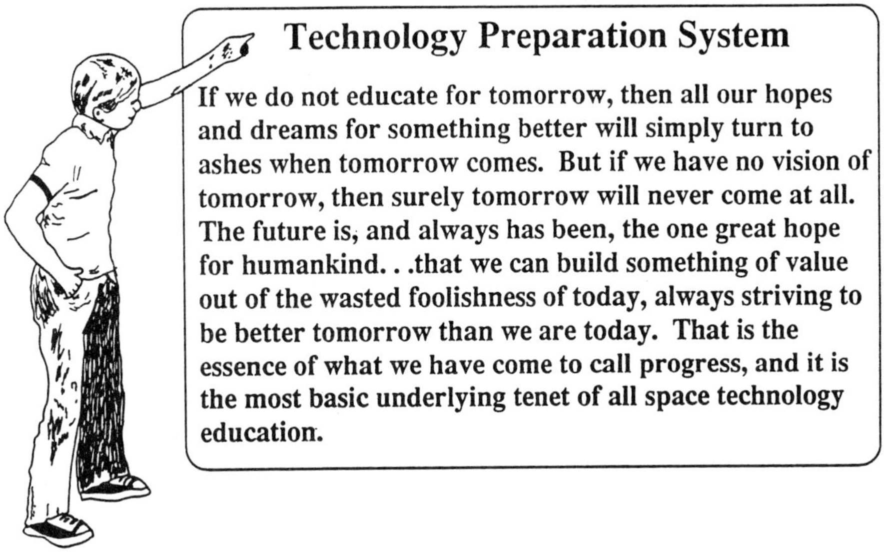

Technology Preparation System

If we do not educate for tomorrow, then all our hopes and dreams for something better will simply turn to ashes when tomorrow comes. But if we have no vision of tomorrow, then surely tomorrow will never come at all. The future is, and always has been, the one great hope for humankind...that we can build something of value out of the wasted foolishness of today, always striving to be better tomorrow than we are today. That is the essence of what we have come to call progress, and it is the most basic underlying tenet of all space technology education.

Figure 4 The technology preparation system is the basic tenet of all space education. Sketch by the author.

The second non-traditional approach is the use of *Advanced Educational Systems* which includes everything from 35mm single lens reflex cameras to computers, camcorders, digital cameras, basic digital data communication, and teaching the technologies of Mars mission. In Advanced Educational Systems:

- Subject matter and materials are not part of today's usual curriculum
- Teachers are required to be more versatile and creative
- Students are led to think and reason as opposed to listening to lectures and filling out "worksheets"
- The systems are computer-related and require more than simple computer literacy, familiarity and word processing.

Figure 5 illustrates three stages in the study and human perception of Mars. The upper level represents an armchair and telescope method; studying only what we can see and understand from ground level. The middle level shows what we first learned and can study based on flyby spacecraft imagery such as the Mariner 4 and 6 missions. The lower stage awaits a human landing of geologists and other scientists, after the Viking and Pathfinder missions, and the final truths that can only be discovered by human contact.

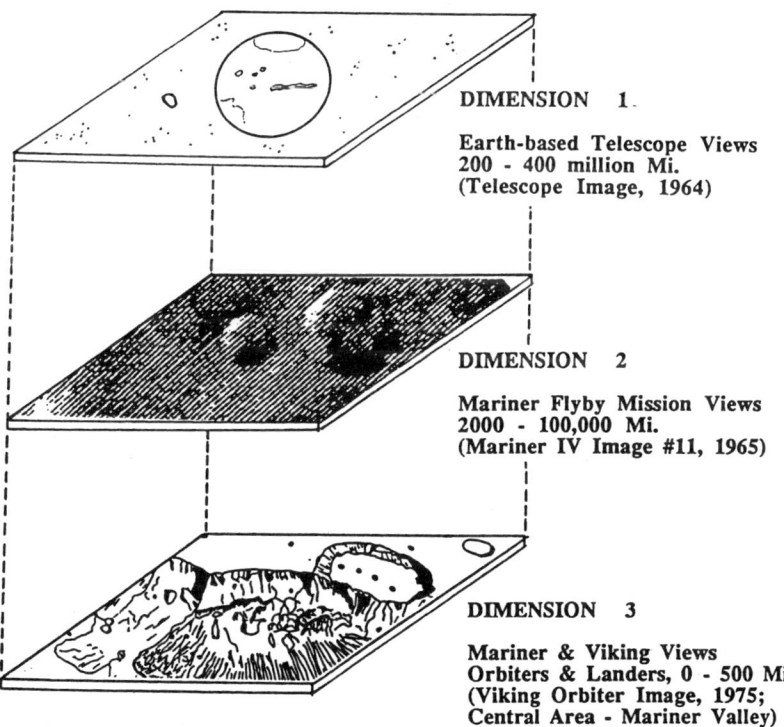

ADVANCED EDUCATIONAL SYSTEMS
Spacecraft Imagery Perspective Of Mars Surface
To Teach Comparative Geography And Geology
(Adding A Third Dimension For Reality)

DIMENSION 1
Earth-based Telescope Views
200 - 400 million Mi.
(Telescope Image, 1964)

DIMENSION 2
Mariner Flyby Mission Views
2000 - 100,000 Mi.
(Mariner IV Image #11, 1965)

DIMENSION 3
Mariner & Viking Views
Orbiters & Landers, 0 - 500 Mi.
(Viking Orbiter Image, 1975; Central Area - Mariner Valley)

Figure 5 Advanced Educational Systems provide greater understanding of remote objects in space education. Sketch by the author.

It is imperative that today's scientists and space community leaders capture the attention of young people at the very time when these students are trying to make career decisions. We must expose high schoolers to as many career options as possible so they can make intelligent choices at this critical time in their lives. These intelligent and curious minds are continually searching for a job field or a professional career line that not only offers security and high job interest, but which also allows them to exercise their creative talents.

Young people today in their 11th year of schooling will graduate in the year 2000 and face all the problems and challenges of a new, intense and much faster-paced millennium. By now they should have a clear view of the kinds of life careers they want to pursue, but for the most part, many have absolutely no idea of their intended job fields even though they are only two years away from entering university. The skills they need to survive in the highly competitive job marketplace to meet the awesome challenges of the next millennium include: abstracting, retrieving, assembling, searching, categorizing, systematizing, comparing, thinking, compiling, transmitting, reasoning and visualizing.

Learning goals should include multidisciplinary approaches to problem-solving, understanding relationships, adopting a global viewpoint, seeing technologies as tools, experiencing rapid cultural change, goal-oriented reasoning, and the ability to think deeply. Arthur Clarke's global village already is a reality and tomorrow's adults will live in it on a much more face-to-face basis than today's adults. But how ready are today's students to learn about new technologies?

A TALE OF TWO COURSES

In 1987 I taught two six-week classes of 47 low-achieving students (only 34 students finished the course) in a suburban high school with a student population of 1800 (Reference 3). Students enrolled in this special two-and-one-half hour summer session were of two specific backgrounds: (1) those who had failed a subject in the regular school year, and (2) those who were interested in simply earning additional high school credit but who had not failed a regular course during the year. Students were tested weekly and also were graded weekly on desk map work. Space technology made up more than 50% of the course.

Students were characterized by poor reading ability, low reading comprehension, poor map-making skills, poor study habits, low organizational skills, and generally low academic achievement. As one might suspect, they had poor self images, low motivation, some behavior problems, low expectations for themselves, and subscribed to a counter-culture value system.

The course, "Technology And Current Events", had a global perspective, strong use of audio visuals, and interdisciplinary subject matter for which they had no background whatever. Topics in the course included the Thames River Flood Barrier, British/French Channel Hovercraft, Channel Tunnel, Landsat and GOES satellites and imagery, French TGV Train, Netherlands Polders, Concorde Jet Aircraft, MIR Space Station, Hubble Space Telescope, and maps from space of the Sinai Peninsula, Europe, Netherlands, Mainland China, Earth's Northern Hemisphere, Chernobyl and Africa's Great Rift Valley.

Final letter grades for the course were A=1, B=9, C=5, D=12, F=7; T=34. Low achieving students can achieve given the right circumstances. Motivating factors in this intense, highly technical course included:

- Studying real-world subject matter
- Studying space technology topics

- Rise in self esteem
- Earning grades they never could earn during regular school
- Students needed a passing grade.

Two years later I was teaching a class of 15 gifted 8th graders for the St. Louis Gifted Resource Council, called "The Space Frontier" which was a walk-in, four hours a day direct contact for six weeks. We studied everything from the Shuttle Orbiter to the European Meteosat weather satellite and the Soviet space program. We also reviewed the political situation of "the space race" in relation to the Cold War. Toward the end of the course, the students built model rockets which they painted and launched on the football field.

One of the students suggested we conduct an experiment to find out how long breathable air would last in a plastic bubble in a swimming pool, and at the same time we could monitor the pH content in the bubble over a period of time. The experiment had some practical application for the Moon and Mars. The students rigged a large plastic sheet with rope and bricks tie-downs, hauled in compressed air tanks, filled the plastic bubble with air, and took turns diving under it to get into the bubble. The experiment was an excellent exercise in design and follow-through, as well as leadership, and a remarkable success considering the students were allowed to spend only one hour a day in preparation, and they received minimal help from me.

These are two examples from opposite ends of the learning bell curve, both highly successful and both rather amazing in their execution. But both were very carefully designed and orchestrated, with motivation coming from within the students themselves. The power of space education is undeniable, and we should put it to work on the study of Mars.

THE MISSOURI SCHOLARS ACADEMY FOR GIFTED STUDENTS

I first taught in the Missouri Scholars Academy at its founding session in 1985 and again in 1986. The in-residence, state legislature-funded Missouri Scholars Academy for rising, gifted 10th graders offered me a third unique opportunity this past summer to put space education concepts into daily practice once more. There were two classes of 18 select students each in *Basic Remote Sensing* (a major course meeting three hours daily) and *Introduction To Mars* (a minor course meeting one hour daily). In addition, the experience presented an opportunity to design a "bare bones" Mars course which now is being expanded and fine-tuned, and could easily become a much larger kind of course. See the MSA scene pictured in Figure 6.

Most of the materials for the course were derived directly off the Internet or obtained from select individuals and agencies working in the space community. A collection of data about Mars with student exercises made up a packet of about 100 pages given to each student. The course was structured to begin with basic facts about Mars as a planet, lead into the planet's major features such as volcanoes, impact cratering, channeling, atmospherics, landscapes, and finish with the application of this knowledge by starting on the design for a science station on the Martian surface. The main emphasis throughout was on geography, geology, atmospherics and comparative planetology. Four presentation methods were used throughout the course.

Audio Visuals/Models

Color slide sets, some made up as far back as my attendance at Viking press conferences and tours of the Viking launch site in 1975, as well as more recent sets obtained from the Lunar and Planetary Institute in Houston, were the backbone of the course. Lively class discussions and short lectures were accomplished using slides as conversation and question-answer starters. Several VCR tapes such as "Made For Mars: Pathfinder", "On Robot Wings" and "The Planet Mars" helped fill in the gaps as did posters, original oil paintings and large pictures. The Mars Global Surveyor and Pathfinder missions were examined in detail and imagery called up on specific computer websites. A working model of the Sojourner rover was demonstrated and the students were encouraged to operate it themselves, as in Figure 7.

Figure 6   Missouri Scholars Academy students work in the computer lab to access Mars website data. Photo by the author.

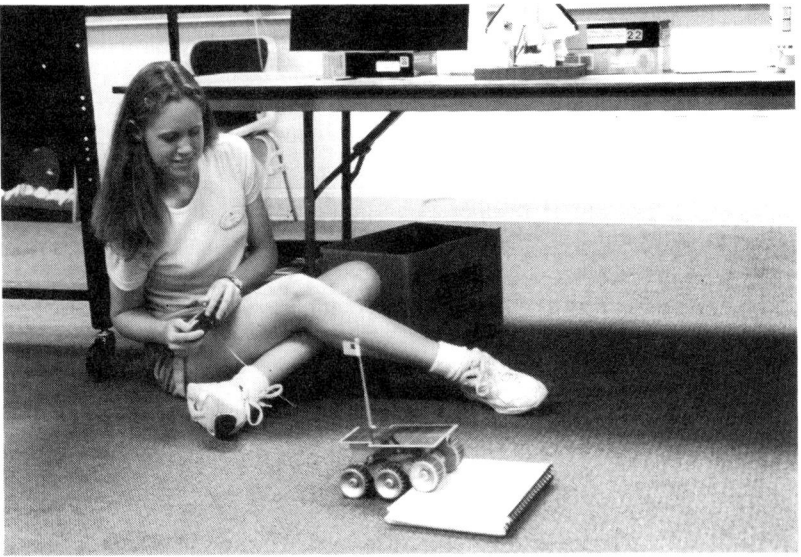

Figure 7   A Missouri Scholars Academy student puts the Mars Sojourner model through its paces. Photo by the author.

Specific Imagery

Imagery of major features such as volcanoes of the Tharsis region (Olympus Mons, Ascraeus Mons, Pavonis Mons and others), Mariner Valley, Argyre Planitia, the Martian poles, etc., gave students an overview of the geography and geology of the planet. Some paper prints of imagery in color from the Viking missions, which students could keep, were distributed to the class for individual study and comment. Other imagery was obtained off the Internet. Occasionally an especially appropriate piece of imagery from the Remote Sensing class was given to students in the Mars class because otherwise they would have no opportunity ever to see the pictures.

Computer Laboratory Research

Using the PDS Planetary Image Atlas online program (more about this later) as a primary access to surface features, students were led to choose specific sites on Mars and to create their own pictures which they could study at the Academy and then take home with them. Students were able to print out as many as several dozen or more pictures off the Internet, allowing them to thoroughly research one or more specific landscapes and features.

Class Discussion/Student Desk Exercises

Mixed with short lectures and studies of surface features, lively discussions were held during which students were helped to analyze imagery, compare geological evidence, and arrive at reasonable conclusions about how certain surface features were formed. Repeated cratering (first, second and third event cratering were distinguished), mass wasting/slumping of canyon walls (see Figures 8 and 9 of the Candor Chasma section of Mariner Valley), evidences of past water flow through extinct riverbeds, outflow channels, and the creation of Olympus Mons and other volcanoes were determined by class discussion and reasoning.

Although only 25 hours length, the course was highly successful for students who had no other access to this kind of structured study of Mars. On the last day of class, I asked students to render an opinion of the course. Students were highly enthusiastic about the course; and a frequent comment was that the course was not long enough. The site they finally chose for the science station was in the terra fossae region in the northern latitudes because of the nearness to water and an outflow channel.

A STUDENT TECHNOLOGY ACCEPTANCE PHENOMENON

Teaching these and other courses over the years, an observable phenomenon has emerged among average students that I term "The Technology Acceptance Curve" (see Figure 10) and which invariably shows up in courses with high technology content.

During the first several weeks of a course there is a gradual climb to a plateau as students experience the excitement of technology education. But then comes a sudden drop in interest, test scores and individual work assignments in the face of subject matter for which students have absolutely no background. A slow climb back up to a new plateau of comprehension occurs as students begin to apply new principles and come to understand the science and technology of space topics.

This phenomenon does not seem to show up in gifted students who already have exploring backgrounds, and are able to self-sustain their interest and efforts for long periods of time out of self curiosity and higher order study skills. Gifted students are considerably self-motivated and goal oriented, and can focus on priority subjects for long periods of time.

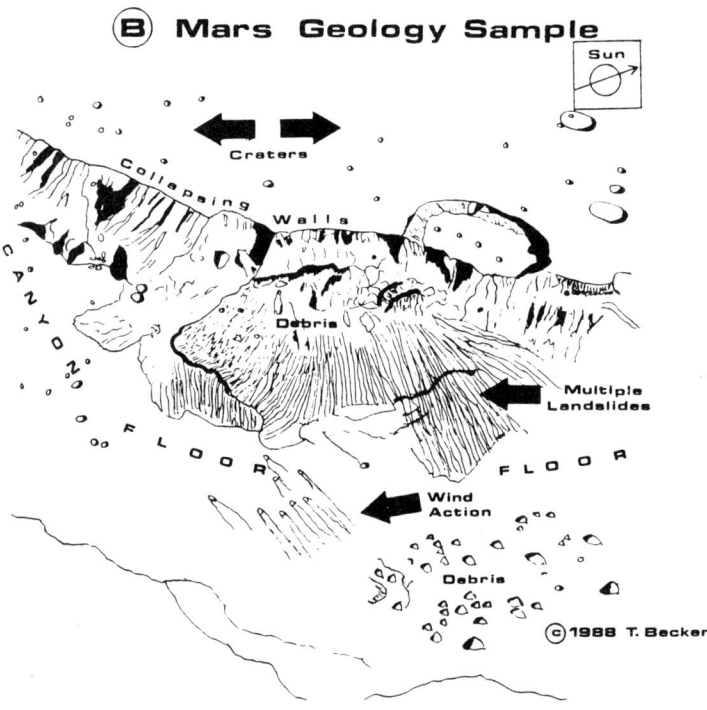

Figures 8, 9 A Missouri Scholars Academy student desk exercise in Mars geology features the slumping event in Mariner Valley. Students used a NASA raw data image to study that had been scanned so it could be Xeroxed. Sketch and computer work by the author.

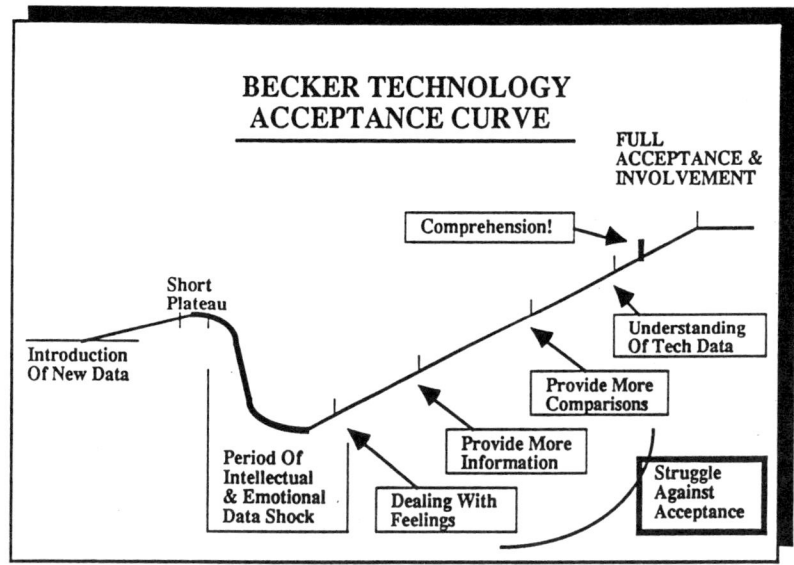

Figure 10 The Technology Acceptance Curve appears during class instruction of non-gifted students. Chart by the author.

SURFING THE SPACE INFORMATION SUPERHIGHWAY

NASA revised its homepage in 1998 so that accessing the basic web site permits the user to reach all other NASA facilities. With the entire NASA organization contributing to the Internet, information about any NASA topic is instantly available to students and teachers alike. This strategy is ideal for preparing lessons, student preparation of special reports, and even larger scale research projects. From Mars and asteroids to the X-33, International Space Station, Shuttle Orbiter and the greenhouse effect, for example, space theme subjects are absolutely endless. In addition, each site offers a number of links from the NASA homepage to private industry.

Pictures, posters, color slides and booklets about space science topics are available via the Internet in amazingly large quantity. Some current examples are:

a) for the **NASA homepage:** *http://www.nasa.gov/*

b) for the **NASA Mars Global Surveyor homepage** and the opportunity to download a series of MGS mission profile slides: *http://mars.jpl.nasa.gov/mgs/home.html*

c) for the **Mars Pathfinder and Sojourner** sites: *http://mpf.www.jpl.nasa.gov*

d) Future Mars robotic missions to the year 2004 can be browsed at: *http:/marsnt3.jpl.nasa.gov/education/table-contents.html*. This site will also gain access to the **NASA Jet Propulsion Laboratory's Mars Exploration Education Program: The Road To Mars**

e) for the **NASA Planetary Photojournal Image Access homepage** and access to pictures and text about the Solar System: *http://photojournal.jpl.nasa.gov/*

f) for additional Solar System and planetary data, the website **Views Of The Solar System** is an excellent and accurate portrayal, including asteroids, comets and meteorites, but be-

ware that some material might be copyrighted
http://bang.lanl.gov/solarsys/eng/homepage.htm

g) for **asteroid and comet impact hazards**, which are important for understanding impact craters on Mars: *http://arc.nasa.gov/index.html*

h) considering Mars' thin atmosphere and distance from the Sun, it might be helpful to learn more about the Sun at this excellent **SOHO (Solar And Heliospheric Observatory)** joint ESA/NASA website: *http://sohowww.nascom.nasa.gov/*

i) **Malin Space Science Systems** (together with the Cal Tech Institute of Technology) built the MGS camera (Mars Orbital Camera), is doing exemplary work studying Mars surface sites. This web is well worth your time and effort. Many of the images are offered free to the public at this website, but show your respect for their request for some copyrighted images and their wish to receive a credit line: *http://www.msss.com*.

These nine "best" examples are sufficient to give an idea of pertinent Mars websites. Of course, there are many, many other websites that offer a large amount of data. Browsing the Net will eventually get you where you want to go. These sites often have public outreach and educational programs of considerable value to the education community, and data and imagery can be printed out.

There are a number of automatic messaging systems for space sciences available on the Internet to which students and teachers can subscribe free of charge simply by applying on the computer. One of the more exciting websites is managed by Linda Porter at the NASA Manned Space Flight Center in Houston, called NASA Science Headline; apply at:
<express-delivery@sslab.msfc.nasa.gov>.
Some interesting recent topics have included the following:
Starquakes: *http://science.msfc.nasa.gov/newhome/headlines/ast09jul98_1.htm*
Urban Heat Island: *http://science.msfc.nasa.gov/newhome/headlines/essd01jul98_1.htm*
Solar 3-D Images: *http://science.msfc.nasa.gov/newhome/headlines/ast22jun98_1.htm*
Gamma Ray Bursts: *http://science.msfc.nasa.gov/newhome/headlines/ast09jun98_1.htm*

If you're interested in the X-33 spaceplane (VentureStar), you can get on the automatic mailing list simply by sending your request to: *Majordomo@treflan.shout.net* and you will receive occasional public information releases. Is this design suitable for a Mars surface transporter, or perhaps you will get some other ideas from the website?

TEACHING MATERIALS OFF THE SHELF

One of the most rewarding and user-friendly online sites is the program **PDS Planetary Image Atlas** developed by Rob Waltz and a development team of the U.S. Geological Survey at Flagstaff, Arizona. The URL is *http://pdsimage.wr.usgs.gov/ATLAS.html*. (PDS = Planetary Data System.) The program, which can be accessed and manipulated online, allows the user to get an image map from which to choose and print any area on Mars with a variety of zoom factors, image sizes, and map projections using data from NASA's Viking missions.

To create an image simply click the area you want on the browse map image and the **MapMaker** program calls up the image which can be enlarged, moved up and down or sideways to select an exact feature, and then printed. Figure 11 is an image taken directly from the program and used by a student for study. Teachers can use the same program to create a variety of posters and/or bulletin board pin-ups, or prepare entire lessons.

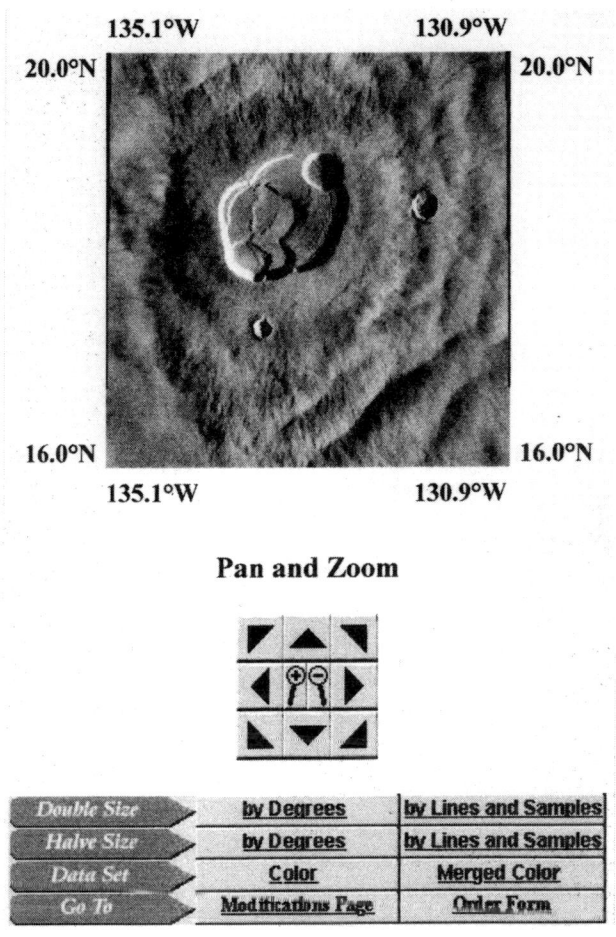

Figure 11 Mapmaker PDS printout called up by a student in the Missouri Scholars Academy computer lab. Artwork by the author.

Another useful online site is the **Astronomy Picture Of The Day (APOD)**. The URL is: *http://antwrp.gsfc.nasa.gov/apod/archivepix.html* which features a different astronomical object each day of the year. While there is a wide variety of cosmic objects to choose from, the site also furnishes a number of Mars images and considerable creative opportunities for the teacher to highlight facts and pictures about Mars. In Figure 12, a student at the Missouri Scholars Academy browses the Astronomy Picture Of The Day information board for new information about Mars. It is worth your while to browse the Catalog and the Index.

The **Lunar And Planetary Institute (LPI)** at Houston, Texas sponsored by the Center For Advanced Space Studies, NASA Johnson Space Center, has four sections devoted to major Mars landscape features, followed by five classroom activities using the image shown in Figures 8 and 9. The website can be found at *http://cass.jsc.gov/expmars/edbrief/edbrief.html* titled *Exploring Mars: 1996*. Activities for examining specific Mars surface features can be found at *http://cass.jsc.gov/expmars/edbrief/classact.html*. Activities can be geared to middle school or (with additional data) to upper high school. The site is excellent for a lead-in to topographic features, and Mariner Valley in particular. Both sites offer innumerable ideas in planetary geology with images for teachers to prepare meaningful lessons, and for students to work through as personal exercises. The **Institute** also offers color slide sets of Mars, Venus, the Moon, etc., each accompanied by a well-written and researched booklet explaining each slide. The sets are affordable and can be ordered by email with a credit card. The slides are ideal for classroom use and for individual study.

Figure 12   A Missouri Scholars Academy student pores over the data board featuring Astronomy Picture Of The Day. Photo by the author.

Finley-Holiday Films located in California probably has the best selection of Video tapes about Mars and other space related topics, at *http:/www.finley-holiday.com*. The company has been in business since 1965 and also offers CD-ROM space programs and color slide sets of a number of space subjects. The company is mentioned here because both service and quality are reliable.

Many **commercial CD-Rom disks** are available under $20.00 featuring Mars. Two of the better ones are:

- Mission To Mars: Exploring The Red Planet, by American MPC Research Inc.
- Mars: Past, Present, Future, by Holiday-Finley Films.

The **National Space Science Data Center** online features a CD_ROM Catalog of 39 titles, at *http://nssdc.gsfc.nasa.gov/cd-rom/set_bd/voyager.html* as well as prices and ordering information. Mars and the Mars spacecraft missions, especially Viking, are featured prominently at this NASA website. Many disks sell for $10.00 and the selection is excellent.

Mars topic CD-ROM disks can be purchased from any of a number of sources including the NSSDC and space activist groups, or by searching through the space magazines. Other disks about exploring the universe or the Solar System invariably have a section about Mars. For example, a CD-ROM about "The Universe" is going to include the Solar System and the planets (and Mars) as topics.

Final Frontier magazine (tel. 1-800-24-LUNAR for U.S.; international calls at 626-932-1033), aside from being a very good magazine, has a product section in each issue. The products are often quite expensive but they also are usually scarce or unique at those prices. The magazine offers authentic and sometimes fascinating space-related products; material about Mars is excellent.

HIGH SCHOOL SPACE COURSES BY DISTANCE LEARNING: UNIVERSITY OF MISSOURI

The **Center For Independent Study** at the University of Missouri-Columbia scored a major breakthrough in space technology education recently by offering two high school for-credit science correspondence courses in space technology that are on the Internet. Written for 9th grade students and up, the courses can be completed right on the computer after gaining access to the website and paying the enrollment fee (*http://indepstudy.ext.missouri.edu*). However, the courses are not for the faint-hearted but offer academic alternatives for creative and talented high school students. Each course has mid-term and final exams as well as progress evaluations along the way.

A third course is in the writing stage about major unmanned space exploration achievements such as the Hubble Space Telescope, the assault on Mars, Antarctica from space, the planet that got knocked on its side, and other exciting subjects. Completion of the course is projected for early next year.

Other courses are in the planning stage, especially a detailed survey of Mars as a special single course which will be created during 1999. Written by this author, all the courses together will form a series of upper secondary school correspondence courses on space technologies designed to take Missouri into the next millennium as a leader in space technology education and distance learning by correspondence, at the cutting edge of space exploration. All the space courses will be on the world wide web.

For the two courses already on the web, I took great care to obtain photographs and illustrations that were appropriate and accurate. Both courses are the result of some 30-40 years researching and teaching space technology. Many foreign space agencies contributed to each course in order to make them as authentic as possible.

a) *Aerospace: Crossing The Space Frontier* - a political and technological history of space exploration from the Treaty of Versailles that ended World War I and helped create the Nazi V2 rocket to the International Space Station of today. The course explains the many scientific principles of space technology, reviewing each decade of space technology. The former Soviet space program and the European Space Agency are reviewed in detail. The course is copiously illustrated with authentic photographs and illustrations.

b) *Studying Planet Earth: The Satellite Connection* - this unique course about satellite remote sensing uses up-to-the-minute satellite images and airborne photography from the world's leading space agencies and allows students to study Earth from the vantage point of space. The course comes with a special packet of color images for study and has lessons on topics such as basic remote sensing, hurricanes, spies in the skies, midwest flood of 1993, Chesapeake Bay, Netherlands Polders, and the Northeast, killer volcanoes, synthetic aperture radar, and Mission To Planet Earth. Each lesson in this course is accompanied by a special study-exercise.

THE KEYS TO MARS SPACE EDUCATION

There are several critical keys to designing and teaching courses about Mars. I've always believed that nothing can really be accomplished in an hour; designing a course for a time allotment of less than 12 hours is non-productive.

An Earth Comparison - Earth has been and always will be an obvious comparison to Mars or any other planet. If we don't compare Mars to familiar landscapes and features on Earth, we just can't relate to the Martian environment.

Understanding Remote Sensing - The basis of all Mars imagery is remote sensing and digital data transmission. Students must learn the technology and how it works in order to understand how the imagery is created and then enhanced or manipulated.

Computer Driven - Because of the enormous amount of information about Mars, courses must be computer based so that imagery can be obtained accurately and quickly.

Image Study - To understand how Mars behaves, students must have access to as much Mars imagery as possible. Aside from pictures in books, a Mars course should include a packet of features, landscapes, and image comparisons.

Technology Development - Each successive Mars imaging mission was accomplished by a corresponding increase in the sophistication of the mission spacecraft. Mariner 4 was the beginning; Mariners 6 & 7 were improved craft; Mariner 9 was more than adequate to do the job entrusted to it; and of course Viking and Mars Global Surveyor as well as Pathfinder are the current stages in technology improvement.

National Mars Curriculum - In the long term, if America is interested in sending human teams to the Martian surface, then Mars courses for young people should be the same across the country. This approach, over time, will establish an authentic national Mars curriculum that is the same for everyone. Aside from the motivating factor, students in Texas will be learning the same material as students in Minnesota or Massachusetts.

A Mars Society Course - The new Mars Society is in a perfect position to initiate and spearhead a national Mars course. In America today, the space education community is too splintered and generally uncooperative to bring about a serious kind of national Mars course (Reference 3). Regional and local politics, defending the turf, trying to look good in the public view, are all unproductive and cannot be tolerated in the face of such a serious undertaking as a national course. A national course must reach far beyond these territorial issues if something of value is to be accomplished. Designing and implementing a national Mars course cannot be left to chance or to the personal whims of space education organizations which have amply proved their inability to cooperate with each other or to function adequately.

MARS BASIC SUBJECT MATTER: STAND AND DELIVER

The general study pattern for understanding Mars as a planet and as a future target for the next step in space is a logical group of subjects, as follows (References 23 thru 28):

Basic Statistics - understanding such basics as comparative sizes, diameter, moons, tilt, distances from Earth and Sun, seasonal changes, etc.

Climate And Climatology - climatic change, factors affecting climate, seasonal variations

Geological Driving Forces - impact cratering, mass wasting, faulting, absence of tectonics, permafrost, temperature as deformation processes, transport of soil material

Location Geology - types of rocks and structures according to location

Landscape Features (Geography) - volcanoes, canyonlands, basins, channels, chaotic lands, poles, place names

Mars Mission Technologies - Mariners 4, 6, 7, 9; Vikings 1 & 2; Mars Global Surveyor, Pathfinder, (future missions)

Atmospherics - temperatures at various altitudes, haze and frost, clouds, movement of the polar ices, aeolian transport features

History Patterns - birth of the Solar System, Mars evolution and current development, impact cratering, water and water flow, Mars development compared to the other planets and their satellites.

THE NEXT FRONTIER AND THE FUTURE

Time and time again the plea from the public sector has fallen on deaf ears in the Congress. The time for the Congress and the President to listen is now - the message is loud and clear. Recent Mars Society Special Reports are nothing short of a call to arms for the membership to stand and be counted and to once again remind the Congress of the power of the constituency.

If a human mission to Mars is to be undertaken, it will have to be America that makes it happen. No other culture has the resources, the technology, or the public will, but the task will not be an easy one and it will require the best thinking of our best creative scientists to bring it about.

Methods to reach and establish a foothold presently are under serious discussion by serious people. Alternative propulsion systems, en route survival systems, landing site choices, surface excursions by foot, scientific experiments on Mars - all are being examined in a search for the best systems. Figures 13, 14, and 15, a Mars Mission proposed by the author and technology artist Darren J. Gillett of London, England suggest a colonization system that might take place some time after the first foothold on Mars is accomplished. I met Darren about 8 years ago in London and, after several months of talking about how to get to Mars, he began drawing more than 30 engineering sketches based on ideas we had discussed. Only a couple of his marvelous sketches are included here.

By the time configurations such as these might be undertaken, already there will be human explorers at work on Mars. An armada of four ships are led by another ion propelled spacecraft shown in Figure 13 and operated by two astronauts. The craft carries supply modules to be set placed in Mars orbit and retrieved at will by the colonists. The main control center, the Bridge, is shown in overall detail consisting of two seats for Spacecraft Commander and Co-Pilot.

Figure 14, top, represents a multi-role Mars Lander capable of ferrying about two dozen passengers and a limited cargo of spare parts, foodstuffs, science equipment, etc. At bottom is an orbital forklift vehicle which retrieves cargo containers and ferries them down to the surface of the planet for unloading, and a "bundle" of cargo containers.

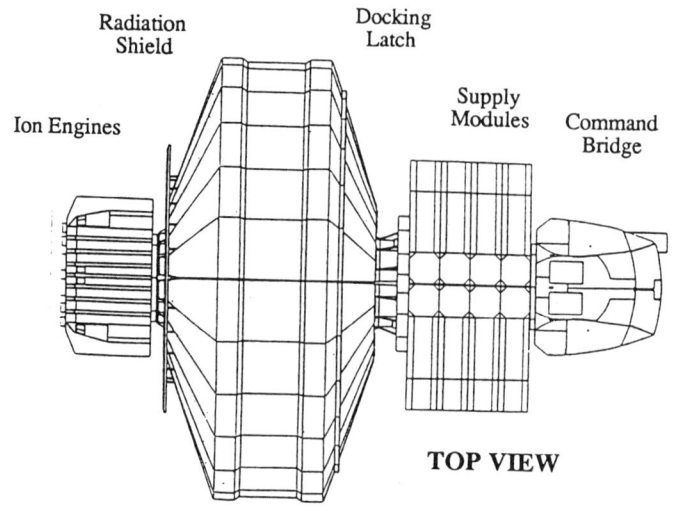

Command Control (Bridge)

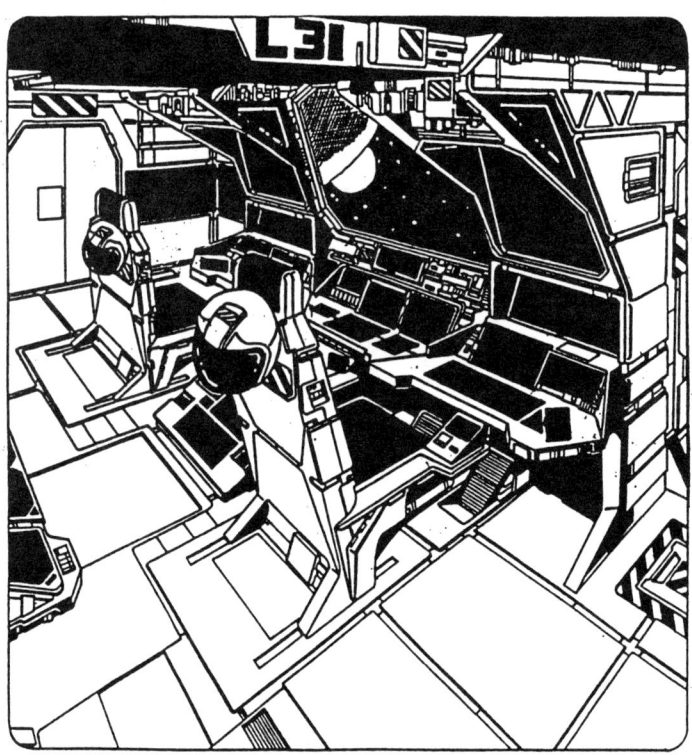

Figure 13 Ion propelled spacecraft operated by two astronauts. Below, the main control center (Bridge). Sketches Copyright 1993 by Darren Gillett.

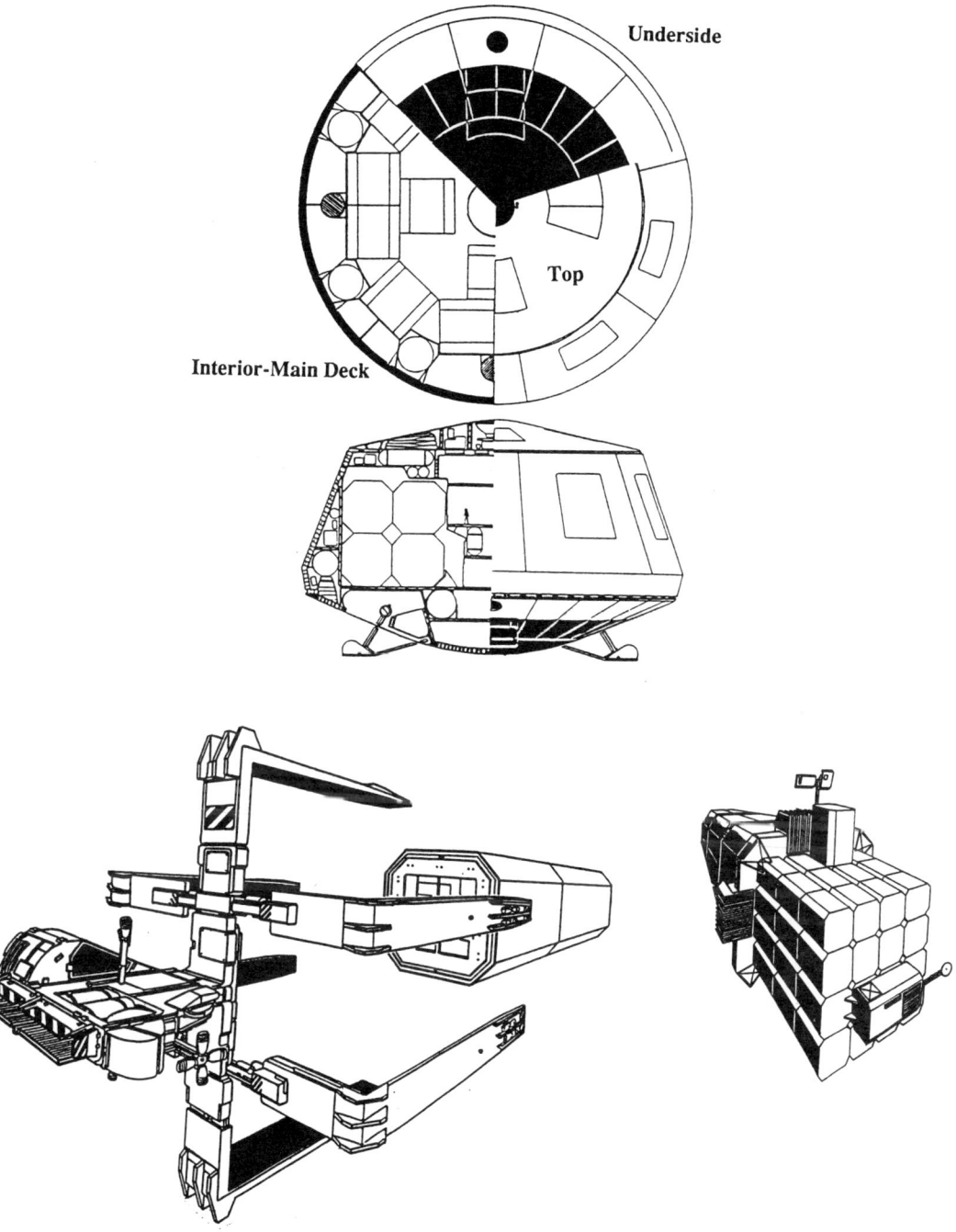

Figure 14 Multi-role Mars Lander (top) to ferry passengers and limited cargo. Bottom an orbital forklift vehicle retrieves cargo containers from a "bundle." Copyright 1993 by Darren Gillett.

Figure 15 is a diagram of an armored mining environment suit used for mining operations on the open landscape. The suit features robotic hands for using tools, picking at rocks and soil, and lifting small objects. The inside of the helmet has a "heads-up" display which gives the worker messages about the direction back to the habitat, distance away from the habitat, heart rate, radiation readings, terrain map, time of day, and air/water survival supply. The life support and communication gear are carried on the worker's back with a monitor unit located on the chest.

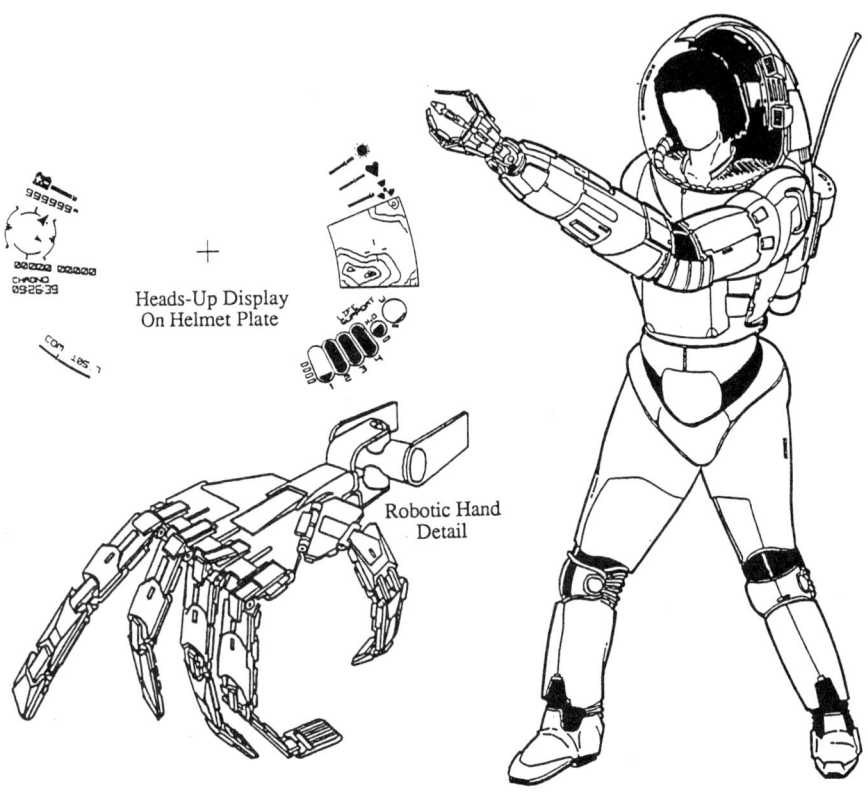

Figure 15 Armored mining environment suit for mining operations on the open landscape, featuring robotic hands and "heads up" display inside helmet Sketch Copyright 1993 by Darren Gillett.

While concepts of setting up Mars bases and colonies are in the thinking stage at the moment, they represent various approaches to solving everyday living and working problems in the Martian environment (References 9, 11, 14, 17, 18, 19). It is important for us to begin to solve these problems now, before we ever reach the Red Planet. There is no way of knowing what new directions the human future on Mars might take.

The settlers at Jamestown, Virginia and Plymouth, Massachusetts arrived with only what their ships could carry. As a result, they lived off the land and made the best use of the natural resources around them. Sickness claimed many lives; there were seemingly insurmountable problems to be solved; starting a New World was not an easy job. But these settlers not only survived - they endured, and raised up on the beaches of North America one of the most awesome civilizations in the history of planet Earth. There is no reason that kind of determination and commitment cannot take place on Mars.

Most of the books being written today about Mars have to do with the past and the present; histories of the public view of Mars, or the Mariner and Viking Missions, etc. These books are important and have their place in society and in the classroom, but now it is time to look ahead to Mars as the next step in human exploration and possibly human evolution. We now need to focus on the future and all that it holds for us. The public needs to be educated about the great potential of Mars settlement, and one of the surest ways to bring that about is by teaching teachers to teach young people, so that these young people grow into adulthood with the concept of settling Mars already fixed in their minds.

A system should exist for training teachers during their college years, before they graduate as teachers, so that they have the background to work with concepts of space education and Mars in particular when they begin teaching. The business of Higher Education is still education - it is the business of learning, and preparing teachers and young people for places of leadership in America's future. Unfortunately the business of most colleges and universities is business - making money and surviving.

The current structure for creating teachers is outdated and, for the most part, does not allow for reality technology training. Higher Education must set the pace for secondary school graduates if we are to build a brighter future and return standards to their previous normal levels. This will be Higher Education's most difficult role in the face of economic pressures of the 1990s and after the turn of the century.

Standardized teacher training seminars about Mars are needed so that Mars can be taught in the classrooms of America. The answer once more is a *standardized teacher training curriculum*. In many cases, young people know more about space than their teachers. In other instances, many young people's perception of space technology is grounded in the Saturday morning television cartoons, and science fiction films made in Hollywood. If teacher training is to be done properly, teachers should be taught before students are taught, and we should not leave these efforts to chance.

If we are to achieve something worthwhile, there must be a consensus among the industry, government and education communities working in concert. In this *triad partnership* (Reference 4), the role of education is to (1) shape the public view and attitude toward a Mars human landing through an understanding of the importance of such an endeavor, (2) educate young people for their places in industry and government, and (3) help establish Mars as a reachable goal in the national will. This kind of cooperation and commitment begins with state and local Boards of Education and local school administrators.

There are only two references to education in the United States Constitution; evidently our founding fathers believed education should be left to the discretion of the individual states. But the development of our country has shown education to be a major factor in cultural affairs and a highly important segment of state government. State mandated curriculum begins with state Boards of Education, and it is those political arms that will have to lead the continued quest for technology education at the grassroots level.

Everyone involved in any way with space exploration is a learner because usually there are no precedents for what we are trying to accomplish. For these educational efforts, then, we may have to invoke what I have come to call the "Leonardo Connection" when DaVinci commented at the height of his career, "Learning is the only thing the mind never fears, never exhausts, and never regrets; and it is the only thing that never fails us."

Neither space nor space education should be viewed as an all-encompassing cure for the frustrations of humankind, but rather as a golden moment of rare opportunity to seek and encounter new dimensions of fulfillment for the human mind. In the final analysis, there can be no alternative for the age we live in, and no greater promise for centuries to come.

Every time I enter a classroom, I'm mindful of the fact that I may be teaching Martians that day. Perhaps someone in that classroom will be the first to set foot on Mars, or the first to make the journey to start a new colony on the Martian landscape. In the long term, then, when we speak of going to Mars we're talking about human evolution.

The colonization of Mars is the next stage in the ongoing evolution of the human species. America IS a spacefaring nation (Figure 16), and although the process of evolving is slow and cautious, evolution is taking place right now and we are all us a part of it. For this reason it is even more important that space education across the nation must show up on our list of priorities.

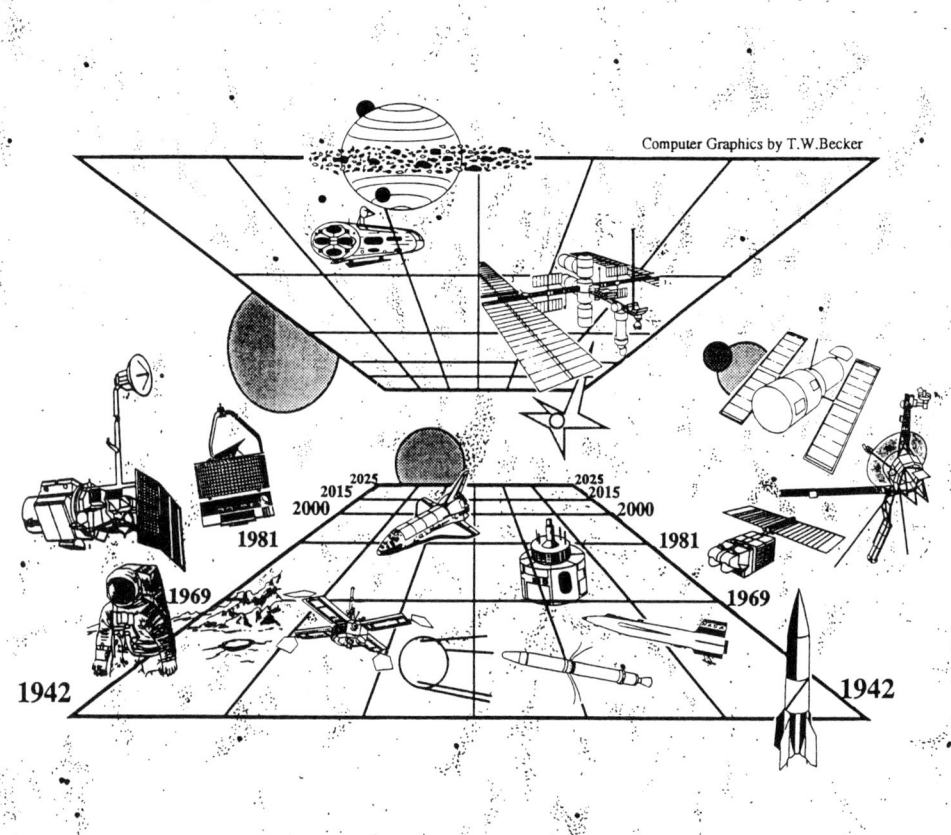

Figure 16 The human species has committed itself to the exploration of space, and in that commitment, America is a leader in the human quest for newer footholds in the universe - and we will go wherever we need to go. Graphics by the author.

We must explore. . .it is our destiny to reach out for the blinking lights in the sky, and to plant our feet on the beaches of faraway shores. If we ever lose that instinct, we are indeed lost to the centuries that lie before us and which could have made of us a far grander civilization. *Make no mistake about it - the Manifest Destiny of the 21st century is on the cold, hostile shores of Mars where creative human activities will be tested to their limits. We dare not fail in meeting that challenge, because it is not only the greatest test of the present but the greatest hope for the future of humankind in the universe.*

Economist Gale Baker Stanton said it another way: "If we are to achieve all that is possible, we must attempt the impossible; if we are to become all that we can be, we must dream of being more."

REFERENCES

1. Becker, Thomas W. "Space Education Policy - A Global Imperative", *SPACE POLICY*, pp. 60-71, Butterworth-Heinemann Ltd., Feb 1991.
2. Becker, Thomas W. "Comparing Earth And Mars: Teaching Applied Space Education" in *SPACE: The Next Renaissance, Proceedings of the Seventh Annual International Space Development Conference*, pp. 87-104. San Diego: Univelt, Inc., 1991.
3. Becker, Thomas W. *Technology And Current Events: The St. Louis Study*. Unpublished study report, 1988, St. Louis, Missouri, 25pp.
4. Becker, Thomas W. "Global Space Education For Secondary Schools: Formulating New Attitudes And Policies", *SPACE POLICY*, pp. 57-72, Butterworth-Heinemann Ltd., Feb 1994.
5. Becker, Thomas W. "Mariner 3 and 4" in *Magill's Survey Of Science: Space Exploration Series*, Frank N. Magill ed., Salem Press, March 1989, pp. 842-848.
6. Collins, Michael. "Mission To Mars", *National Geographic*, November 1988, Vol. 174, No. 5, pp. 732-764.
7. Grieve, Richard A. F. "Impact Cratering On The Earth", *Scientific American*, Vol. 262, No. 4, April 1990, pp. 66-73.
8. Kasting, James F., et al. "How Climate Evolved On The Terrestrial Planets", *Scientific American*, February 1988, Vol. 258, No. 2, pp. 90-97.
9. Robinson, Mark S. "Surveying The Scars Of Ancient Martian Floods", *Astronomy*, October 1989, Vol. 17, No 10, pp. 38-45.
10. Squyres, Steven W. "Searching For The Water Of Mars", *Astronomy*, August 1989, Vol. 17, No. 8, pp. 20-27.
11. Zubrin, Robert M. "The Key To Mars, Titan And Beyond?", *The Planetary Report*, Vol. X, No. 3, May/June 1990, pp. 9-13.
12. A series of basic Pathfinder articles in *Science*, 5 December 1997, Vol. 278, is in the "must read" category.

Useful Books About Mars And Technology

13. Barbree, Jay and Martin Caidin. *Destination Mars In Art, Myth And Science*. New York: Penguin Putnam inc., 1997.
14. Booth, Nicholas. *SPACE: The Next 100 Years*. London: Mitchell Beazley Publishers, 1990.
15. Clarke, Arthur C. *The Snows Of Olympus*. New York: W.W. Norton & Company, 1994.
16. Fraknoi, Andrew, David Morrison and Sidney Wolff. *Voyages Through The Universe*. New York: Saunders/Harcourt Brace College Publishing, 1997.
17. Miles, Frank and Nicholas Booth, Editors. *Race To Mars: The ITN Mars Flight Atlas*. London: Macmillan London Ltd., 1988.
18. National Commission On Space. *Pioneering The Space Frontier*. New York: Bantam Books, 1986.
19. Ordway, Frederick I. and Randy Liebermann. *Blueprint For Space*. Washington DC: Smithsonian Institution Press, 1992.
20. Smith, Arthur E. *Mars The Next Step*. New York: IOP Publishing Ltd., 1989.
21. Spencer, John R. and Jacqueline Mitton, Editors. *The Great Comet Crash*. New York: Cambridge University Press, 1995.
22. Verschuur, Gerrit L. *Impact! The Threat Of Comets & Asteroids*. New York: Oxford University Press, 1996.

Required Reading Basic Books

23. NASA-SP263 *The Mariner 6 And 7 Pictures Of Mars*, Stewart A. Collins, 1871.
24. NASA-SP329 *Mars As Viewed By Mariner 9*, Mariner 9 Television Team And Planetology Program Principal Investigators, 1976.
25. NASA-SP337 *The New Mars: The Discoveries Of Mariner 9*, Hartmann and Raper, 1974.
26. NASA-SP425 *The Martian Landscape*, Viking Lander Imaging Team, 1978.
27. NASA-SP441 *Viking Orbiter Views Of Mars*, Viking Orbiter Imaging Team, 1980.
28. NASA EP-179 *Activities In Planetary Geology*, 1982.
29. *Scientific American*, Special Issue 1990. *Exploring Space*. The entire issue is devoted to space science, especially Voyager missions to the planets, written by the scientists themselves.

GROWING THE FUTURE

Charmin P. Gerardy

*"It is not WE who will go to Mars, it is THEY—
and THEY will be very different from us."*
Jesco von Puttkamer

INTRODUCTION

The generation of a cultural paradigm to support the tremendous endeavor of sustained exploration of Mars need not, in fact, cannot wait until the members of that culture have reached the maturity to vote and pay taxes. The attitudes and skills needed must be implanted and nourished throughout the developing years, intrinsically linked to the interests, abilities, and priorities of the members of that supporting culture.

Acknowledging the significance of education to the development of the Martian endeavor is only the beginning. It remains for the educator to find ways to communicate the vital issues in meaningful ways to the students. In addition, even educators who are enthusiastic and knowledgeable about space exploration must address the ever present challenges of low budgets, administrative limitations, logistical obstacles, and standards requirements. To meet those challenges, my second through fifth grade students and I planned a detailed Mars Mission. To accomplish this feat with extreme limits on funds, materials and time, we borrowed techniques from clever scientists everywhere and the space program itself. We used the best information we could find, utilized what resources and skills we had, and adapted findings to our own purposes. We "stood on the shoulders of giants" and reached for a new world.

All of the projects were carried out using easily accessible materials. Most of it was recycled "junk" collected from everyday sources. (The operating budget allowed for this project was $0.) Projects and experiments were geared to the varying ages and abilities of the students. For information resources, we used the most specific and up-to-date sources we could find, relying heavily on the work being done by real aerospace experts. Utilizing the news of current events and ongoing missions helped generate enthusiasm and gave students a sense of involvement in the larger world. Making real-life connections with working experts provided powerful role models and gave us a chance to get feedback on our work. Dividing the work into smaller groups which focused on specific aspects of the mission allowed us to generate greater depth and produce better results. Communicating their work to others reinforced learning for the students and intensified their experience of having really accomplished a powerful goal. Sharing their work also inspired the other students to desire involvement and ask questions. Adapting the program to the needs of the clients and acknowledging their personal perspectives generated intense involvement and creative solutions to problem solving. Extending what they had learned into a second application they were already enthusiastic about provided even more reinforcement and involvement.

Using the mission to Mars as the focus, this project involved the study of microbiology, planetary evolution, environmental and health issues, physics, engineering, and cultural issues. Like the real Mars endeavor, it became a vehicle for my students to explore their own future. The project became more complex as it grew, and provided a paradigm for meeting standards at many levels. It became a powerful experience for all involved.

PRELIMINARY STUDIES :
MICROBIOLOGY AND THE MARTIAN METEORITE

With the news of the Martian meteorite so vividly available, I began a preliminary study of microbiological topics with the 4th and 5th grade students. We explored the theories of the origins of terrestrial life and the evidence offered by stromatolites. We used microscopes and examined microbial life in many different habitats. We examined the theories concerned with the effects of life on planetary atmospheres and climate, and explored the evolutionary development of living systems on earth.

Then, using the techniques developed by the Microcosms Curriculum group, we built "Stinky Little Planets". These were varied soil samples collected by the students and placed with water, newspaper strips, and raw eggs in modified two-liter beverage bottles. The chambers were sealed, and exposed to sunlight. Over a period of weeks, the little "planets" developed bacterial colonies. There were four types: anaerobic, sulfur bacteria; anaerobic non-sulfur bacteria; aerobic sulfur bacteria; and aerobic non-sulfur bacteria. By observing the development of the colonies and relating it to environmental variables, students gained understanding of photosynthesis and fermentation as energy obtaining strategies. By correlating changes in the colonies with environmental variables and substrate variables, students developed understanding of the effects of climate and resources on living systems. Discussions of the processes of decomposition and gas production led to understanding of the importance of bacterial life in the ecosphere and why the evidence of AH84001 was so significant.

A SURVEY OF THE SOLAR SYSTEM: GETTING TO KNOW THE NEIGHBORHOOD

Having developed some understanding of planetary formation and biological development on Earth, the study then turned to the local neighborhood of other planets. We examined each in turn, each of the students in each class researching one planet or satellite, and presenting their findings to their class. Using a system presented by Robert Heinlein in *Have Space Suit-Will Travel*, students learned average distances of planets in Astronomical Units. This system equates A.U.s to dollars (a decimal system most children are familiar with fairly early). As we researched the planets, we discussed the climate and life possibilities of each. These could now be related by the students to distance from the sun and energy supplies. We practiced problem solving "mail runs" on "space scooters" using A.U.s to compare time and fuel requirements. It also gave them preliminary understanding of communication time limitations over such a large area. Comparisons between planets led to discussions of global environmental issues like global warming and greenhouse effects.

Resources used to help the students research planets were varied. There are many children's books available, and these we supplemented with findings from the Mariners, the Pioneers, the Voyagers, Magellan, astronomy texts, and Internet sources. We also took a field trip to Fiske Planetarium to view a presentation on Mars and to actually walk the Scale Model Solar System.

BACKGROUND STUDIES: FOCUS ON MARS

Having examined known conditions in the rest of the system, students now were ready to examine the characteristics of Mars and its climate. We examined precursor missions and the findings of Mariner 9 and Viking I and II. We used a video presentation of JPL compiled imagery that allowed them to "fly" Valles Marineris. The video also demonstrated the problems associated with exaggerated scale in computer imaging. Students compiled an awareness of the climate conditions on Mars as well as the challenges of colonization. Discussions of temperature extremes and gravitational differences as well as resource availability were held, and continually cross checked with available sources. Of particular importance were some copies of artwork donated to us by Carter Emmart. The visual impact of the pictures, showing humans working and living on Mars, was inspiring. The students were ready to begin planning their own "mission" of exploration and colonization.

STUDENT MISSION TO MARS

As students began investigating the challenges and requirements of human exploration of Mars, they also began discussions of why the endeavor should be undertaken. As these developed, it became clear that, though they were very young people, their thoughts on the matter were not trivial. Some students were worried about hardship and expense, others compared it to the development of the American West. In all cases, even the youngest, the students were taking it very personally. They clearly saw that the endeavor would have a strong effect on their own lives, and on the rest humanity. They wondered about leaving families and homes behind, and also about Earth and its population problems. They questioned if it was right for humans to take over another planet. They debated who would own Mars if people from many countries went there. They argued about whether they would need weapons.

The first step was for every student to design a habitat for Mars, to enable them to focus on the requirements for living there and give them a personal investment in the process. Connecting with their own physical space is critical for young students, and visualizing where they would live allowed the mission to become more real for them. Every student presented their own plan to their class, then as a group, each class formulated a "packing list". This generated intense discussion about the criteria for "necessities vs. luxuries", and helped formulate group cooperation and communication.

The next phase of our efforts, actually creating the detailed plans, posed our first major obstacle and led to a decision that became critical to the overall project. The immensity of the endeavor was quickly apparent, as were the differences in ability and interests of the students. It was necessary to split the project into smaller units, with different groups becoming responsible for smaller aspects of the overall mission. In this way, students could concentrate on specific tasks and assimilate more in depth understanding of that topic instead of only gaining a general knowledge. It also allowed the needs and abilities of each student to be better addressed. Many students worked on more than one topic. Individual student desires and skills were included as a valid criteria for choosing their own tasks, which, in my opinion, was critical to the phenomenal success they achieved.

One group of second graders really wanted to perform experiments, so they built miniature hydroponic greenhouses and monitored plant growth and water consumption. Some of the students, who were veterans of my rocketry program, were used to designing and building their own vehicles and wanted to concentrate on spacecraft and propulsion systems. One class of particularly kinetic third grade boys was interested mostly in sports, so they invented sports suited to the reduced gravity of Mars. As part of their endeavors, they produced a sports page for a newspaper called, of course, "The Martian Chronicle". It included advertisements for sports equipment designed for Mars, and a huge controversy about the legality of "cloned" basketball teams. A group of fifth grade students that had worked with a LEGO-Logo setup designed and built a remote-activated, computer-controlled rover. The acceptance of student choices, both as groups and individuals enabled them to devote more effort to their work and give it more meaning as part of their own preparation for the future.

As students began to work on specific aspects of the mission, they continued to explore the research materials that were available, checking their own plans with the "experts" whenever they could. They became familiar with not only the work that was happening, but also aware of the people who are doing it. For example, one girl wrote, "Chris McKay has been studying life on Mars, so he went to Antarctica. Why? Because they are both very cold!" The quest for Mars was no longer science fiction, but the endeavor of real people they could recognize.

Students turned their own best skills to the task. Artistically talented ones made lovely mission patches and maps. Writers generated biographies of the chosen crews. Builders made models of wind turbines, greenhouses, rovers, ships, bases and habitats and water recyclers. Some of them were working prototypes powered by wind, electricity or solar panels. One musician was worried that the crew might become bored, so he designed some instruments for them to build out of "ship junk" to entertain themselves with.

As student efforts continued, and their work began to accumulate, I was astounded by the results. These very young students had addressed most of the necessary parameters of a Mars mission. A detailed description being prohibitive, here is a brief list of their products:

Martian habitats, designs and models
"Packing lists"
Martian bases, designs and models
Crew, selection criteria and biographies
Flags
Maps and landing sites
Mission patches
Musical instruments, designs and models
Martian calendars
"Martian Chronicle" newspaper sports page
Nutritional requirements and diets
Martian resource utilization report:
    analysis of water and oxygen sources and use of indigenous fuels.
Hydroponic greenhouses; designs, working models and experiments
Water recyclers; designs, experiments, and working models
Waste disposal strategies; design and model
Solar power plants; working models and experiments
Wind turbines; working models
Shuttles (for base to orbit); designs and models
Surface craft (manned and unmanned):
    Rovers; designs and working models, solar, wind, and electrical powered
    Robot walker; design and model
    Mars plane; design
    Mars "copter"; design
Computer controlled, remote activated LEGO-Logo Rover; working model
Spacecraft (Earth to Mars); designs and models,
Propulsion systems; designs and comparative proposals
Orbital trajectories; comparative proposals
Spacesuits; designs
Biomedical requirements and equipment lists
Communications equipment requirements; designs for satellites and ground equipment
Mission Parameters:
    experiment designs for the Martian surface (there were over 30)
    exploration strategies and itineraries
    data collection strategies; balloons and indigenous fuels

MARSFEST: PRESENTING THE MISSION

Groups shared information whenever possible, to coordinate their efforts, often at the lunch table or on the playground, since the overall school schedule prohibited larger gatherings. Students shared discoveries, information, and technical details. However, the incredible scope of the project was not apparent and it became obvious that this tremendous effort needed to be witnessed by more than just one teacher, me. I asked my administrator if I could arrange to display the materials so that other students and parents could view it. I was granted the lunch room for one hour at 3:30 on a Monday afternoon, with a margin of "maybe fifteen minutes" to set it up and take it down.

To meet the tight constraints on physical and temporal space was a significant challenge, but one that we met, and well. From the local appliance store, I collected large cardboard boxes. These were cut down one side, and used as self-standing "accordion" folded screens. They were covered in red craft paper, and student works were mounted on them, with photos taken during the project. Labels and other information for the models were also mounted on these, with space left to display the models on the floor or table at the base of the screen. The necessary cabling for the computer controlled robot rover was tethered on a string suspended from an overhead water pipe.

We added a special display of some of our resources, including posters and visual art to help create atmosphere for the exhibit. We had sent home notices announcing the event to parents, and on the day of the exhibit, I put up notices mounted on red backing, near the entry as reminders.

The allotted time was already scheduled for the weekly meeting of my Young Astronauts Chapter. So, since they had all worked on the project, we made the presentation an official function of the Chapter. The members became the "Official Hosts and Tour Guides" for the exhibit, and they performed marvelously. As parents arrived to pick up students or see the exhibit, they were greeted by welcoming "posters" that also informed them that "The Mission is Already Underway" (a useful phrase from Case For Mars III.) Then the student "Tour Guides" would lead them in, answer questions, and explain their work.

The preschool students were given a special guided tour. And the "older" students were very professional and gentle in their explanations to the "little ones". Both groups benefited from this encounter. They were already "sharing the spark".

As a special treat, Carter Emmart and Rene Munoz also came down from NCAR to be part of our Marsfest. Carter, who had provided our inspiration with his artwork and video, showed slides of his IMAX work on Mars, shared his expertise, and discussed technical topics with the students. Rene Munoz, Education and Tour Director from the National Center for Atmospheric Research, and a long-time sponsor of our annual egg drop event appraised the exhibit and discussed the students work with them. She generously invited us to put the exhibit on display in the main lobby of NCAR for the summer, where it was viewed by thousands of visitors to this national laboratory. In addition, Larry Esposito, Randy Davis, and Michael Shull, all of the Laboratory for Atmospheric and Space Physics of Colorado University, (and most importantly, parents) were also present and conversed with students about their work.

In only about an hour, this event became a phenomenally exciting occasion. Students were animated and invested while discussing their efforts. Parents and other students were amazed by the scope of the project and the effort displayed. Visiting scientists were besieged for autographs by the young "Martians". The event was surpassed only by the two follow-up events that occurred as extensions of the project.

EXTENDING THE SCOPE OF THE PROJECT : THE EGG DROP

Applying the knowledge students had gathered about the Mars missions beyond the original project was easy in this case. Every year, my classes joined NCAR staff for an Egg-Drop contest. It was clearly the single most popular and creative event of the year for students, and we chose to dedicate our efforts that year to honor the Pathfinder mission which was underway.

As work on their own Mars Mission Projects were finished, students began to design egg-droppers for the "Mars Egg-Stravaganza", to be held at NCAR. Students designed and built their own craft to deliver a raw egg payload safely through a six story drop onto concrete, and many reflected the theme by creating models of real Mars exploration craft. There were Mariners, Vikings, Pathfinders, and Sojourners. Some students made other rovers, robots, and Mars base models to be dropped. One young fellow made a very large version of a design called "Ganymede" that was shown at Case For Mars I, in honor of its creator Carter Emmart, who had become an inspirational figure for the students. Carter was on the roof to help deploy the droppers and, when he appeared beyond the edge, students recognized him and spontaneously began chanting "Carter!, Carter!", in the manner normally reserved for rock stars and sports figures.

The finale of this event, constructed by NCAR staff in honor of the students, was a model of the Pathfinder craft covered with helium balloons that bounced very like the real thing when dropped from the tower, and opened to emit a homemade radio-controlled model of Sojourner that rolled around the plaza to the delight of everyone. Coverage of the event was included in the *NCAR Staff Notes*, of June 1997, and videos were distributed as Quick Time movies on the NCAR web page.

Student involvement and excitement was obvious at this event. They were having fun, but they had also mastered knowledge and applied it well. As they left, I heard repeated comments about the Mars landing that was still upcoming for them. It was a real event that they were looking forward to, and planning to follow.

VALIDATION AND REWARDS: THE RECEPTION AT NCAR

Rene Munoz of NCAR had requested the Marsfest exhibit be put on display in the main lobby of the Mesa Laboratory facility. When it was set up, she thoughtfully arranged a "formal reception" for the creators and their parents at the laboratory. Punch and cookies were served and the students got to see their work prominently on exhibit in one of the most beautiful buildings on the planet. As they walked in, they were clearly in awe of their surroundings and very proud to see their work given such serious consideration.

In addition, there were experts on hand for them to "mingle with". At one point, there were four young ladies sitting on the floor around Mars terrain maps, discussing the relative advantages of various landing sites with a planetary scientist whose specialty is Martian geology. This group was in turn surrounded by a circle of parents who were too mesmerized to speak, as if afraid to disrupt the moment they were witnessing.

I will never forget the look on the face of one young lady as she reached out to touch with her own fingers a fragment of Martian meteorite. She was later observed looking around the lobby with a friend, and proclaiming, "This place is COOL! I think I might want to be a scientist!" Formerly, she had displayed ambitions to be a prominent member of the fashion police.

Earlier at Marsfest, Rene had been guided around the display by one particular student, and she had later notified me of his thoughtful efforts and detailed explanations. He sought her out again at the reception, and made sure she was again well escorted. I later explained to her that this young fellow was extremely bright, normally had dreadful social skills, and had endured a tragic home situation that made it very hard for him to trust adult females. At this event, he was, able to connect with someone new in a very positive way, receiving acknowledgement for his intellectual prowess at the same time.

After months of work, to have such significant validation of their efforts was a pivotal experience for these students. Although they did not produce any ground-breaking technology, and all their models were made of recycled junk and LEGO, their effort and creative ideas were still valuable. Each of their discoveries were original to them, if not to the world, and were no less precious for having been made before. They were very young, and did the very best they could with what they had available to tackle a complex endeavor. In the process they increased their own skills and gained appreciation for the efforts of those who have gone before. They became aware of the significant effect that technological issues, including the exploration of Mars will have on their own future. They became invested in actively taking part in the process of directing that future. In my opinion, we can ask for nothing more, and I am in awe of their accomplishment.

THE FRONTIER OF MARS AS AN AGRISCIENCE CLASSROOM: TERRAFARMING MARS

Larry Payne[*]

INTRODUCTION

When we have reached the red rock, whom will we have brought along? Better yet, whom will we have left behind? In the presentations and discussions at this, the founding convention of the Mars Society, I think it appropriate to request you consider an active role for high school students. Specifically, high school agriscience students would make a most excellent partner to work along with you. A quality team effort is necessary to identify, address and overcome such challenges and issues as may confront us in our explorations outside the planet of origin.

PURPOSE

It is the intent of the Oroville High School Center for Astrobiology and Agriscience to be the world class leader in terrafarming education at the high school level. We will act as a stimulus and catalyst to form innovative partnerships with NASA Ames' Astrobiology Institute, the Mars Society, academia, the National FFA and the private sector. Partnerships will direct efforts toward curriculum development, implementation and cooperative research as the state-of-the-art and mission needs of secondary agriscience programs change. This work will guide resource investments for producing and integrating cutting-edge instructional technologies in agriscience/terrafarming education.

EMPHASIS OF PROGRAM

The centerpiece of the Oroville High School agriscience program will include a multidisciplinary emphasis centered around the field of astrobiology and agriculture. We envision working in conjunction and cooperation with anthropologists, astrophysicists, biologists, chemists, geologists, mechanical engineers, physicists, planetologists, and psychologists to conduct research and apply findings in terrafarming Mars. We will also examine the cosmic implications of plant and animal life elsewhere.

Emphasis of the program will be placed upon topics related to the formation, development and habitability of other worlds. Such a focus is well suited for study by students enrolled in an agriscience program. The formation of habitable worlds and their evolution can be readily incorporated into such components of the agriscience curriculum as principles of technology, geology and soil science. Emergence of living systems readily matches such areas of study as plant and animal science, botany, and chemistry. Biosphere and ecosystem changes fit well into the study of ecology, resource management and environmental horticulture. These are merely a

---

[*] Principal, Oroville High School. E-mail: lpayne@bcoe.butte.k12.ca.us. © 1998 all rights reserved.

few examples of the ways in which linkages can readily occur. As a means to effectively establish an interest in life beyond this planet, programs such as this will generate immense appeal.

CURRENT INTEREST IN LIFE BEYOND EARTH

The current interest in exploration and settlement of other habitable worlds and the need for a pioneering spirit is being used to create high interest levels in high school students. Using CONTACT's Cultures of the Imagination model, teachers at our site are developing a unique thematic series in courses for students beginning in grade 9. The project has caught the interest of students and will expand each year. By encouraging the participation of such students in the process of scientific research and application, a connection will be established to the next generation of workers and researchers. We are now beginning to design an entire program to focus on terraforming Mars. This, as you could well imagine, has the attention of many students especially those interested in agriscience classes.

A mission-essential strategy for those of you involved in emerging forms of science and technology, is strategic knowledge engineering. Value-added knowledge exchanges between professionals such as yourselves and high school agriscience students should be one of the most important components of your work. The most effective way to convey knowledge about life beyond planet Earth is through an interactive experiential process.

WHY AGRICULTURAL EDUCATION STUDENTS ARE MOST APPROPRIATE

Agricultural education involves three components: (1) classroom and laboratory, (2) Future Farmers of America (FFA) and (3) supervised occupational experience. All three components are interrelated. Each compliments the other. Science education in this country has been led by "ag" teachers. Ag teachers, of all the various high school disciplines, are the best-organized and most effective in teaching hands-on science. For the best in hands-on, practical and theoretical learning, you need only look inside an agriscience department.

Teaching science through agriculture has resulted in some of the most impressive advancements known to date. Cutting-edge technologies, innovative practices and world leaders are regular products of agriscience education. Students enrolled in high school agriscience courses are most always National FFA members. Being a member or officer in the FFA is where they learn leadership skills. These leadership skills are the most useful and longest lasting of skills, which carry over into all aspects of life. Supervised occupational experiences blend opportunities for success in the management of individual projects or employment in occupations related to what has been learned. It is this mix of concepts, practices and leadership that provides the ideal working partner for you in terraforming Mars.

THIS CALL IS FOR YOU

There is a need for scientists to become much more cognizant and supportive of inclusion of young people in the process of discovery. A crucial issue that continues to elude pensive thought relates to the effect of the rate of change in knowledge on young students. We know that the velocity, complexity and magnitude of information are being compressed into shorter and shorter life cycles. This has an as yet to be addressed impact on young learners. Our efforts at the high school level can now easily result in sharing historical information with students. Unless we are very careful and place our emphasis on teaching underlying principles, we risk creating a class of citizens whose skills and knowledge are outdated from the git-go.

Contrast this with the real-time knowledge acquisition and applications occurring in today's basic and applied technology and science studies. In other words, you who are working on the creation of knowledge are so far from your future protégés that, with the lack of timely and relevant information, these students end up receding faster and further from a common knowledge base essential to the future. Failing to recognize the fact that we are absolutely and irreversibly interconnected as creators of knowledge and wisdom are to, in the end, deny young learners real-life opportunities to learn. The result of this is to create a chasm of alienation and ignorance for the sake of "club" science. There is, however, a possible way out of this dilemma.

LEARNING ASTROBIOLOGY BY DOING ASTROBIOLOGY

The learning of astrobiology by doing it makes the practical application of science and technology a reality in which students experience their own learning. Here, students can be put in touch with your knowledge as you create it. Into this dynamic web of interrelated events must be drawn younger generations. These students must be provided opportunities to study the nuances and implications of the inter-relatedness of various disciplines.

Classroom and laboratory study of theories and practices will provide the patterns of knowledge and skills to access the future. In this manner, we will enable learners to become more effective in adapting, evolving and utilizing tools and resources. Cooperative, real-time work with you will enable them to experience the effect of what they have learned and to experience present-moment study and discovery. This "school without walls" learning is precisely what will have a tremendous impact on every class a high schooler takes.

KIDS AND PARENTS - YOUR BEST SUPPORT FOR TERRAFORMING

If you choose to have broader support and interest in space exploration and terraforming other planets, you need to hook the kid who can go home at night and exclaim to his parents, "You know what we did in school today? We tested Martian soil for evidence of microfossils!" To get this kind of support does not take huge sums of money. The investment that needs to be made is measured in terms of real-time. Our agriscience classrooms would be directly tied in to your work site and our students would work along with you - in a reasonable manner, of course. Who better to lead this grand adventure than Mars colonization proponents, visionary educators, and high school students?

Still, there are numerous high school principals and agriscience teachers who need to create a broader vision of the possible. Agriscience on the frontier of Mars is the same agriculture which has forever epitomized the development of mankind. Civilizations, except for hunters and gatherers, require ecosystem and biosphere management to survive. Vast numbers of technologies for the advancement of village societies have been wrought from the soil workers. So-called modern societies owe most of their development to changes in agricultural practices. There are now incredibly challenging and unique possibilities awaiting the terrafarming of Mars. No, the last frontier has not been settled. Questions have yet to be posed about cultures, life, intelligence and cosmology. Who better to seek answers to such questions with you than our youth?

QUESTIONS ABOUT LIFE BEYOND EARTH

Astrobiology seeks to answer questions regarding the formation of life, its existence beyond Earth and whether humans can survive in environments other than Earth's. There appears to be no sound reason, other than traditional practice, to limit such studies to scientists.

Multi-age efforts to identify and solve problems or resolve issues are known to be most effective. Oroville High School agriscience students are willing to participate in efforts with NASA astrobiologists, university researchers, the private sector and others who are attempting to:

1. Identify any natural processes which may spread life from one planet to another;
2. Define the minimal ecosystem required for organism survival and adaptation beyond Earth;
3. Identify and understand these adaptive mechanisms and strategies;
4. Determine ethical principles for cultivating life elsewhere in the solar system; and
5. Demonstrate a new kind of multidisciplinary learning about space and the future.

Students will also seek to know how habitable worlds form and how they evolve. They will study the emergence of living systems. Determining how other spheres which show life can be recognized will present an excellent challenge. Students will examine the mutual influencing of the Earth and its biosphere. They will look at how rapid changes in the environment affect emergent ecosystem properties and their evolution. Also to be considered will be the potential for survival and biological evolution beyond the planet of origin. Contact with other cultures and the means to establish mutual trust, understanding, respect and friendship will be studied.

DISCIPLINE-BASED TECHNOLOGIES

Through the use of various discipline-based technologies, the Oroville High School Mars terrafarming program will identify and/or conceive concepts, instruction and applications of an emerging field of study related to terrafarming education. Technologies to be used will be much the same as are used by astrobiologist and others. These include:

Modeling and Simulation
- virtual environments and artificial life
- simulation-based design
- interdisciplinary design of simulated/synthetic systems
- living systems and physics-based models

Data-Based Management
- data mining
- information interconnectivity
- pattern recognition
- discovery of knowledge

Human/Computer Interaction
- knowledge from information/data
- scientific visualization methods
- learning from intelligent agents and embodied intelligence
- immersive technologies

Automated Decision Making
- computational intelligence (neural nets, fuzzy logic, etc.)
- knowledge-based (expert) systems
- model-based reasoning (for design, diagnosis, control, etc.)
- reasoning under uncertainty.

SUMMARY

Most adult professionals don't know high school students. Due to a lack of direct contact, most adults are apprehensive about teenagers. Most adult professionals need to know that high school students are capable of doing wonderful and incredible things. Many of the most technologically-literate workers and post-secondary students of today came out of high school agriscience programs.

When those of you who work in such occupations as anthropologist, biologist, explorer, historian, mathematician, philosopher, scientist, and writer begin to engage high school students directly in your everyday work, you will find that work more fulfilling. I can think of no better place to begin than in designing, developing and implementing this model with highly motivated students and teachers choosing to create the future. Let's not play a half-court game. Let us open the frontier of Mars to a uniquely creative cross-generational team of pioneers who will do the work of a quality team.

MAR 98-027

TEACHING FROM MARS

Gabriel F. Rshaid*

A few months ago I attended an educational pre-conference that took place before a NASA workshop. One of the sessions dealt with how to organize teacher workshops and implement successful space related educational activities. Although some of the points were well made (activities had to be complete, easily integrated, enjoyable, involving experimentation, everybody likes quality handouts and classroom posters, etc.) I was somewhat shocked to see explicitly listed in the presentation some of the general principles that I had managed to detect in space education during my many years of browsing through educational brochures and guides. "Successful" activities should take no longer than one or two class periods, be low cost and easy to set up, and definitely not include homework, technical details about missions and spacecraft, or involve multi-week curricula. They should be designed bearing in mind that teachers have a weak math background and in some cases math phobia altogether, possess few computer skills, cannot assimilate too much science too fast, are easily lost by technical jargon and have no research experience.

Beyond the mere anecdote of this workshop, and understanding that any generalization is intrinsically unfair, after working in space education for more than five years, participating in several conferences and being in contact with many teachers that have introduced space in the classroom, I have to admit that the great majority of space related educational activities are designed with these principles in mind. With some exceptions, most space education projects target elementary and middle school students and can easily be implemented by the kind of teacher described above.

Hold on. I am not here to defend teachers or challenge the above profile of an average teacher. Like in any other profession, there are good, mediocre and bad teachers, and it would certainly be hard for me to tell whether the majority of them respond to those characteristics. And I do not deny the validity or usefulness of short, easy to set up, understandable activities that can help introduce space in the classroom at the elementary and lower middle school level. But I think we have to ask ourselves what our goal is in trying to introduce space in the classroom. It depends on which way you look at it. From the perspective of a space agency or space corporation, instructing the upcoming generations helps establish a conscience of what space exploration means, and thus ensures sustained support from years to come.

From my own point of view as a teacher, I am not so much interested in space exploration per se, but in what it does to my students. Through the study of space I want them to see humankind attempting the almost impossible, teaming up to solve the most difficult problems; I want them to witness the adventure and the drama of space exploration, to encourage them to pursue their own challenging careers in the light of the inspiration of our contemporary space

---

* Mars Academy Coordinator (http://www.marsacademy.com). E-mail: grshaid@marsacademy.com.

pioneers. I like to see them amazed at the complexity and order of the Universe, perplexed at its mysteries, their imagination stretched to the limits.

And it is hard to do that with single lesson "successful" activities that have no Math or computers. Simply because I cannot fool my students: they know that space is just not like that. A successful activity is perhaps not the most popular, the one that every student participates in. In my own quite unauthorized opinion, a successful educational activity is one in which some students are willing to go the extra mile on their own. When you tell them to stay after class or come in during the holidays and they do not complain. When they study seemingly difficult topics from a new perspective without any prompting.

What kind of activities are those? I have found them to be quite the opposite of the prescribed paradigm. Semester long activities, that involve intensive use of computers and other high tech equipment, generally requiring long hours of work and extensive research on the part of both teachers and students. Sometimes they are not even almost zero cost. Just a few examples of such projects are high fidelity mission simulations, design and construction of spacecraft simulators, analysis of real time data from spacecraft, design of space colonies and settlements, interactive Web sites, etc.

Mars Education: the key area in space education today. The choice of topic is also essential to the success of a project. Most space related topics are interesting and attractive to students, but not all of them engage them in the same way and ensure continuity. There are many options available to the educator, starting with current systems (Space Shuttle, International Space Station), and going on into the near foreseeable future (manned Mars exploration, lunar and Martian colonies) or directly to the realm of quasi science fiction (terraforming, giant orbital space settlements, intergalactic travel) Speaking again from my own experience, at this point in time, Mars is the ideal subject of choice.

I have worked extensively in Space Shuttle simulations and, sure enough, the Shuttle is quite a sexy machine and has the undoubted flair of a current transportation system. But after an initial tantalizing effect on students it is difficult to keep them hooked up in follow up activities. Their imagination is not captured by Earth orbit space operations and they know that it is, to say the least, a controversial issue in the space community as to its effectiveness as a spacecraft system and it cannot compare to the deeds of the past (namely the Apollo program). Lunar missions and lunar bases are more attractive, but again to a young mind there is always the stigma of we-did-it-before that somehow undermines their incentive. On the other hand, the almost futuristic projects like orbital space settlements or terraforming are exciting and charming for students in that the lack of technical constraints imposed by the limits of current and foreseeable technology helps materialize the vision that everybody has of a space faring civilization and makes them more free to dream their projects before designing them. But the drawback of this kind of activity stems from its own virtue. The fact that they will materialize so much ahead in time, probably not even in the lifetime of the students makes them more science fiction than science. Onward to Mars. Manned exploration of the Red Planet is foreseeable, and what we always say never fails to ring a note in their young minds: that they are the Martians, that their generation will be the one to physically make the trip. The road to Mars is challenging enough to compare and even surpass the Apollo days, providing the adventure and drama that is always needed to make the topic exciting, and it is current in that manned missions to Mars are currently being designed. The study of human exploration of Mars is also ideal in that it will remain as the main focus of the space program for decades to come and in that it is not limited in any way. Students can start by designing pioneer missions, then regular transportation systems, Mars colonies, Martian cities and even terraform the planet. So the (pink) sky is the limit.

The Mars Society: teaching from Mars. This convention will give birth to a new and unique organization whose focus is to promote exploration of Mars. A strong Mars education program is, in my opinion, one of the ways in which it can achieve its objectives. At a general level, such an education program should be based on activities that have the following characteristics:

* Officially sponsored programs: Whatever the project or activity, it must be officially sponsored and if possible bear an attractive, self explanatory name that says everything about it. The life cycle of an educational project at the school level, especially if it involves spending money on it, starts with the interested teacher who wants to participate because of the educational value of the project itself. But when the educator needs to ask for administrative support and perhaps even funding, it becomes crucial to be associated with a prestigious space organization. In that way the school administration can in turn justify the support given in terms of the impact on the community of such activities. In many cases, high tech space projects in which local students participate generate favorable media repercussions.

* Internet based projects: It is unnecessary to stress here the fundamental impact that the Internet has had on education. However, it is still difficult to find applications that make full use of the Internet's potential for communication. Internet based collaborative projects where students interact with their peers from all across the world add the excitement of the multicultural environment and mutual enrichment as well as provide an ideal media for research and collaboration. In the coming years, more and more schools will be connected and an increased bandwidth available to share video and audio.

* Provide opportunities for exceptional students and teachers: Although not all activities have to be of a competitive nature, it is always a good incentive, especially for teachers who are the ones to initiate participation in Mars programs at the school level, to recognize exceptional educators and student projects by means of internships, field trips, privately sponsored awards, etc. Again the element of prestige is a powerful driver for administrative support and funding.

* Active participation: It is essential, even if the task involved is somewhat trivial, that the participating schools be active in some way in the project. It makes a world of difference for the students if they feel that even in some minor way they are contributing actively to the task and not just reading or receiving information.

All this is easier to say than to do, but here is a list of suggested activities for such a program, targeted to fill in the mentioned void for this type of projects:

* High fidelity mission simulations: One of the most exciting, fun and challenging activities for students and teachers alike is performing high fidelity mission simulations. There are various ways to do this, either via the use of a specially designed hardware simulator or by running software simulations. The emerging programming tools that enable almost full fledged networked applications over the Internet can help develop an online collaborative mission simulation where students can control a Mars mission and role play astronauts and mission controllers. An example of a similar web based Space Shuttle simulation was developed by the author at *http://www.sptschool.com/sim/pub* and we are currently in the process of designing a Java based simulation of a manned mission to Mars through the Mars Academy *(http://www.Marsacademy.com)*. The use of the Internet for collaborative mission simulations can add to the

natural excitement of a role playing adventure by enabling students to interact in real time with their peers from all over the world.

* Mission outreach activities: Another very interesting possibility is the relaying of data directly from spacecraft for analysis by the participating school. This has been done already in the Moonlink program by Space Explorers, Inc. together with NASA Ames Research Center following the progress of Lunar Prospector. On top of performing a 2 hour launch simulation with 13 computers simultaneously connected to the Internet, each of the schools was able to select a 150 km square on the Moon and receive data for all the spacecraft's instruments. For this type of program to be effective, however, comprehensive tutorials and technical support must be readily available for participants to help them sort out a complicated real life problem.

* Assistance from the experts: Many scientists, engineers and other space experts generously volunteer their time to participate in web chats with students and teachers. These chats are invaluable opportunities for students to interact with them and gain insight into high tech careers. But many times, due to the general nature of the chats, the questions and consequently the experts' answers cover the whole range in terms of complexity and it is difficult to extract any content learning from them. It would be much better to organize these interactions within the context of a specific project so that the students and teachers can ask questions directly related to a certain problem or area. Our own experience in this matter has been very positive. During the design stage of our Mars Academy project we held several web chats with experts who helped our students gain insight in, for example, how to adequately formulate the mission goals, choose the landing site or define the number and functions of crewmembers.

* Contests: Prizes (especially if they involve field trips) and prestige are very powerful drivers for participation. Existing contests, like NSTA's SSIP program and the two ongoing space settlement design competitions have every year an increased number of entries.

* A Mars teacher workshop: There are a good number of teacher training programs in space education but most of them, in order to be open to all levels of teachers with diverse backgrounds are not specific or intensive enough to provide a solid foundation for the kind of projects suggested. An annual one week intensive Mars teacher workshop can be organized to instruct participating educators in Martian geology, spacecraft systems, orbital mechanics and various other related topics. It can also provide them with hands-on experience and even certification for on-going simulations and other projects as well as printed materials, curriculum guides and direct contact with experts, professionals and fellow teachers interested in Mars education.

The preceding list of activities does not obviously preclude simpler projects of the type described initially targeted at lower grade levels (although most of the above activities can be implemented at any grade level if the teacher is able to scale down the technical content appropriately).

For most of the above proposals an affordable fee should be charged to participants, in order to help the organizing institution cover costs and thus improve quality and service and also to discourage discontinuous participation. Schools are generally able to pay a few hundred dollars for a semester long program and there are always ways in which an educational community can raise funds.

Can we get there?

Mission design papers generally conclude with a succinct feasibility study of the proposal. Although our educational mission will not physically get us anywhere in space, it is nevertheless pertinent to perform a similar analysis in this case. This radical approach to Mars education with a stress on high tech medium and long term activities can undoubtedly be carried out without too much strain taking advantage of emerging technologies, widespread use of the Internet and faster and cheaper communications. In particular, the advent of Java as a cross platform Internet programming language can make possible what only a few years ago was considered little more than science fiction: real time graphical applications that can be executed coherently regardless of operating systems and system architectures. The flaw in the plan is self evident: the majority of the suggested programs involve a great deal of dedication, some technical knowledge and essentially the desire to study on the part of the teacher. This obviously limits participation to educators willing to accept the challenge. But our goal in teaching space is precisely the challenge itself. In the same way as surely almost nobody working in space exploration would be doing so if it were commonplace and routine, I would not be interested in teaching Mars if it was not challenging and difficult. I have said before that the young generations will be the Martians, and that any of them, even my own daughter, might one day be making the trip. However, whether they go or not is not important to me. What I really care about is that through learning Mars they will realize, especially in their teens; at that age when their eagerness and natural curiosity are slowly and subtly eroded by doubts and skepticism; that whenever we set ourselves any goal in life and work hard to achieve it, even if it is as seemingly impossible as traveling to another planet, nothing can stop us from making our dreams come true. And I do not think we can teach this particular lesson in a short time or without a great deal of effort. It is up to us to communicate this vision and inspire educators to reach out to Mars. It is quite far out, but very much worth the trip because, at least in Education, the road is much more important than the destination.

PREPARING FOR THE JOURNEY: AN INTRODUCTION TO MARS EDUCATION*

Donald M. Scott[†]

This paper is about Mars educational resources, both of the nuts-and-bolts type, and of the higher country of the mind.

During the past 25 years, much work has been done to prepare humankind for Mars. In technical, scientific, and economic terms, we are pretty well prepared to go, to explore, and to return. But a major Mars preparation task needs more attention. This is the education of the "Mars Kids"; the children now in school, who will design, fund, and conduct the human missions to Mars.

Those first Martians currently attend school here on Earth. A good education will be the foundation for their successful missions to Mars. This is even more true if we seek to develop the most efficient and cost-effective human missions. Teachers who are teaching the Mars Generation need resources to help them with the job.

One such resource is the set of nuts-and-bolts materials available to teachers who would prepare the Mars Kids. Another is a new way of thinking, along the lines described by NASA's Jesco von Puttkamer: "... a new frame of mind that shifts the emphasis from individual subjects to the interactions and relationships between them."

ON PREPARATION

During the past 25 years, much work has been done to prepare humankind to journey to Mars. In technical, scientific, and economic terms, we are pretty well prepared to go, to explore, and to safely return - repeatedly.

But a major task, while underway, needs more attention. The preparation of the "Mars Kids": Children now in school who will design, fund, and conduct the human missions to Mars.

Teachers are teaching the Mars Generation today; but they need lots of resources to help them with the job. This paper provides information about Mars Education resources to those teachers.

Since teachers have an immediate need for nuts-and-bolts resources - teaching activities, films and slides, news about space - the paper begins by describing such resources.

There is also a need for new ways of teaching, based on new ways of thinking. As NASA's Jesco von Puttkamer, writes:

"...we need a new frame of mind that shifts the emphasis from individual subjects to the interactions and relationships between them."

Jesco von Puttkamer,
Spaceflight and the New Enlightenment

---

* 1998 Donald M. Scott (all rights reserved).
† Oklahoma State University / NASA Ames Research Center.

So here, also, is a description of such a new frame of mind.

ON EXPLORATION

July 4, 1997—

A small rover begins to examine rocks on the surface of Mars. Playfully, the scientists name one rock Yogi.

On the World Wide Web, students of all ages observe this event in near-real time. Through their selection of images, they interact with Mars. In the weeks between July 4, and August 1, 1997, NASA web sites receive over a half a *billion* visits! During the coming months, through downloading images and examining them with public-domain image processing software, students will be able to do real descriptive science simultaneously with Mars scientists.

But working through the computer networks of Earth is not the same as walking Mars (or Earth). Students can benefit from studying data from Mars or images of Earth from space. But students also need fieldwork - to study real things in real places. They need places like parks and science museums to learn how to do the fieldwork. They need the *context* of knowledge that can be provided in classrooms, in discovery centers, parks, and museums.

Classrooms are important in their own right. Major outreach science programs like Nasals Aerospace Education Services Program are important. Equally important are the Mars education programs developed by organizations like the Mars Society.

Finally, in this age of systemic reform of education, this era of the Millennium, this time of Mars, it is prudent to take some time to consider new ways of thinking about research and education.

THE TOOLKIT: Resources for Educators

NASA offers educational support to formal and informal educators. Support includes national and Center-based workshops, educator activity guides, audiovisual materials, an educational television network, educator resource centers, World Wide Web sites, and personal services from NASA Educators.

The goal in all that we do is to enhance literacy in science, mathematics, technology, geography, and all education. We actively seek to work in partnership with all educators with a similar goal.

Some important information about NASA:

NASA is an independent agency, whose director reports directly to the President. NASA is therefore not in any cabinet department.

NASA works within the framework of 4 strategic enterprises: Aeronautics and Space Technology, Space Science, Human Exploration and Development of Space, and Earth System Science. Mars Exploration actually involves all four enterprises.

Some NASA resources are directed by the Washington NASA Education Office; others, by various missions or offices within the four NASA enterprises. This allows our agency to offer some resources which pull from all the enterprises, and others which go into great depth in a particular area of knowledge. Since so many missions currently focus on Mars, there is a lot of Mars material available to teachers and other educators.

Resources from NASA's Washington Education Office:

AESP

It has been said that one year on the road as an Aerospace Education Specialist (or Spacemobiler) gives more hands-on, direct knowledge about the state of American education than most educators get in a lifetime.

The Aerospace Education Services Program (AESP) has existed for more than 30 years. It is run under contract by Oklahoma State University in Stillwater, Oklahoma. At the request of any school, public or private, the program will schedule an Aerospace Education Specialist to visit the school or District. There are some variations in program operation from region to region, but all Specialists provide educator workshops, and student programs. Informal science educators, like interpreters, can request support of specialists at public program or interpretive skills workshops.

AESP Specialists are *teachers*. Qualifications state that they possess three years of full-time classroom teaching experience, a Master's Degree, and teaching credentials. Specialists must also have the ability to interpret science educational concepts to a wide variety of groups, and to a new group every day. Within a given day, a Specialist may present programs for students from kindergarten to eighth grade, and to the adult educators after school.

Specialists are *educators*. That is, they have a deep knowledge of educational methods and materials. Specialists attend national educational conferences of science, math, technology, geography and other disciplines to present workshops for educators, and to consult with educators.

AESP Specialists are *connectors*. They connect different disciplines involved in Space and Earth science in an interdisciplinary fashion. They connect the lives of individual students and educators to NASA's work. And they connect the educators of America to scientists and educational planners.

Specialists present workshops. They work with science museums, schools, parks, and museums to present educator workshops. There is no charge to the requesting organization for our participation in these workshops.

Check out the AESP home page, listed at the end of the paper, for more information. Make sure to follow the link to additional resources for educators, which will connect you with dozens of sites.

EDUCATOR RESOURCE CENTER NETWORK

To provide continuing support to educators, NASA established a national network of Educator Resource Centers (ERCs). Each NASA Center has a major ERC. Other ERCs - Regional ERCs - are established in conjunction with universities or non-profit educational groups. In some states, ERCs are located in planetariae.

Materials available in the ERCs varies, but each ERC does have the same basic items: NASA public domain films on VHS cassette and equipment to duplicate them, sets of slides for duplication, a small curriculum library, and educational publications for educators. A educator can visit a ERC, duplicate a film about Earth from space, and obtain an educator's activity guide and posters on Earth System Science - all in one trip, and at no cost for services.

ERCs stock many Mars Education materials. Lithographs of Mars help students and educators become familiar with the landforms of the planet. Posters like that of the Pathfinder landing can become centerpieces of a classroom.

NASA TV

This programming service regularly schedules NASA public domain films, so that educators can view or copy them. NASA TV also schedules live national educational conferences. These conferences focus on different topics - recently, Earth System Science, Robotics, and Space Life Sciences. NASA TV is accessible to every school or museum which has satellite ground station connectivity. In addition, many local cable systems carry it.

NASA TV is a window into the history of science. Major announcements - For example, the discovery of possible Martian fossils - are presented live. All Shuttle Missions are broadcast in their entirety. Other missions are also broadcast live on NASA TV, and science museums take advantage of this fact: When the Pathfinder landed on Mars in July of 97, the Oregon Museum of Science and Industry projected the entire event live - and had to stay open extra hours to meet the public's interest.

ON THE WEB: SPACELINK, and other resources

NASA has a plethora of world-wide-web electronic resources. These include many Earth pages. Sites often contain links to curriculum materials: Lesson plans, educational activity guides, and pictures for educators to use. Most pages also point to dozens of sites for Space and Earth Exploration.

The major page is Spacelink, the electronic entry point for educators. It will direct you to other NASA pages, curriculum materials, and images. The "cool" and "hot" sections contain late-breaking information about NASA missions. Other pages of note are listed at the end of the paper.

EDUCATOR WORKSHOPS

For educators who wish to become more directly involved with NASA, there are many On-Center workshops available. These include national workshops like NEWEST (NASA Educational Workshop for Elementary Science educators) and Center-based programs like NASA-Ame's Space Down To Earth. Note: NASA is currently offering a NEW (NASA Educators' Workshop) for informal science educators such as those in science centers, parks, and museums. NASA also co-sponsors other workshops, offered at locations around the country. The GLOBE Workshops, for example, train educators in student collection of Earth System Data; this information is then entered by students into the Earth-wide GLOBE database. Mars scientists will often present workshops on Mars Exploration at national conferences.

RESOURCES FROM MISSIONS

Other organizations also provide materials. NASA now requires researchers to use a percentage of funded research budgets for education. The result has been an explosion of educational materials.

Mars Missions have produced a plethora of materials - web pages and CD's, for example. JPL's web pages on Mars exploration include news, images, and educational materials. Ames Research Center's Center for Mars Exploration - CMEX - has produced the excellent CMEX CD's, which include NIH image and a complete atlas of Mars. Contractors working with

NASA produce posters and other materials. For example, The SETI Institute and the Planetary Society co-sponsored the production of the CMEX Educational CD ROMs.

IN CONCLUSION

NASA tools for Educators - publications, films, Web resources, or personal services - can help science educators like yourselves to do good work.

The challenge is not to find NASA materials that you can use in your park, or science center; but to find the *best* materials. Good starting points are the local Educator Resource Center, Spacelink, or the AESP program. AESP Specialists will help you plan programs to enhance scientific, geographic or other literacy.

But now it is time to consider grander educational tools for Mars exploration - new paradigms of thought, a necessary educational revolution.

A Wander ... Into the high country of the mind

We live in the dawn of a new Enlightenment, a new Renaissance, born of Apollo and the first photographs of Earthrise over the Moon. We need this Enlightenment, since it will help us to think in new ways, and we need to do that. The nature of our tasks for the next few decades demands it. But how do we get there?

Science, they say, always starts with question. And questions come from wandering and wondering.

Robert Pirsig says that we should enter the high country of the mind, to enjoy a kind of intellectual chatauqua. Since Boulder is home to a Chatauqua, and borders the high country, let us follow Pirsig's example in "Zen and The Art of Motorcycle Maintenance", and wander up to the high country chatauqua to consider some questions about education.

It makes sense to me. I began to consider these ideas when I was a Ranger, wandering around large or small wild places. And I note that new paradigms of the nature I offer here are developed in conjunction with wilderness or wild places.

Paradigm shifts, according to Kuhn, are not usually the work of teams, or committees. Individuals see new ways to consider the universe, and begin to apply them. As these new paradigms are adopted, it then requires teams of specialists to apply them. But in the early stages, loners develop the paradigm, and small informal groups begin to apply it.

Perhaps wilderness settings, or rural areas, allow more time and inspiration for the birth of such revolutionary ideas, and more freedom of communication necessary for them to be nourished and to grow.

One of the qualities of the Corporate/scientific State in which we live is its specialization. Doing everything in teams of specialists does permit drawing upon diverse strengths to achieve mega-projects like a Mars mission. The problem with this approach is that it often prevents the kind of thinking that leads to breakthroughs.

Let us consider for a moment the Scottish Enlightenment. From Edinburgh, Scotland, in the mid-18th Century, the entire modern world emerged. In the space of a few years, Edinburgh produced Adam Smith, David Hume, Robert Burns, John Locke, James Watt, Michael Faraday, Sir Walter Scott, James Hutton, Joseph Black, even Philadelphia's Dr. Benjamin Rush (who went to school at the University during the Enlightenment).

Smith gave us Capitalism, Hume the theoretical basis for nuclear physics, Burns vernacular (i.e. the common man's) poetry, Locke the theory of the Social Contract, Watt the steam engine, Faraday the dynamo, Scott a literature of heroism and character and the land, Hutton geology, Black Carbon Dioxide, and Dr. Rush helped to give us the Declaration Of Independence and The United States Constitution.

Why Edinburgh? It was a backwater capital of sheep and highlanders, where the Stuarts had just been defeated in their attempted return to rule. Yet, the modern world was born here. The paradigm under which we live was created here.

How did these Scots think, to create such structures of the mind and of life? David McCullough, in a keynote speech to the National Trust for Historic Preservation, 1991, commented on this phenomenon, and suggested an answer:

> "Edinburgh, that wonderful Edinburgh described by Robert Louis Stevenson, gave forth to what was essentially an English-Scottish-Western Renaissance of its own in the Eighteenth and early Nineteenth Century. In medicine, philosophy - people like David Hume - the origins, the beginnings of the whole idea of an Encyclopedia Britannica, not to say Sir Walter Scott, Robert Louis Stevenson.
>
> And why did it happen in that tiny, little, northern, bleak, town?
>
> Why did it happen in a place smaller than New Haven Connecticut?
>
> Well, we don't know. But one of the things we do know is that in Edinburgh in that day everybody saw everybody. There were clubs, societies, and associations. They met almost every night. Lawyers, doctors, engineers, poets, people of all professions and all persuasions, saw each other, talked, worked, imagined. Together. No walls, no barriers, no status order.
>
> We must encourage that."

Indeed we must, if we are to create the paradigm for the explorations of a New World, as did the Scottish Enlightenment or the Renaissance.

One club such as the one described by McCullough was the Oyster. Here, many of these men met, regardless of social station, to interact socially with each other and each other's ideas. Members included Joseph Black, Adam Smith, James Watt, David Hume, even Benjamin Franklin (when he was in town), and the father of Geology, James Hutton.

We need to learn again to share ideas as people in those places and times did. More than that, though, we need a model of thought that also has nuances appropriate to the coming millennium.

In the spirit of the Oyster Club, and 18th Century Scotland, we need a paradigm which allows for developments of interdisciplinary thought, encourages the interchange of such developments; or, as von Puttkamer indicates, one which emphasizes *"...that shifts the emphasis from individual subjects to the interactions and relationships between them."*

It seems wise also to have this paradigm begin close to nature, as did the Scottish paradigm. Is there a model for such a new way of thinking?

George R. Stewart and Geo.S

George R. Stewart was a professor of English at the University of California, Berkeley for most of his life. He was also the author of books which are described as "regional" or "geographic" or as "works of Place". Two of these books were best sellers and Book-Of-The-Month club selections.

But Stewart's work was much more profound that "best-seller" status might imply. Over the years, his works have become simultaneously cult works, and works with profound influence on all humankind. For example, few people know Stewart's name, but most people know that we name storms. George Stewart was the one who popularized that practice in his scientific novel STORM, written in 1941.

As Pulitzer Prize winning author Wallace Stegner writes, "(Stewart) was a much more important author than the general public knew."

Stegner writes, "Of George Stewart's twenty-eight books, I find that I have seventeen on my shelves à Three or four of them I read all the time, and refer to, and quote, and steal from, and couldn't get along without."

The Geo.S Paradigm:
A Model For Mars Education, and Other Thought in the Next Millennium

In his works, fiction and non-fiction, George R. Stewart presented a paradigm for the organization of knowledge - useful for both research and education - of great value for teachers and students seeking to find new and better ways to study and to think about things.

Equally as important, Stewart's paradigm is applied in his works to "place". It can be applied to any domain, physical or intellectual, but it is clearly presented by Stewart in such a way that it can be easily applied to learning about a place. As "Sheep Rock" is a place, Mars is a place.

What are the characteristics of this educational paradigm?

Like the thought of the Scottish Enlightenment, it is interdisciplinary. Stewart's works always combine art with various sciences, social sciences, details of industrial or social organization, the human mind, heart, body, and soul.

But Stewart also always considers a world from two perspectives, of ground and space. Here is the key to the unique usefulness of Stewart's paradigm, for the coming Millennium - its two perspectives on worlds. "It is fine and important to know a place from the perspective of all the disciplines and arts," Stewart seems to say in his work. "How else would you know it?"

"But," we might continue to paraphrase his thought," you must also know it from the two perspectives of ground and space." You must study Earth and other places from the two perspectives of the Astronaut or Cosmonaut, and the Ranger/Poet."

Some key passages from Stewart's work illustrate the paradigm:

TWO PERSPECTIVES

"The reader must for a moment imagine himself raised in space some hundreds of miles above a spot near the center of the state of Nevada. Far to his left, westward, the onlooker from the sky just catches the glint of the Pacific Ocean; far to his right, on the eastern horizon, high peaks of the Rockies forming the Continental Divide cut off his view. Between horizons lie thirteen degrees of longitude, a thousand miles from east to west. The only mark of civilization is a tenuous trace winding from east to west, a faint pair of lines, the California trail."

... and ...

"Probably the best way to feel the actuality of the story is to travel through its setting. For the country is tangible and solid, now as then. I have in the telling often stressed the scene until the reader has, I hope, come to feel the land itself as one of the chief characters of the tale."

George R. Stewart
"Ordeal by Hunger", 1936

INTERDISCIPLINARY THOUGHT

"I, George Stewart, did this work.
I have looked into the blue and green depths of the spring, and have climbed the rock, and gazed out across the desert. That first night, the grim fascination of the place rose within me and I thought of this book.
That time I was with Charlie. I was there again — with Jack, with Selar, with Carl and Parker and Starker, with Brig and Roy. I said to myself, 'I shall know more about this place than anyone knows of any place in the world.' So I took the others there, and one looked at the beaches and the hills, and another at the grass and the shrubs, and another at the stone-work among the hummocks, and so it went, until at least each had shared with me what he knew. Besides, I read the books.
But if you ask me, 'What is true, and what is not? Is there really such a place?' I can only say, 'It is all mingled! What does it matter? In the end, is what-is-seen any truer than what-is-imagined?' Yet, if you should look hard enough, you might find a black rock and a spring—and of the other things too, more than you might suspect. ...
And he who will brave the desert and come after me to the spring, he may find my traces, where I wandered wondering, and thus wandering wondered, and he too may take to himself a little of me as I in my time took to myself a little of those who had been there before me."

George R. Stewart
"Sheep Rock", 1951

THE RELATIONSHIPS BETWEEN DISCIPLINES

In his Paradigm, Stewart realized a truth about knowledge that other great minds have also realized. (John Muir wrote, "Whenever you try to pick out anything by itself you realize it is hitched to everything else in the universe.")

Stewart knew that artificial divisions of space or time or the human heart help us to organize the world. But he also knew these artificial divisions could keep us from seeing its truth. He knew also that what was true of space was equally true of knowledge. Divide knowledge, and lose truth:

"...Certainly, every place is only a part of long continuous space, and every place therefore leads the way into every other. Like so much else in the world, then a place turns out to be nothing more than an abstraction fathered by the human mind, setting up artificial barriers, pretending that something ends at a hard and fast line, and that it is not in actuality running on continuously into something else..."

George R. Stewart
"Sheep Rock"

So Stewart's work truly reflects the interdisciplinary nature of the Edinburgh paradigm. As the first quotation from "Sheep Rock" shows, he did work in small collegial groups to learn about places.

In addition, Stewart's work has that emphasis on seeking the meeting place of disciplines that von Puttkamer supports.

So it would seem that the body of Stewart's work may give us the foundation for the way of thinking, learning and educating necessary for the tasks of the next millennium. Tasks like the exploration of Mars.

THUS, GEO.S

Geo.S is the abbreviation for Stewart's name. And Geos is Greek for Earth, which symbolizes his concern with planetary place. So call his paradigm of thought the Geo.S Paradigm, if you will.

Does Geo.S Apply to Space Exploration? I would only note in passing that Konstantin Tsiolkovsky: grew up as the son of a Forest Ranger; learned from libraries; began by producing works of art - science fiction; and from these Geo.S type foundations, developed the basis for the astronautics which will someday carry us to Mars.

IN CONCLUSION. FROM GEO.S TO MARS

Time to return from the high mountains of the mind, and end our chatauqua. And to summarize:

To teachers of the Mars Generation, of the Mars Kids, NASA offers an extensive toolkit for the nuts-and bolts of teaching about Mars Exploration. From personal educational programs in schools, to interactive Internet projects, to Mars Experiment Design competitions for students, to workshops on NASA Centers - and all the rest - there are resources.

To those same teachers, George R. Stewart offers his paradigm. To help them help their students to understand the world and other worlds in new ways.

Nuts-and-bolts toolkits and a new way of thinking, together, will move the Mars Kids onto the surfaces of other worlds, and into a millennium which promises to be the most exciting in the history of humankind.

Our exploration of Mars will be the start of a successful Millennium of journeying throughout the cosmos. And the continuation of a great human adventure, in which we have seen the best of ourselves:

> "... Our vessels consisted of six small canoes and two large pirogues. This little fleet, although not quite so respectable as that of Columbus or Captain Cook, was still viewed by us with as much pleasure as those deservedly famed adventurers ever beheld theirs, and, I daresay, with quite as much anxiety for their safety and preservation. We were now about to penetrate a country at least two thousand miles in width, on which the foot of civilized man had never trod. The good or evil it had in store for us was for experiment yet to determine, and these little vessels contained every article by which we were to expect to subsist or defend ourselves. However, as the state of mind in which we are, generally gives the coloring to events, when the imagination is suffered to wander into futurity, the picture which now presented itself to me was a most pleasing one. . . ."
>
> **Captain Meriwether Lewis, "The Journals**
> **(Setting out from the Mandan Villages**
> **Toward Terra Incognita)"**

BIBLIOGRAPHY

Anonymous, "The Center For Mars Exploration". Booklet. *The Center For Mars Exploration*, Dr. Geoffrey Briggs, Director. MS 245-1, NASA-Ames Research Center, Moffett Field, California, 94035-1000.

Lytkin, Valdimir V, and Madry, Scott, "The Biography of Konstantin Tsiolkovsky", at The Konstantin E. Tsiolkovsky State Museum of the History of Cosmonautics. Http://deathstar.rutgers.edu/museum/tsiol.html.

Bakeless, John, "The Journals of Lewis and Clark", New York, Penguin Books Mentor Edition, 1964.

Kuhn, Thomas S., "The Structure of Scientific Revolutions" (Third Edition). Chicago, The University of Chicago Press, 1996.

McCullough, David. Keynote address, "National Trust for Historic Preservation", 1991, San Francisco.

Pirsig, Robert M. "Zen and the Art of Motorcycle Maintenance : an Inquiry into Values", by Robert M. Pirsig; with a new introduction by the author. New York: Morrow, 1984, c1974. 412 p.; 22 cm.

Scott, Donald, "One Planet From Two Perspectives: The NASA-NPS Project of Wider Focus". Presented at the National Commission on Space, San Francisco, California, 1985.

Stegner, Wallace, "George R. Stewart And The American Land", in "Where The Bluebird Sings To The Lemonade Springs", New York, Random House, 1992.

Stewart, George R., "Ordeal by Hunger", New York, Henry Holt and Company. 1936.

Stewart, George R., "Sheep Rock". New York: Random House, 1951.

von Puttkamer, Jesco, "Spaceflight And The New Enlightenment", NASA Magazine, March 1992.

FROM BRADBURY TO BLAMONT: THE SCIENCE OF MARS IN THE ARTS

Michael Carroll

A BIT OF HISTORY

Throughout history, a curious feedback loop has existed between the science of Mars and the culture of our civilization. Mars has inspired the way we think, write, paint and sing. The arts, in turn, influence and inspire the next generation of scientists and engineers, who make exploration happen. This exploration leads to more data and an increased awareness on the part of the general population. Advances in our understanding of the nature of Mars affect the quality and content of the arts, which inspire the next generation of scientists and engineers.

In the sixteenth century, Albrect Durer did an accurate engraving of a rhinoceros. He had never seen one. However, he had first-hand reports and eye-witness accounts from travelers who had recently returned from North Africa. With the aid of a rhino ear and horn (some reports also mention a tail), Durer was able to reconstruct the strange beast. A few years later, a live specimen of a white rhino was brought to Germany. Today, we have travelers' tales from distant lands as well. Some are eyewitness accounts brought back by astronauts, while others are returned robotically. These tales tell us of worlds we will one day explore, work and live upon.

Our tales of Mars were not always so clear. Early telescopic views were fuzzy glimpses through foggy lenses. Astronomers like Schiaparelli and Lowell struggled to understand the seemingly geometric patterns they saw. These first views gave rise to the supposition that Mars had canals. Lowell's maps and writings, in particular, captured the imagination of the public. Lowell envisioned a dying race of Martians trying desperately to irrigate their parched homeland with water from the shrinking poles [1]. If there were canals, many reasoned, why not Martians? And if Martians, why not kings and princesses and great cities and strange creatures? Edgar Rice Burroughs—most famous for his *Tarzan of the Apes* books—penned the *John Carter of Mars* series. Burroughs portrayed Mars as a dry world with great canals and Martians who had quite human qualities. Some were noble, while others were villains and scoundrels.

H. G. Wells wrote the classic *War of the Worlds* [2], in which a dying race of Martians tried to conquer the Earth for their own devious purposes. While the book was a commentary on Victorian society, it also reflected societal ideas about the Red Planet. A few years later, the *Buck Rogers* comic strips [3] also depicted Martians as evil threats to our civilization.

At the beginning of this century, the director of Paris' Meudon Observatory, Lucien Rudaux, painted the first scientifically accurate portrayals of Mars. Rudaux took advantage of his skills at the telescope to paint scenes of Martian dust storms as seen from the surface, and several vistas of Mars from the vantage point of its two moons, Phobos and Deimos [4].

Science progressed, and we began to sense that Mars was even colder than originally thought, with a very thin atmosphere. Paintings done for magazines such as *Life* and *National Geographic* in the 1950s began to show a thinning blue sky with creatures armored against the cold and solar radiation. Werner Von Braun and Willey Ley wrote a series of articles for *Collier's* Magazine [5] which described the first piloted missions to Mars. The articles were beautifully illustrated by architect Chesley Bonestell. Bonestell's paintings were to become the most famous space art of all time, inspiring many to pursue scientific careers in the upcoming space age.

In the meantime, the romance of Mars continued to grasp our social consciousness. Ray Bradbury wrote the fine collection of short stories *The Martian Chronicles* [6]. Bradbury's presentation of Martians as an innocent race of peaceful beings indicates a subtle paradigm shift on the part of society. Mars is no longer a hostile place to be feared, but a real world with great promise.

The movie *Conquest of Space* [7] attempted to show Mars as accurately as possible, as did such classics as *Robinson Crusoe on Mars*. Others, like *Angry Red Planet* (1959) and *Mars Needs Women* (1966) contented themselves with Mars as a vehicle for a bizarre story. George Pal teamed his special effects with Barry Lyndon's adaptation of Wells *War of the Worlds* to create Paramount's fine 1952 film of the same title.

THE SPACE AGE

With the advent of planetary exploration, Mars became a dead world virtually overnight in the minds of the general public. Mariner IV radioed back images of cratered, moon-like terrain in 1964, and the impression seemed to be bolstered by Mariners VI and VII at the next encounter opportunity. But with the arrival of Mariner 9 and the Vikings, Mars became a world that we could comprehend, and not nearly as dead as we once thought. It is an inspiring world of soaring volcanoes, billowing dust storms, plunging canyons and deserts much like the deserts of our own world. It is a world that has been scarred by ancient floods and ice sheets, a world that had a much different past than the desolate planet we see today. Science writer and musician Jonathan Eberhardt was inspired to write the song *Blues for a Red Planet* [8], in which he says:

> *Ten thousand times a hundred thousand dusty years ago,*
> *where now extends the Plain of Gold did once my river flow.*
> *It stroked the stones and spoke in tongues and*
> *splashed against my face,*
> *'Till ages rolled; the sun shone cold on this unholy place...*
>
> *For the water of my river*
> *and the air that was my wind,*
> *though bound in rocks and wintry wastes I pray may*
> *flow again.*

ANATOMY OF A MARS PAINTING

Armed with the new science of these modern missions, along with fresh information from Pathfinder and Mars Global Surveyor, space artists can now depict Mars with greater accuracy than at any other time in history. Comparative planetology enables the artist to study Mars analogs on Earth for unprecedented detail. Many desert areas bear strong geological and morphological resemblance to Mars. Several glaciers in Iceland exhibit layered terrain similar to that found at the edge of the Martian polar caps. Patterned ground like that discovered in mid latitudes on Mars can also be found at a different scale in many subarctic and arctic regions [9] [10]. Thanks to space exploration, scientifically accurate space painting is now possible using comparative planetology and creativity constrained by the laws of science. Paintings range from the hyper realistic forms of Miller, Flynn, Eggleton, Emmart, Hartmann and others to the visionary works of Bob McCall and John Berkey, to the impressionistic styles of such artists as Pesek, Avary and Sokolov.

MARS AND THE CULTURE OF TODAY

The tabloids spread wild stories about the Allen Hills SNC meteorite. Arnold Swartzneggar terraforms Mars instantly in the movie *Total Recall*. The Mattel toy company markets a Hot-Wheels Pathfinder, which becomes one of the most successful and sought-after toys in its history. Popular novels like Kim Stanley Robinson's *Red Mars* [11] reflect advanced technical concepts for populating and terraforming Mars. All of these phenomena are indicative of a society which is still enamored of Mars, and becoming more knowledgeable about the planet.

Elton John's song Rocket Man [12] states:

Mars ain't the kind of place to raise your kids.
In fact, it's cold as hell,
And there's no one there to raise them if you did.

It is significant to note that Mars is, in fact, cold as hell, and that at this point in time there is no one there to raise your kids. However, it is my conclusion that, as indicated by the sophistication of artistic depictions of Mars and their reflection on society's level of awareness, there may be someone raising kids on Mars very soon.

REFERENCES

1. Lowell, Percival; *Mars as the Abode of Life*, c. 1909 The Macmillan Company.
2. Wells, H. G.; *War of the Worlds*, c. 1899.
3. *Buck Rogers in the Twenty Fifth Century*, beginning c. 1929 National Newspaper Syndicate, Inc. Initial story by Philip Nowlan, illustrated by Dick Calkins.
4. Rudaux, Lucien; *Sur Les Autres Mondes*, c. 1937 Hollier-Larousse, Paris.
5. *Collier's* Magazine, 1950.
6. Bradbury, Ray, *The Martian Chronicles*, c. 1950 Doubleday.
7. *Conquest of Space*, Paramount studios, 1954.
8. Blues for a Red Planet, c. Jonathan Eberhardt, 1984.
9. Lachenbruch, A. H. 1962. *Mechanics of thermal contraction cracks and ice wedge polygons in permafrost.* Geol. Soc. Am. Spec. Paper 70.
10. Lucchitta, B. K. *Martian outflow channels sculpted by glaciers.* Lunar and Planetary Science XI, pp. 634-636.
11. Robinson, Kim Stanley, *Red Mars*, c. 1993 Bantam/Spectra.
12. Rocket Man lyrics by Bernie Taupin, 1972, from the album *Honky Chateau*, Dick James, Ltd.

Chapter 7
PRIVATE FUNDING FOR MARS MISSIONS

Book sales were brisk in the busy hall. Rachel Zubrin (age 6) makes sure everyone has a "Mars or Bust" button.

MAR 98-030

A SPONSORING CONCEPT FOR MANNED MISSIONS TO MARS

Michael Bosch[*]

After the first lunar landing in July of 1969, public interest in manned space flight has continued to decrease. In contrast to this phenomenon, other comparable high technology events, such as international automobile racing, are growing in popularity. Formula 1 racing, in particular, has developed into a highly profitable, privately sponsored economic enterprise. Because of its tremendous financial power, Formula 1 racing is able to recruit top personnel in the fields of technology, sports, marketing and management. As a result, it's possible to achieve challenging project goals very quickly. Project successes lead to an increase in public interest. Thus, additional private sponsors are willing to put their money into Formula 1 events.

This paper starts by pointing out the critical success factors of Formula 1 sponsoring. Next, a number of possibilities are discussed on how to increase public interest in manned space. On this basis, a sponsoring concept for manned missions to Mars is developed.

1. INTRODUCTION

Over the last 25 years, public interest in manned space flight has continually decreased. This trend has continued in spite of the scientific and technological achievements of manned space projects in the post-Apollo era, such as Skylab, ASTP, STS/Spacelab, STS/Mir.

Diminished public interest has led to a subsequent decrease in congressional funding. The ambitious plans of the 60's for manned space flight to other planets have yet to receive public funding. Presently, not a single space agency has concrete plans for a manned mission to Mars. Although NASA has the largest budget of any space agency worldwide, and has always been proud of its pioneering spirit, it, too, has recently canceled all plans for manned missions beyond LEO (Raumfahrt Journal 1/98, p. 21).

These projects were canceled against the advice of well-respected aerospace engineers, who assert that a manned flight to Mars would be readily achievable with current technology. Development of untried, 21st century, *'Battlestar Galactica'* technology would not be necessary (Zubrin, R., The Case for Mars, 1996, p. 2). The main obstacle to manned flight to Mars remains the financing of such an endeavor.

Since public funding is out of the question, both currently as well as in the near future, this paper explores the possibility of private sponsor-financing.

2. HOW TO REAWAKEN BROAD PUBLIC INTEREST

Inspiration of broad public interest is an important prerequisite for private sponsoring models.

---

[*] University of Regensburg, Faculty of Business Administration, D-93040 Regensburg, Germany.

During the Apollo program, the main motivation for political and public interest in space flight was competition with the U.S.S.R., rather than science. The public found the "Race to the Moon" both thrilling and fascinating. As a result, citizens were more than willing to spend their tax dollars to support "their" space program.

Today, manned space missions are conducted as international cooperative efforts (e.g. the International Space Station), without any competition involved. The public is unfortunately not interested in microgravity research, LEO missions or the cost advantages of international cooperation. Public interest would be much greater in the following areas:

- thrilling entertainment provided by spectacular competition
- new mission destinations, such as Mars ('to boldly go where no man has gone before')
- real-time participation in mission events via Internet (e.g. Pathfinder)
- extreme speed, acceleration and exciting flight maneuvers
- the danger and adventure of manned space flight
- the personal appeal of astronauts and other space officials as heroes or role models.

Thus, a *Race to Mars* could plan a critical role in winning back public interest in manned space flight.

In contrast to the present manned space flight, other comparable high technology spectator events are growing in popularity. International auto racing, such as Formula-1, is driven by competition between drivers and teams sponsored by industry. Formula 1 racing has developed into a highly profitable, technologically-driven enterprise, which generates billions of dollars in revenue each year.

Manned space flight has much in common with Formula-1 racing and could therefore appeal to a broad public. These similarities include:

- high goals
- high speed for their respective conditions
- high requirements for both man and machine
- application of the newest technologies
- test pilots subject to permanent test situations under dangerous conditions
- use of light and highly durable materials
- heroic status of astronauts and Formula-1 drivers.

If it proves feasible to make use of the competitive spirit, organizational and management structures of Formula-1 racing in manned space missions, then it should also be possible to take advantage of similar commercial sponsoring mechanisms to finance these missions. The principal determinants of success in Formula-1 sponsoring are discussed below.

3. SUCCESS FACTORS IN THE SPONSORING OF FORMULA-1 RACING

3.1. Competition

Empirical observations show that public interest in Formula-1 racing is primarily due to competition between drivers and teams sponsored by industry. Viewers are only interested in one thing: who will win. If all teams combined their efforts and built a single car in order to save on development costs, viewer interest would disappear.

Financial success in a market economy is based on competition. Competition is the sole force which motivates a company to develop technological concepts more quickly and cost efficiently than its competitors. The rate of technical innovation slows down as soon as competition is replaced by international cooperation. The same principle holds true for the establishment of monopolies through fusion or acquisition of competitors.

In Formula-1 racing, it is quite evident that extreme competition between teams drives technological development. Figure 1 illustrates the chain reaction effect of competition in Formula-1 racing.

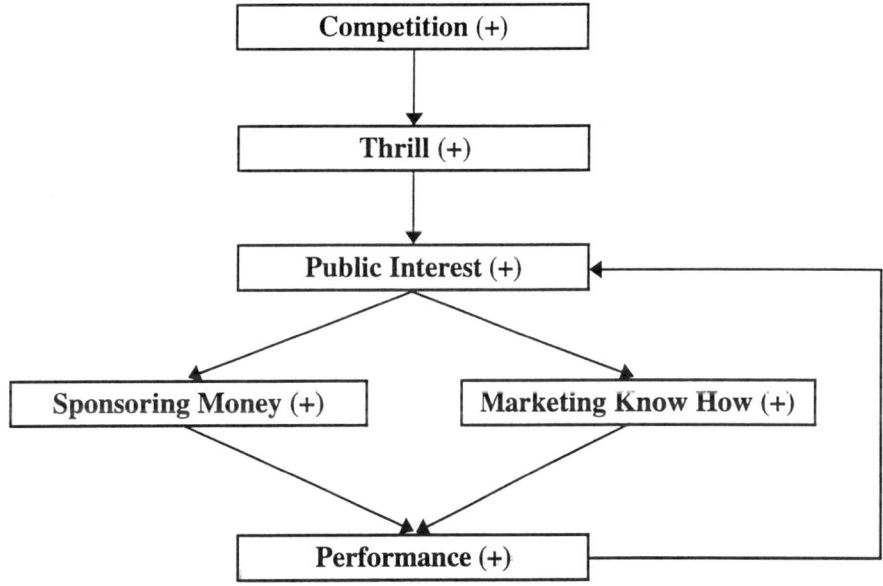

Figure 1   Cycle of Success.

Competition between different drivers and teams raises the thrill-level for the public. Public interest attracts sponsors, who advertise their products and their corporate image. Sponsors contribute large sums of money as well as personnel with marketing and sponsoring expertise.

The budget for each team can exceed $300 million per year. This makes it possible to hire top personnel, who are responsible for superior technological achievements. This increase in achievement leads to an increase in public interest in winning drivers and teams. For example the recruitment of aerodynamics expert Adrian Newey from Williams Renault to McLaren Mercedes contributed to the success of the McLaren team in 1998.

3.2. Broad Public Interest in Drivers

The Formula-1 race is billed as the world championship for drivers. It often develops into a title bout between two top competitors. For example, between

- Prost and Senna, even though they were on the same team (McLaren)
- Hill and Schumacher
- Schumacher and Villeneuve
- Hakkinen and Schumacher.

The broad public is especially interested in the successes and failures of individual drivers. This means that the drivers themselves are the main attraction in the Formula-1 business. No one would bother watching an auto race with remote-controlled cars.

The public is not only interested in the professional achievements of a driver, but also in the driver as a person. The result is, that sponsors want the drivers to market their products or corporate image. It is in the sponsors' interest to sign the best drivers.

3.3. Marketing of all Aspects

The large number of viewers is the key to the success of the Formula-1 sponsoring concept. This makes Formula-1 racing very attractive to sponsors. All aspects of Formula-1 racing have been successfully commercialized:

- races (TV-rights, admission, VIP-services)
- drivers
- all hardware and software used in the races
- sponsors' products and corporate images
- other products, which are advertised during commercial breaks
- merchandising.

4. TRANSFER OF SUCCESS FACTORS TO MANNED SPACE FLIGHT

4.1. Introduction of Competition

Up until now, conventional wisdom has held that a manned mission to Mars would require the combined efforts of all countries with space programs. For the International Space Station program, all competition has been completely excluded. International cooperation has been elevated to the highest goal. Since the project began in 1984, public interest has been low and progress on the project has moved very slowly.

The introduction of competition is the most important prerequisite to increase public interest. The only remaining question is, „**Who will be the competitors in the Race to Mars?**"

Competition between different countries, as in the Apollo program, is not an option today due to the changed political situation. Government agencies have neither the will, nor the capability to carry out a manned mission to Mars.

On the other hand, global mergers have led to the fusion of huge conglomerates in the aerospace industry. Due to their technical and managerial competence, the following corporations would be capable of planning and carrying out a manned mission to Mars:

- Boeing (USA)
- Daimler-Benz Aerospace (DASA) in Daimler-Chrysler conglomerate (Germany, U.S.A.) or
- European Aerospace and Defense Company (EADC) as a currently proposed future merger of Aerospaciale, BAe, DASA and other European Space Corporations (Europe)
- Lockheed Martin (USA).

These corporations could compete against each other in a Race to Mars. Such a race between different teams sponsored by industry would guarantee an increase in public interest. Sending the first person to Mars would be the best advertising for each industrial team. This event would deliver enough material for advertising claims for the next century. Both the winning team as well as all other participating teams would be assured a place in the history of space exploration.

Financing such a mission still remains a problem. Even a conglomerate as large as Boeing would not be able to absorb the entire costs for a manned mission to Mars. The costs for the hardware alone would be prohibitive.

The gap between total costs and the amount each corporation would be willing to contribute could be filled by sponsors. These sponsors could advertise their companies and products as part of the manned mission to Mars. As soon as one of the aerospace conglomerates decides to support a manned mission to Mars, public interest would be aroused. Other sponsors could be approached immediately thereafter. The following headline could appear in newspapers in 1999:

EADC to Send People to Mars in 6 Years
Sensational Announcement Made by the European Aerospace Conglomerate
Project to be Financed Exclusively by Private Sponsors

Once plans for a private mission to Mars are announced, other aerospace corporations, or even NASA itself, will also decide to sponsor their own missions to avoid falling behind their competitors. This would open the Race to Mars.

The public would attentively follow the Race to Mars. As a result, an increasing number of sponsors would be willing to invest their money in the industrial teams. The cycle of success shown in Figure 1 would develop its own dynamics. During the planning, development and test phases, public interest can be raised by regular reports on:

- project advancements
- selection of the astronauts
- flight crew /backup crew decisions
- astronaut training
- test flights and other hardware tests
- alternative launch windows and trajectories.

Money from the sponsors finances project advancements. As shown in Figure 2, two complementary effects occur simultaneously. During the project life-cycle, more money is needed as the project progresses. Phase A (Feasibility Studies) is much less expensive than Phase C/D (Design, Development). As the project progresses and success seems imminent, more sponsors will be willing to contribute. Financial requirements and resources increase together as the project life-cycle progresses.

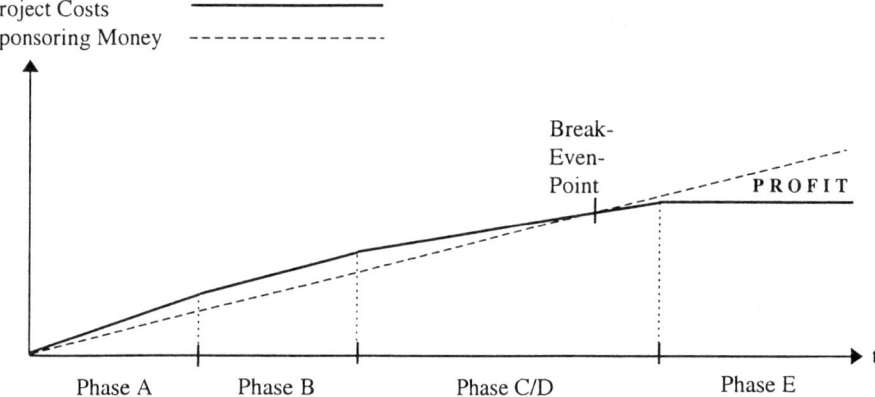

Figure 2   Complementary Correlation between Project Costs and Sponsoring Money.

The first step is to find a way to finance the first phase of the project. Thus, sponsors who come on board at the start of the project should be offered the best conditions.

4.2. Opportunities for Marketing

4.2.1. Advertising Space

The sponsors' logos could be presented on different types of advertising surfaces. The size of each logo would be cost-dependent. Sponsors could reach a wide array of viewers during news reports covering different race events: astronaut training, launch, the flight, landing, reentry. The following types of advertising could be considered:

(a) Flight Hardware

The rocket boosters, the landing vehicle, the reentry module, etc. would provide ample space for sponsors' logos.

(b) Clothing

The astronauts and other representatives of each industry team could wear sponsors' logos on their clothing whenever they are in the public eye. The astronauts' space suits and helmets would be especially effective for advertising.

(c) Billboards

Sponsors could present their products on billboards at the launch site, on shop floors and in aerospace and astronaut training centers.

4.2.2. Advertising in Print and in Electronic Media

Sponsors can advertise both their products as well as their corporate images in print and in electronic media. Taking a cue from the Olympics, a company can buy the rights to call itself an „official sponsor" to the first manned mission to Mars.

Some examples:
- Schweppes - Official Soft Drink Sponsor to the Motorola-EADC-Mars Team
- The new sport utility vehicle from Daimler-Chrysler: Mars-proven technology
- Computer Associates - The quality of our software shines through in critical moments
- Fortis Official Cosmonauts Chronograph.

4.2.3. Merchandising

Merchandising is the marketing of products which are loosely tied to the project by the sponsor's logo. The rights to use certain images, symbols or logos can also be licensed to third parties.

Some examples:
- miniature flight hardware and model-building kits
- posters and flags
- Caps
- T-shirts
- Watches
- Pens
- Sunglasses
- coffee mugs
- Magazines
- Books
- computer games
- food products.

4.2.4. Marketing the Main Events

Important milestones during the preparations (roll out, test flights, astronaut training), the launches, critical flight maneuvers, the landing on Mars, the exploration of the planet, reentry and landing (e.g. splashdown) are spectacular events for the broad public. These events could be subject to more intense marketing.

For example:
- selling the TV rights
- advertising during commercial breaks
- selling tickets for the launches; prices would be based on the distance between the observation platform and the launch site

- special VIP service during critical mission phases

4.2.5. Marketing the Astronauts

As mentioned earlier, public interest focuses on the astronauts themselves as famous personalities. Astronauts are viewed by the public as heroes. This image can be quite valuable to sponsors. Frequent media reports on the astronauts can continually raise their levels of recognition. Astronauts could represent their team sponsors at public events. The names of especially popular astronauts (e.g. the first person on Mars) could be developed into name brands. The advertising rights to use astronauts' names could be licensed to third parties.

4.3. Organizational and Managerial Implications of Sponsoring

The organization of the sponsoring model will be illustrated using a hypothetical team funded by the future European Aerospace and Defense Company (EADC) as an example. The EADC-Team, would develop all of the flight hardware itself and then later buy launch capacity from another firm.

The EADC Team's marketing department would be responsible for finding sponsors to support the Mars project. The proposed organizational structure is illustrated in Figure 3:

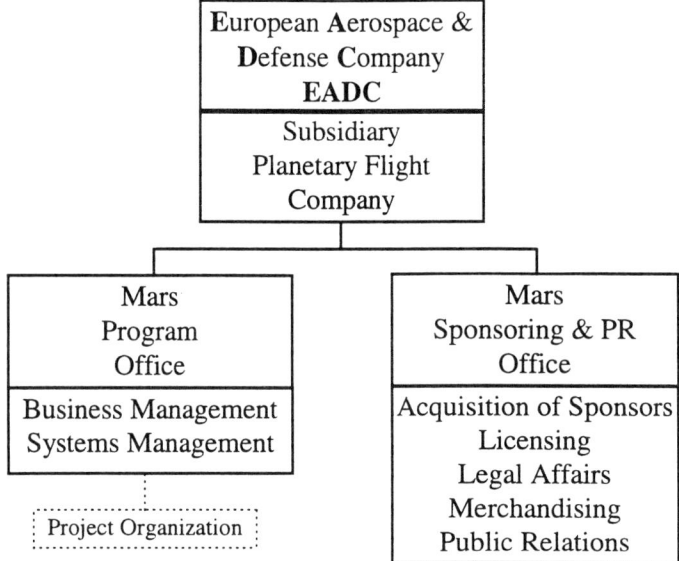

Figure 3   Organizational Structure of a Hypothetical Mars Team.

5. SUMMARY

This paper presents a sponsoring concept for a privately financed, manned mission to Mars. The concept can succeed, provided that the following conditions are met:

(1) Public interest for a manned mission to Mars can be awakened.

(2) Sponsors can be found to support the first critical phase of the project.

(3) One of the major aerospace corporations can be convinced that a manned mission to Mars is worthwhile.

(4) An inexpensive project proposal can be developed, which would convince the sponsors that such a project is financially feasible.

ACKNOWLEDGMENTS

Firstly I would like to give special thanks to my colleague, Dr. Patricia Shiroma Brockmann for the perfect translation of this paper into English as well as for her continued encouragement and thought provoking discussion.

REFERENCES

Raumfahrt Journal 1/98, T. Block Publishing Corp., 1998.

Zubrin, Robert, *The Case for Mars*, 1996.

CONDUCTING MARS EXPLORATION ON A PRIVATE BASIS BY REVIVING THE "EAST INDIA COMPANY" FINANCING MODEL*

John Q. Coston[†]

The fabled East India Company was created as a "share" venture to spread the extraordinary risk of financing early trade voyages among the maximum number of investors.

If a ship on a two or three year trade voyage failed to return or arrived late, the shared loss spread among many investors, would avoid individual bankruptcy. Conversely, if the ship arrived, especially if it arrived before any others carrying similar cargo...

Space exploration is risky. It is also new. So new, risk cannot be actuarially determined. The only large group intellectually prepared to assume this risk is the highly disorganized group of space enthusiasts collectively known as the "space community".

Using the East India Company model and the program called "Power of Ten Clubs", we can organize this incoherent space community into a coherent investment force/vehicle capable of financing the exploration and colonization of Mars, regardless of risk.

This financing vehicle will be named the "MARS CORP." MARS CORP. is a for profit business corporation. The corporate benefit being: Opportunity to bring the widest possible spectrum of space supporters into the Mars exploration program as active profit or colonist motivated participants. This occurs with maximum individual risk reduction.

The great failure of the leadership in the space community is failing to successfully impress upon the consciousness of the community the long-term character of space exploration and the need to soberly commit to that problem.

MARS CORP. combined with the Power of Ten Club program will focus attention on the problem. Focusing by using the "Lens of Time" as the investment tool for successful exploration and colonization.

The Power of Ten Club program was created by Power of Ten Clubs International, Ltd. (PTCIL), a nonprofit State of Delaware corporation chartered in 1985. The primary purpose of the corporation is: "to organize, charter, and direct the activities of an international association of independently governed and operated mutual investment clubs, dedicated to investment in, promotion, and development of, near and deep space exploration".

Over the years, PTCIL has developed an extensive comprehensive program to accomplish this purpose which will not be included in this writing.

---

* Copyright July, 1998, Power of Ten Clubs International, Ltd. This document may be reproduced or quoted in whole or in part with source acknowledgement.

† President, Power of Ten Clubs International, Ltd., 3743 Pulaski Avenue, Philadelphia, Pennsylvania 19140.

Using "time" as an investment tool means using the long lead-time nature of initial exploration mission(s), combined with the Earth/Mars confluence problem, to provide "Demming" just-in-line cash flow programmability and to reduce risk.

"Demming" being W. Edward Demming, who designed this program to provide resources precisely when needed rather than draw-down from frozen assets of prior manufactured inventory. Demming's program demonstrated that frozen assets are losses waiting to be recognized.

When applied to MARS CORP. projects, this means realizing that you don't need all of your project money to start. What you do need is reliable, incremental cash-infusions, based on a reasonable schedule, that you can work your production requirements around.

For example: If a carbon dioxide/hydrogen to methane/water "Zubrin Generator", originally built full scale for U$47,000.00, can be built, Mars proof, with container, controls, and storage for U$500,000.00, less atomic power plant; and lead time is two years, MARS CORP. would need to generate funds at the rate of U$20,833.33 per month to amortize the project in that time span.

If the project was funded by the National Space Society (NSS) membership of 25,000, structured as 2,500 Power of Ten Clubs, total investment for a Zubrin generator, per member, per month, would be U$.833 - less than U$10.00 per year. (Note: 500K is nearly eleven times the original 47k cost.)

Why use an investment club to accomplish this? Simple. You are one person of a ten person investment club. You want to invest in Zubrin generators via MARS CORP. stock and you need $8.33 in new money each month to make the deal. ($.833 per member, per month.)

Solution: You take $17.00 of each members investment each month or $170.00 and buy RPM stock, a NYSE listed company. (RPM pays as cash dividend of $.14 per share, per quarter.) You do this consistently.

At the end of the first quarter, RPM sends the club a $4.20 dividend check on 30 shares. The club has sent MARS CORP. $24.99 for stock purchases. Net club out-of-pocket money to MARS CORP. is $20.79. Second quarter net MARS CORP. investment drops to $16.39. At sixth quarter net MARS CORP. investment drops to zero.

In brief, your club pays for its high risk MARS CORP. stock with dividends/O.P.M. (Other People's Money). End Game: Your club has RPM stock, MARS CORP. stock, and is making the deal with other people's money. Time and planning takes the risk and cost out of high risk investing.

Under this plan, no club may purchase MARS CORP. stock until it has existed for one year and gone through at least twelve monthly investment cycles.

Once the generator(s) is firmly funded, it should be possible to have the atomic unit(s) funded by a manufacturer of atomic units or possibly an Electric Power utility in a joint venture with MARS CORP. If not, MARS CORP. will fund.

Breaking the massive project of Mars exploration/colonization down into layman-scaled tasks, applying the same engineering frugality as was applied to the Zubrin generator, and programming funding on a precise when needed basis, will turn our current enthusiastic space rabble into a disciplined and dedicated space army to carry the day for Mars.

Arranging the tasks chronologically will further inculcate the relentless constraint of time on exploration/colonization into the psyche of the space community.

Does this mean we turn down the Bankers if they show up? It's been about thirty years since "The Eagle has landed". Figure on marching without them.

A prodigious undertaking such as Robert Zubrin's "Mars Direct", organizing its investors from dedicated lay persons, has the potential for running head-on into so-called "Blue Sky Laws". Blue Sky Laws are S.E.C. prohibitions against selling extremely risky, unproven, speculative securities to unsophisticated investors. MARS CORP. fits this description. Blue Sky Laws can be satisfied by investors forming "Power of Ten Club" mutual investment clubs.

The Power of Ten Club method, with its charter-driven requirement of 40% general business stock, 20% energy company stock, and 40% space related stock, precludes excessive investment in high-risk space stocks.

If the value of this program is recognized by this Society and efforts are made to limit pre-Mars settlement investment to Power of Ten Clubs, composed of NSS members and Mars Society members, the unsophisticated small investor is eliminated. Because NSS and Mars Society members represent the most sophisticated and knowledgeable group of persons on space exploration in the U.S.A. and probably the world.

No Blue Sky violation can occur due to this sophisticated prior knowledge, and the restrictive nature of the Club Charter. In addition, investment per "Demming" is spread out over project years. There is no front-end load, no contractual purchase requirement, ongoing current knowledge of space project/problems via NSS, NASA, and the Mars Society. Any club can abandon any investment in MARS CORP. at any time. If the Society and MARS CORP. stumble, the investors take a hike. This makes for prudent management decisions.

If the Mars Society embrace the MARS CORP. concept and the Power of Ten Club program is implemented leading to colonization by members of the space community, the inherent process will probably become the template for Mars government.

Mars could easily become the first corporate state due to the inherent Democratic nature of corporations. When analyzed correctly, the corporate structure closely mimics the democratic process; i.e. one share one vote.

This requirement of stock ownership for a voice in the affairs of the corporation can eliminate the possibility of "politically correct" social engineering which currently plagues these United states. The corporate charter can be written with a restriction prohibiting the officers and directors from engaging in any activity, which would obligate the corporation to provide any goods or services, to any person or group not specifically allowed in the Articles of Incorporation or the Bylaws. The restriction can be included in both. Any bylaw can incorporate a super majority stockholder vote for change.

Immediately upon formation, the tasks that MARS CORP. will relieve Mars Society of become apparent:

- MARS CORP. will build and own the carbon dioxide/hydrogen/methane Zubrin generators shipped to Mars. Upon landing, deployment, and initiating the fuelling process of the Zubrin generators, the MARS CORP. will establish *de facto* ownership of the first privately owned profit motivated Public Utility on Mars. MARS CORP. will, by this act, firmly establish, under all forms of the law and ownership, a commercial "Mars Property Right".

- MARS CORP. should construct and deploy all habitats. There are a number of reasons for this. When deployed and set-up, a functioning saleable private property right on Mars is established. By possession, a real saleable claim to land occurs. A property conveyance mechanism is created.

- MARS CORP. should build and operate the Rovers. Particularly in early exploration, as a transport service and survey company. In this participation/cooperation with the Mars Society, MARS CORP. will establish a non-conflicting, defensible claim to "boundaries" of contiguous property claimed by MARS CORP. This claim will be important as settlement expands. With the survey properly pre-planned, eventual establishment of road right-of ways, strategic settlement points, storage depots, etc., will be easier. It will also provide "land" for sale to investors.

This strategic plan will lead to the establishment of a survey company, a property title company, and a habitation construction company. Eventually these entities can be spun-off as privately owned companies. They will require laborers, support personnel, and support services. This will create a requirement for people on Mars.

If the Rover and the two-story habitats can be engineered with the same degree of frugality as the Zubrin generators, it is conceivable that the current NSS membership, or any significant fraction thereof, structured as Power of Ten Clubs, can fully fund the Mars Direct infrastructure requirements on Mars via MARS CORP. without imposition of economic stress on any participating member.

This is a chronological process taking years to develop. A process admirably suited for application of just-in-time funding using Power of Ten Clubs as the funding source.

In pro forma vision, the Mars Society will probably be "project oriented", dealing with the totality of the problems associated with the exploration and colonization of Mars; Social, Scientific, Political, Logistical, etc.

Contrarily, MARS CORP. is totally "task oriented", engaging in fulfillment of "people needs" for a price. Creating cash flow and assets for the benefit of its' investors; the bottom line purpose of all commercial corporations.

MARS CORP., shouldering the infrastructure problems of Mars Direct, will relieve The Mars Society of the problems it will otherwise have balancing its no doubt nonprofit status, with the requirement for involvement in infrastructure needs which are totally commercial.

There is no conflict between these disparate goals. The solution is in fact elegant in that The Mars Society can be a holder of both stock in MARS CORP. and a directorate position without impinging on its nonprofit status, provided such holdings fit within the guidelines of Sect. 501 of the Internal Revenue Code. A natural symbiotic relationship is thereby maintained. The two organizations will never work at cross-purposes.

The Mars Society, pushing Mars Direct, will be in conflict with those equally passionate members of the space community who view the Moon as the logical choice for first-step to space. The "Moonies" have formidable arguments. In terms of distance, fuel costs, and in-transit life support costs, the Moon is a no-contest winner. The Moon's disadvantage is the long-term and possibly perpetual requirement of infrastructure subsidy.

Once a Mars base is has been established, on-site life support produced lowers each add-on unit cost. On a Moon base, add-on unit costs remain static, or go up. Inflation is a cost driver.

If MARS CORP. take trip fuel cost, in-transit life support costs, and on Mars infrastructure costs out of the equation, Mars is a better investment for NASA than is the Moon. The Moon loses all of its advantages based upon cost.

According to Zubrin, the same lifters can be used for both efforts. This is a point of agreement upon which Mars and Moon boosters can work in concert, Some serious "horse trading" could lead to a pragmatic coalition on distribution of space shots, The preceding points provide a basis for strong negotiation.

Turning now to the question of how a stockholder in the highly speculative MARS CORP. profits from the investment. A profit-driven investor profits from MARS CORP. income generated via:

1. Branded Mars merchandise sold by MARS CORP.
2. Martian souvenirs.
3. Documentaries and docu-dramas for television.
4. Contract experiments carried out or monitored by MARS CORP. personnel on Mars for other entities.
5. Leasing of Martian facilities owned by MARS CORP. to science projects and industrial exploration teams.

Profit for a colony-driven investor is different. Recognizing that "money" will probably circulate in the form of electronic "credits", this investor can profit by the accumulation of "Mars credits" paid as stockholder dividends into a MARS CORP. account on Mars.

Credits to be exchanged for:
1. A cumulative credit/kilowatt energy account
2. Establishment of a "home site" on Mars, surveyed, sold, deeded, recorded, and conveyed by MARS CORP.
3. Accumulation for purpose of buying, financing and erecting a private habitation on Mars after acquisition per 2.
4. A credit/utility account for other services such as water, sewer, air, waste disposal, health care, earth freight charges, etc.

Differentiation in investor returns can be accomplished by issuance of two forms of common stock. Each form limited to either credit or cash dividends. The MARS CORP. "credit" unit may become the *de facto* medium of exchange for all private and business transactions on Mars. "Credits" are administered using standard credit card procedures.

If the space community is truly dedicated to space exploration/colonization, attention must be focused on "People Need" market requirements. Until there is a real space market, meaning; people there who require goods and services, nothing will happen to aid us, as members at large, to reach our goal of exploration participation.

People "needs" creates markets. Markets generate profit. Profit attracts capital. Capital will not flow into space exploration until a "people need" market is established in space. The Mars Society and MARS CORP. can create that market.

To accomplish the foregoing, MARS CORP.'s policy will be: Everything needed for the exploration/colonization of Mars has already been designed, proven, and produced. Indeed, the need to invent will demonstrate a failure of management. Costs will fall.

MARS CORP.'s credo will be: Find, adapt, standardize. This uncomplicated philosophy requires application of the shop-worn admonition of; "thinking outside of the box". Oft quoted, rarely initiated. I envision a truly amazing reduction in mission cost by simply not reinventing the wheel.

The point being made here is that it is time to stop thinking like engineers and start thinking like financiers. We should look inward first. The solution to Mars Direct already exists within the space community. We must stop regarding time as an enemy and learn to use time as a friend.

With The Mars Society, let us structure Mars Direct, via MARS CORP., to satisfy the most vicarious dreams, of any member of the space community. The twin problems of time and money/survival needs, will tend to moderate the dreams of the most unrealistic space booster, provided they get honest, mathematically verifiable information to substantiate the obstacle/solutions. None of the problems facing the United States when Kennedy announced the Moon program apply to Mars.

If the space community, collectively, has the intestinal fortitude to take up the challenge, there is absolutely no engineering reason why permanent ongoing colonization of Mars should not be a reality in ten years.

MAR 98-032

PRIVATE SECTOR MARS
WAKE UP CALL FOR NON-GOVERNMENT PARTICIPATION*

Thomas A. Gunn†

SETTLE OR VISIT?

It is indeed fitting to be discussing how the first Mars pioneers will travel a great distance through a turmoil of unknowns while the rest of society basks in day to day normalcy. It is equally interesting that such pioneering missions accomplished by non-government backed sectors are little known by the general public. Just 600 years ago to this month, a nobleman named Prince Henry Sinclair, Sir James Gunn, the Zeno brothers and a contingent of 200-300 settler-warriors landed near Plymouth Rock Massachusetts. The year was 1398, nearly *100 years before Columbus* and his government supported mission.‡

The Sinclair group came to establish permanent settlements. Columbus came to visit, explore as a personal agenda, and occasionally find social valuables to return to the supporting government in order to continue financing his personal life agenda. Some characterized the government funded group as having little regard for the local environment or the human race as a whole. We will designate settlement groups like the Sinclair expedition as non-government-related (NGR). Government supported and government dependent business (GDB) visiting groups like Columbus's voyages are still characterized today as seeking individual agendas, using government funds, and making claims of land ownership with little or no regard for settlement.

The private sector group activity, whether Sinclair's or today's, is historically closer aligned to a permanent settlement scenario than it is to the hit or miss government (GDB) exploration scenario. Therefore, we will address only the settlement of Mars as having any financial importance to the private sector.

WHO CAN FUND MARS SETTLEMENT?

The government (taxpayer) dominated space effort will have some key importance to overall future success, but costs and public sector tax support has major drawbacks in maintaining consistency, content, timeliness, and mission purpose. Because big space businesses are government fund-dependent, they must be considered to be in the government sector.

In addition to the government sector as in past exploration and settlement efforts, there exist large, uncommitted, non-government related (NGR) segments of the population can make significant contributions to the Mars effort. These NGR sectors historically have gone unno-

---

* Copyright © 1998 by Thomas A. Gunn, All Rights Reserved.
† 414 West Brookside, Bryan, Texas 77801. E-mail: tgunn@myriad.net.
‡ For information on the Sinclair-Gunn voyage:
    http://www.uwf.edu/~coehelp/studentaccounts/dsutton/webdoc6.htm
    http://www.mids.org/sinclair/600/960915.html.
For information contained in this paper: http://personalwebs.myriad.net/gunn.

ticed. Government and government-dominated big business (GDB) have been the sole managers of what should rather be a global Human effort.

Any ONE of these NGR groups could, by themselves, produce the funding, manpower, technology and effort to design, build and launch a permanent Mars Human colony. These NGR groups are:

1. Students & Schools: 7-12, University, Community College
2. Non-Space Related (NSR) Businesses
3. Adult Supporters from all walks of life
4. Institutes, Societies & non-space related Agencies
5. Wealthy Individuals

The size, scope, and cost-effectiveness of their efforts might vary, fall short, or even exceed the slower, costlier tourist approach built into the current GDB-only Mars effort. Pressures to organize these NGR groups is increasing. More of their constituents want to participate in this effort but find themselves outside the select few running the currently government dominated effort. It is therefore important to expand the effort to include, promote, coordinate, and foster cooperation between all available resources and not continue the narrow path of the past.

There are also smaller government entities that fit historically in the above NGR settlement sector groups and not in the government GDB sector. These resource groups are:

6. State government (and their equivalents in other Countries) including State Commissions, and
7. Municipalities.

They also have a role to play and potentially, they too, could accomplish their own Mars settlements. These groups are a rapidly emerging sector in space transportation, funding, education, and economic growth and are grouped together below as the "State/City" sector.

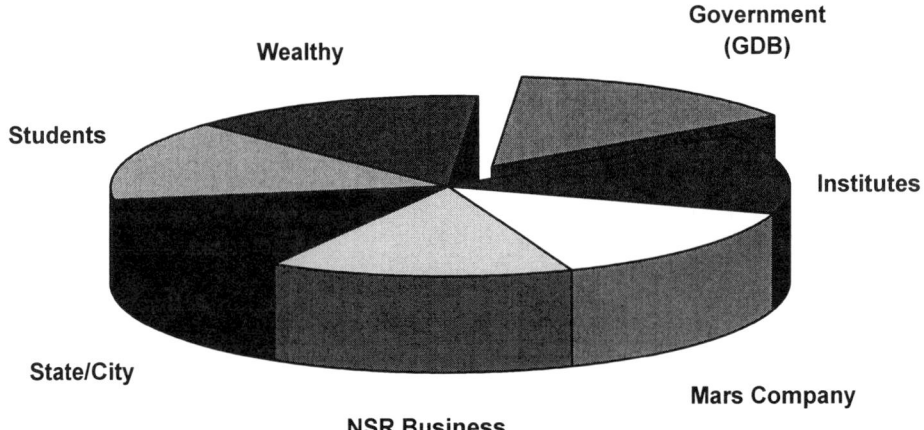

Mars Resource Sectors
Worldwide Potential Funding/Launch Capability

If you take out the "Government" pie slice, the remainder of the pie chart is the non-government related (NGR) private sector. One slice, "NSR Business" contains all the Non-Space Related (NSR) businesses from software companies to oil conglomerates. Containing all the interested adults from all walks of life not in any other sector is a hypothetical "Mars Company" as a publicly held stock company similar to the Hudson's Bay Company. The "Institutes" sector contains both profit and non-profit groups including the Mars Society, and the "State/City" sector represents all non-federal agencies and groups in any country.

This "Mars Resource Sectors" chart, therefore, represents the **totality** of the emergent Mars society demographic. It remains for future historians to write of the role each sector will play in Human outreach. Organizations like the new Mars Society can be instrumental as focal points for the interplay of these resource sectors, their success, and the direction of the final outcome. The success of the Mars Society will depend in large measure on its ability to embrace all sectors rather than become a tool of the currently dominant GDB sector.

WHY THE GOVERNMENT DOMINATION?

It is easy to understand why fundamental business requirements (listed below) have not been addressed by the government dominated industry to date. By definition, their source of funding is largely the taxpaying public through Congressional or other government action not unlike the Columbus mission. Their focus, and rightly so, must be to promote media events and appeal for the social support that educational activity and hype can produce in order to garner the maximum pressure for continued financial support from Congress.

That funding is increasingly difficult compared to the public interest of the 1960's is everywhere evident. Unfortunately, it is this need to expend a greater effort to raise funds that has crippled the basic purpose of federal agencies. Furthermore, it has never been the purpose of NASA and government-dependent big business to address how you would go about conventional business funding for a Mars project, and it would certainly be counterproductive from their viewpoint to promote space related efforts of the NGR private sector. This is why there is a desperate need for a forum that embracing the agendas of all Mars resource sectors. The Mars Society could help fill this need.

WHY HAS NSR BUSINESS REMAINED SILENT?

From the CEO point of view, all that was broadcast for the last 40 years is:

1. Launching people to Mars entails unbelievable **costs**,

2. There is nothing in the effort that could show a normal business **ROI** in normal **timeframes**, and

3. The required **Technology** to accomplish a Mars shot is beyond the average company's ability on its own without government help.

Since all these images are **not true** if business were running the project, there has been great damage done to the business perspective of an entire generation of business and financial leaders.

For the most part, this damage result is just fallout from projects that are publicly run from a taxpayer financially driven viewpoint. It is now time to provide business-specific information to the special interest business groups that have been inadvertent victims of misinformation, lack of information, and propaganda for too many years. The impact could be dramatic.

WAKE UP CALL TO CEO'S

Major and minor NGR businesses have been virtually non-participants in the Mars effort to date. Some would say they have been blind to the obvious potential gains that can be realized from taking part in a new frontier. This is not the case. Business of any type, Mars related or not, must work within traditional rules that make the economic machine run smoothly. Until the Mars effort can effectively communicate that we understand and can provide the necessary business factors for conventional success, we cannot expect to awaken the private sector giant. It is now time to do so, to set aside for a moment the media events needed to appease the GDB sector legislative gods, to issue the wake up call, to address business specifics.

ELEMENTAL BUSINESS FACTORS

To foster the interest of historically non-space related businesses, the following conventional elemental factors must be presented and satisfactorily answered for funding participation:

1. Any participation must provide a **breakeven point within 18 months**
2. Return on Investment **(ROI) must occur within 5 years**
3. Price/Earning **(P/E) ratios above 15** should not be required
4. Profits/Investment ratio beyond ROI should be in **multiples** of investment.

Even if these factors can be shown to be met, additional aspects for involvement must also be present to assure *stockholder acceptance, corporate image enhancement, and mission critical cost-effectiveness*:

1. Positive Company promotion from any media news coverage
2. Alliance with socially acceptable image issues only
3. Reinforcement of Corporate trustworthiness in long range planning
4. Positive Images of future corporate financial successes
5. Insure solid, sustainable traditional corporate growth.

Any perceived risks must be positively embraced by Wall Street and stockholders alike. This can result from solid planning and clearly achievable corporate product mission goals through successful short-term demonstration projects. All of the potential gains from future Mars exploitation cannot occur until we satisfy short-term and sustainable business requirements.

PACKAGING

If participation in the Mars effort can be packaged to satisfy the above conditions, we will have the attention of this entire private sector for the first time. *How* this attention is then addressed and incorporated into the effort will determine:

1. *The extent* to which businesses will participate,
2. *Which business* can and will participate, and
3. *How many businesses* in this sector will participate.

These three areas are quite different and all must be addressed successfully to involve the business sector in the Mars settlement effort. The Mars Society should take a leadership position in working on these areas.

Remember we are ***not*** addressing existing contractors, subcontractors, and would-be subcontractors in the space related industries. The non-space related (NSR) group consists of the other 90% of large and small business that see no current relationship of their missions and products to a Mars effort, nor have they seen evidence that any aspect of this effort can meet the conventional economic factor requirements listed above.

WORK NEEDED ON EMPLOYEE-EMPLOYER LEVEL

The successes of NASA and other agencies has gone a long way to pave the way for a business social attitude change throughout all of the above sectors. The non-space related business sector is poised, ready and participation is long overdue. However, the extent of the *personal interest* of management and employees' without regard of the business's product line, is not clear.

To what extent personal interest can interact and reinforce a *newly stated business interest* is also unknown. Would more employees be interested in space work if the company embraced such discussion? Would management become more interested if existing employee's interest in Mars work were made known? Can historical employer and general social recriminations of being a "space cadet" and "buck rogers" thinking be turned into serious, positive support and open discussion?

Possibly. All the business sector needs to get on board is just proof that the Mars effort work can, indeed, stand on its own *economic merit*, have some *relevancy* to their business needs for success, and meet traditional business *ROI* requirements.

FOCUS ON PROFIT

For those who have not thought about it recently, permanent Mars settlement embodies all that is necessary for serious, straightforward profit making. The *three fundamental profit areas* to explore that characterize **Mars settlement** from strictly a business viewpoint are:

THE THREE FUNDAMENTAL PROFIT AREAS
(1) Ownership or control of large LAND areas will always be profitable.
(2) Marketing and supply of unique INFORMATION is always a solid investment, and
(3) Multidisciplinary design efforts (such as Mars settlement effort) always produce New Product Manufacturing and Licensing "SPINOFF" benefits.

All three fundamental profit areas have many historical earthly examples of financial success, and all three represent solid, sustainable, long-term growth characteristics and opportunities.

Media and space proponents have never approached space with an effective profit focus. Unlike the above, the current predominant targets for profit have been:

1. Make money from government contracts; and,
2. Make money from delivering goods into orbit close to Earth.

Contracts (1) result in little risk and marginal profits. Delivery transport systems (2) involve very high risk and little profit. To a limited degree, these profit areas work for some. However, they are closely related to the near-Earth environment - not *the* space frontier. **The moon, asteroids, and Mars are an entirely different profit scenario and focus.** That far from Earth focus centers on the above three fundamental profit areas to the exclusion of all other considerations.

The first two profit focus areas, ownership of land (1) and information marketing (2), will yield financial benefits **during and after** settlement of Mars - little wonder that pragmatic, day-to-day business is not interested. The third area however, "spinoff" exploitation, will yield profits **before** Mars settlement. This profit focus area is therefore the *prime profit focus area* in which to begin any discussion, analysis, and understanding of realistic and attainable business goals.

The focus of space business, then, is reduced to the question "Can the elemental business factors for solid, sustainable, profit and growth be met from multidisciplinary product spinoffs from design of a permanent human Mars settlement?"

The details of how to do this are outside to scope of this paper. However, the answer to the above question for most business ventures is "yes." There is little need to dwell upon what follows this step one spinoff profit focus area. As profits from mission design development spinoff production begin to occur, the settlement of Mars begins and the other two profit focus areas *automatically* kick in.

Following any other path would not be productive, as witnessed by the failure of Lunar business development over the past 30 years. Basing business funding on some fuzzy economic item to be discovered on an asteroid or on Mars, and convincing stockholders that its transportation over millions of miles will be cost-effective is pure folly. Yet these goals seem to be the only ones generally discussed. Unless and until you can make profits every day, any discussion of the profit "potential" of Mars settlement and asteroid mining will continue to be pure science fiction. The above spinoff profit focus is viable, clear, and simple and will reap solid rewards for those who participate now. Satisfying the elemental business factors for spinoff manufacturing and marketing is well understood.

However, **deliberately creating spinoffs for profit** has not and will not be done by government and represents an untapped business area. The fact that a few spinoffs have managed to find their way into our business sector and to the public is pure happenstance. Making the *creation of spinoffs* our prime economic focus would have a dramatic outcome. It is now time to do so.

The job and rewards belong entirely to the private sector...That we also happen to populate another planet in the process is simply the inevitable byproduct of business expansion.

NEEDED TECHNOLOGY IN PERSPECTIVE

The world has spent too much time on technology and too little time on mission. Neither Columbus nor Sinclair worried about the "technology" to get to the new world. Both had access to boat building and operating experts. The *mission of settlement* rather than *mission of exploration* was the real substance of 400 years of effort Such is the current difference in projects between government GDB rocket programs and NGR private sector groups today.

The defining difference is in (1) purpose; and, (2) the efforts required. Sinclair needed specific and reliable information about where he was going in order to successfully establish settlements. Columbus (possibly married to a descendant of Sinclair)* had some advance information, but remained mute to obtain funding since prevailing government information was that he was going to come the Eastern shore of India. That leaders of old world governments knew nothing of America was of little consequence to his personal exploration mission agenda.

Successful settlement missions by Sinclair, however, required input of scouting missions throughout the previous 400 years by the Viking relatives of Sinclair and Gunn. Such is the major difference today. The American government could have gone to Mars to visit in the 1960's with a crewed mission - all the technology needed seemed to exist for the purpose of a visit similar to a Moon landing. Waiting for the technology of a first-class million dollar spacecraft bathroom may not be needed by some. The Vikings went coach, but achieved a European settlement in America first, nevertheless.

Compared to information required for a "visit," technology information needed for "long term settlement" is legion. To provide a settlement with social and trade success required then and requires today: (1) solid scouting and (2) dedicated individuals who know they will never return to the old life. They went one way to stay. In addition, they had to provide for both their own survival in a hostile unknown environment, as well as send reports back that would be useful to those following them. The reports had to contain specific data that would continue to compel others to commit resources to the settlement effort, even though these backers would not participate in the settlement themselves.

The same resource scouting is required today. Some of the immediate questions that must be answered are appended in a "to do list." Some items cannot be adequately accomplished by government GDB interests. Many can be accomplished with robotics by government while the NGR sectors prepare for their human settlement missions.

The "to do list" is grouped into five sections: 1) Solicitation of non-monetary contributions, 2) Funding source organization and promotion, 3) Remote control Mars on-site resource identification, 4) On-site equipment needed, and 5) Specific special projects needed now. It is now time for the CEO to revisit the down-to-earth profit potential that the Mars venue can provide. We have seen government GDB "to do lists." Private sector NSR business and the other NRG sectors must now explore what must be on their own "to do lists."

SUMMARY

In short, human settlement has never been, is not now, nor ever will be solely the result of government. By involving the larger NGR private sector in the Mars settlement arena now, we force:

1. Needed but still unasked questions to be answered,

2. Solid elemental business conditions to be satisfied; and,

3. The tapping of a global, nearly unlimited, NGR resource for sustained Mars settlement participation and social-economic growth.

These precursor items are all impossible to achieve if left to government and government-dominated business (GDB) participation alone. This is the wake up call to non-government related (NGR) and non-space related (NSR) people and groups. The message

---

* For information on the Sinclair-Gunn voyage:
   http://www.uwf.edu/~coehelp/studentaccounts/dsutton/webdoc6.htm
   http://www.mids.org/sinclair/600/960915.html.

is that the long wait is over; and, that a Mars settlement thrust is not only achievable, but also a timely and viable business proposition.

APPENDIX

TO DO LIST

We could build the rocket and fly people to Mars right now. However, there would be much guessing about what we would need to take. The more we can find out now, the more we can do on Mars and the less weight we would need to take. Exactly the same as if we were moving to an unknown (to us) new home - we would take more than needed.

This list is a framework to help organize our "move" list. Keep in mind these items are all **now** - not things we could do "someday" or that might be nice to have "later". **We can never get all the data we would like by move time** no matter how long we put off the move. Reality is that someone will go soon, totally ready or not. Here are some items we can be working on up to that launch.

1. **Solicit contributions (for a permanent human Mars settlement) to provide:**
   Research (all areas)
   Ideas (all areas)
   Services (labor & skills)
   Equipment (launch center, rocket, settlement)
   Testing (services & equipment)
   Facilities (launch center, staging, warehousing, testing, assembly)
   Legal Issues (ownerships, regulations for private sector public stock group)
   Promotion (writers & media)

2. **Organize and promote funding from all of the following:**
   Direct TV/Internet Advertisements (example: $49.95 for your name on a memorial erected on Mars)
   Part of Fees from Memberships
   Part of Profits from Spinoff products and Private Labels
   Stock in New Private Sector business (direct funding of facilities, equipment and operations)
   Advertising Fees from Endorsement Projects (example: books, Mars Society, Mars bars, M&Ms etc.)
   Fee Surcharges (on stamps or $3 checkbox on 1040 etc.)
   Traditional Taxes
   Direct Contributions from Wealthy

3. **Document detailed data from Mars to identify resources (or lack thereof):**
   Food (nitrates, phosphates, water, etc.)
   Atmosphere & Liquids
   Local Environment variations
   Caverns/Caves (initial habitat)
   Minerals (indicates prior geological water existence & mining sites)
   Construction Materials (for concrete, soil bricks)
   Manufacturing Materials (organics N_2, O_2, C, H_2 to make plastics, etc)
   Soils & preparation effects

Magnetic Anomalies (iron, nickel, platinum (for glass making), etc.)
Hot Spots (possible geothermal & settlement site)
Mining Materials (sulfur, saltpeter, carbon to make gunpowder, bauxite for aluminum etc.)

4. **Equipment to take or create on site:**
   Mining (identification, extraction & processing)
   Communications (local and Earthlink, Mars Web Server)
   Power (in situ all types)
   Drilling (seismic & core)
   Construction (to make bricks, fabricate plastics, work with stone)
   Heating
   Sealing (for domes and cave entrances)
   Medical
   Sports & Entertainment
   Environmental (habitat & farm atmosphere and liquid cycle)

5. **Specific special projects needed:***

   Government Policy
   1.1 Delete Politically Correct SAFETY Constraints
       1.1.1 From Environmental Design/Fabrication
       1.1.2 Remove "Safety Gag" for Atomics
   1.2 End U.S. Government Interference in Private Sector
       1.2.1 Launch Location Bans
       1.2.2 Launch FINES
       1.2.3 Pass Spin-Off Mfg SEC Rule 507
           (which can finance permanent Mars work)
   1.3 Create Private Sector Solid Fueling Access
   1.4 Fund Private & Education Development Tasks (HLLV,etc.)
       (this is ONLY government funding - balance outside)
   1.5 Reaffirm Moratorium Banning Returns from MARS
       (Zero Risk rather than some risk no matter how small)
   1.6 IAA Strong Public Statement of Support & Objective
       (ref. "Why Space", attached)

   Social
   2.1 Handle Publicity Right
       2.1.1 Media & Internet Hero Excitement Creation
       2.1.2 Home Robotics Interactive Telemetry for Fee
   2.2 Begin Training "Martian MacGyver" Crews
       (Don't try to solve everything before going)
   2.3 Develop Public/Private Donations System
       (includes "your name on a plaque on Mars" project

   Science
   3.1 Certify 30+yr Algae/Fuel-Cell System

---

* Gunn, T. A., "Mars One Way To Stay: Economics, Commercialization, Public Policy and Technology Needs Required for Mission Success", AAS 96-321, in *The Case for Mars VI*, K. R. McMillen, ed., AAS *Science and Technology Series*, Vol. 98, pp.223-234, 1999 (Conference Paper presented at the Case for Mars Conference, held in Boulder, Colorado, July 17-20, 1996).

3.2 LEO Human Gestation Experiment
3.3 Precursor Shopping List of Mars Materials
      for immediate atmos, const, and food survivability)
3.4 Investigate "Cristofv Effect" 4.5x Lighspeed Comm
      for faster Earth-Mars Links

Technology
4.1 Report on PRA-Gunn Direct Mars Boost HLLV & Facility
4.2 Create Substantive World Planetary Agency (ref.PRA)
4.3 Benchmark Tensor Calculus PC Based Transit Mission
      (all command control/orbital-xfer mechanics on-board)
4.4 Benchmark Capsule Solar Radiation Hall-Effect
      Reactive Magnetosphere Shielding.
4.5 Establish Permanent Stay Reference Mission Specs

THINKING ABOUT MARTIAN ECONOMICS

Edward L. Hudgins[*]

Why, four decades after men first ventured into space are there no regularly scheduled commercial flights into orbit? Some 35 years after the Wright brothers' 1903 flight the commercially viable DC-3 was flying. But today the cost of placing payloads into orbit on the Shuttle is perhaps a magnitude more than on Apollo. By contrast, in the past twenty years the cost of airline tickets per passenger mile have dropped by 30 percent, with twice as many people now flying. The costs of other goods and services have declined as well. The cost of shipping oil, for example, has dropped 80 percent.

For too long space enthusiasts have ignored economics at the peril of their passion. If men are to journey to Mars, markets are the means.

THE GOAL

In this discussion I shall address two questions. The first is "Can a mission to Mars be privately funded?" My answer will be "Yes." The second question is "Can development and exploitation of Mars be sustained beyond the initial landings?" Here I answer is unclear.

PRIVATE VS. GOVERNMENT SUPPLIERS

It is useful first to review some economic truths that should be obvious to most people today. The first is that the free market system is the only known way that, on a large scale, resources can be used in the most efficient manner to satisfy the needs of individuals

What constitutes a market? First, individuals own property and, especially, the means of production. Second, individuals are free to exchange their property or services with others. Third, all exchanges are based on mutual consent. This means that the prices of all goods, services and labor are based on contract. Fourth, the purpose of government is to protect private property and enforce contracts.

The market system results in the following:

First, each individual has a strong incentive to act in an entrepreneurial manner, offer the goods and services that will bring the greatest return or profits or wages for one's efforts.

Second, in a market system division of labor occurs. Individuals who are most efficient at performing some task or producing some good or service will prosper and those who are less efficient some particular area will have to find work elsewhere.

Third, the market is a discovery process. No individual or group of individuals can know for certain what goods or services, produced in what manner, by what individuals or enter-

---

[*] Dr. Hudgins is Director of Regulatory Studies, The Cato Institute, 1000 Massachusetts Avenue, N.W., Washington, D.C. 20001.

prises, will best serve consumers. It is only in market exchanges that the system "discovers" who is best at what. This is why central planning does not work.

And fourth, the market process allows for the creation of new wealth, the invention of new products and services.

By contrast, governments generally do a poor job of allocating resources and delivering goods and services. To begin with, there is little incentive for governments to do so. After all, revenue is not does not result from satisfying customers; it comes from taxes.

The Space Station, for example, originally was expected to cost $8 billion. The price through the 1980s escalated to $16 billion then to $32 billion. When it reached $39 billion in the early 1990s, a stripped down version was offered that was suppose to cost only $29 billion. But the cost continues to rise. We saw, for example, that when asked how much it would cost to go to Mars, NASA's "90 day Report" in 1991 put the price at $450 billion. Remember, it is governments that give us $900 toilet seats and hammers, and $700,000 outhouses.

A second reason why governments cannot deliver goods and services efficiently is that they are subject to political pressures. Indeed, if decisions are not made on a basis of market demand, then politics is the other alternative. In the case of government spending on technology, businesses compete for corporate pork. For example, recipients of taxpayers' money through the Department of Commerce's Advanced Technology Program include Xerox, IBM, and Dupont, all no doubt fine companies but hardly poor, struggling newcomers. Political pull is their to the federal pocketbook.

BETTER GOVERNMENT APPROACHES

The most economically and politically viable approach to a Mars mission, based on market principles, is the Mars Prize approach, supported by House Speaker Gingrich. A version of this approach was considered in 1988 by a Commerce Department led group looking at ways to return to the Moon. For a Mars mission a $20 billion prize, with an actual mission cost of only $5 billion, indeed creates an incentive for the private sector to find the best way to the Red Planet.

But these funds still will prove difficult to secure. Thus Mars enthusiasts might support a radical approach: A way to make such a prize-based approach politically and economically practical might be for advocates of a Mars mission to back the elimination of the planned Space Station and privatization of the Shuttle as a means to turn over civilian space activities to the private, civilian sector where it belongs. Some of the savings could be used for the Mars Prize.

The General Accounting Office estimates that the actual cost of building the station will be around $48 billion with a cost of $49 billion to operate it between 2002 and 2012. The station will never pay for itself by selling space to the private sector. Space on the station will continue to be given away for below cost.

Further, there is little useful science that will be performed on the station that could not be done with unmanned, expendable launch vehicles. In 1991 a special Presidential Advisory Commission, chaired by Martin Marietta Corporation CEO Norman Augustine, stated that "We do not believe that the space station ... can be justified solely on the basis of the (non-biological) science it can perform, much of which can be conducted on Earth or by unmanned robots."

As part of the transition, some of the savings from NASA's $13 billion annual budget could be set aside for a Mars price. Two billion dollars per year would produce a $20 billion prize over a decade.

PRIVATE APPROACHES

But a better approach would be for the Mars mission to be totally financed by the private sector. There is certainly historical precedent for such an approach.

For example, between 1903 and 1919 the Carnegie Foundation spent $2.38 million to build the Mount Wilson Observatory. That would be $32 million in current dollars. But a better comparison is to the federal budget. Between 1903 and 1919 Washington spent $11.1 billion (adjusting for World War I expenditures). This might seem a small amount. But the same portion of the cumulative federal budgets for the seventeen year period between from 1982 through 1998 Carnegie would be $4.45 billion, an amount that could put a entrepreneurial company well on its way to a trip to Mars.

Carnegie spent $2.29 billion between 1920 and 1929 on Mt. Wilson ($20.4 million in 1996 dollars), $2.38 million from 1930 to 1939 ($26.37 million in current dollars), and $2.15 million between 1940 and 1949 ($18 million in current dollars).

The Rockefeller Foundation, starting in 1929, paid out $6 million to build the Mount Palomar Observatory, which saw first light in 1948. That's about $60 million in 1996 dollars.

Between 1925 and 1929 the Guggenheim Foundation operated a $3 million Fund for the Promotion of Aeronautics. That's $26 million in 1996 dollars. (Total federal expenditures during that period were $14.8 billion.) And in the 1930s the Guggenheim Foundation paid about $100,000 to finance Robert Goddard's pioneering rocket work.

If the actual cost of a Mars mission is between $5 billion and $10 billion, a consortium of enterprises and educational institutions could fund a Mars mission. Yet without taxpayers funds, what would be the incentive to go? Many institutes and foundations no doubt would invest in a Mars mission for the scientific knowledge that it would yield. For example, the National Geographic Society took in $401 million in 1996, and spent $4.7 million in research grants. It would no doubt be part of such a consortium.

The prospect of property rights and owning Martian assets would be a strong incentive. Also a consortium might earn money and develop technology for a Mars mission by taking on other tasks for profit. If NASA stops subsidizing cargo on the shuttle, there would be a larger market for launch services and companies that might provide such services and use part of the profits to fund a Mars mission.

The entertainment industry, no doubt, would have an incentive to be part of a Mars consortium. Consider the Disney Company. It might put up several hundred million dollars to put camera-equipped rovers on the Moon to provide holodeck-type virtual reality entertainment on Earth. It might use some of the profit to fund similar activities on Mars. Disney, by the way, took in $22.47 billion in revenue in 1997. Its pre-tax net revenue was $3.387 billion. With a tax bill of $1.42 billion, the after tax profit was $1.966 billion. If Disney were given a tax credit for money used to finance a Mars mission, it alone could pay for that mission in 14 years. The road to Mars does not have to go through government territory.

IS THE TIME ECONOMICALLY RIGHT?

Mars enthusiasts must also ask the question "Are the economic conditions right for sustained development of Mars?" Many correctly worry that initial landing on Mars will be followed by decades in which the Red Planet sees no human visits, as happened with the Moon after Apollo. The promise of profits for investors on Earth who finance missions to Mars could cover some of the costs. This is why the initial consortium contracts will be important.

But it is often the case that what is technologically possible at a given time is economically impossible. An analogy might be made here with technology itself. Leonardo diVinci designed a flying machine 500 years ago. Let us assume that he actually designed a model that could work. Of course, without materials with certain properties, electric motors, fuels, etc., while the design might work in theory, it would not in practice.

Or consider computers. In the 1950s it might have been possible to string together enough Univacs or early IBMs to do what a good desktop can do today to perform certain functions. But would that be the best way to perform certain tasks? For example, much of the special effects work in the movie *Jurassic Park* was done on desktop computers. in the 1950s it might have been possible to create such effects on a series of mainframes. But it would have cost a substantial portion of the country's GDP. It only became both technologically and economically practical to produce such a movie, or, for example, design an aircraft such as the Boeing 777, on computers in the past decade.

Let us assume that the first Mars landing will cost around $20 billion. And let us take the optimistic view that the cost of each flight will drop after the first landing. Based on a flight every year and a half, the costs will drop by 80 percent over twenty years, the same as the drop in the costs of shipping oil over the past two decades. That's fourteen flights with the first $20 billion and the last $4 billion. That still means that 14 flights will cost a total of $168 billion. And let's assume that by the eighth flight, a crew of eight can be sent to Mars than just four, but for the same cost. That means that the cost of sending 84 individuals to Mars, with 14 flights, over 20 years, is $2 billion each. That's a hefty airfare!

Sustaining activities on Mars will require considerable capital. In his enjoyable book *Red Mars*, Kim Stanley Robinson fails to appreciate the costs of the capital equipment needed for decades to bring that planet to self-sufficiency. A lot a material will have to be paid for and transported to Mars. This will add to the total costs.

But there is still cause for hope for sustainable settlement of Mars. First, the costs of technology needed for a Mars mission would fall much faster than, say, the 80 percent figure that I used. This has certainly happened with the costs of computers.

Second, strong GDP growth could carry us to Mars. Between 1982 and 1989 the economy grew at a real rate of about 4 percent annually. Today the GDP is over $8 trillion, with inflation very low, though growth rates are not as high as in the 1980s. If deregulation and tax cuts spur 4 percent annual growth, a $8 trillion economy will double its size in about 18 years. That means that $168 billion is less of a price.

CONCLUSION

The lunar landings were great human and engineering achievements, but they unfortunately created the mistaken impression that the government must take the lead in such projects. The exploration of Mars will also manifest humans fulfilling their destiny but also will show that free men and women working without government can achieve great things as well.

MAR 98-034

THE BUSINESS OF COMMERCIALIZING SPACE

David M. Livingston[*]

This paper stems from my research connected with my doctoral dissertation in the School of Business at Golden Gate University in San Francisco, CA. My dissertation, *Evaluating the Business Potential, Costs, and Risks of Opportunities Resulting From the Commercialization of Outer Space*, presents a business perspective approach to this topic. It is with this background that I examine both the principal issues and the merit of space based businesses.

Based on my research to date, the general non-space business community does not have a high level of interest in commercializing space except for those businesses currently operating in space with proven records. Nor is there a strong belief that space can or should be commercialized. There is even a lack of understanding as to how profits can be made in space, and there is considerable uncertainty with regards to the laws, regulations, and the role of government in space businesses.

Uncertainty is a concept which business does not like. When risks are taken, usually the risk/reward ratio is significant enough to warrant taking the risk. Investors and management understand the business and its risks, and have experience in these matters. Venture capital companies and wildcat oil drillers that finance businesses when they are most vulnerable for success or failure are examples of business taking risks. However, these risks are usually confined to business risks, i.e. financial issues, sales, market acceptability, or the management team. Other risks such as taxes, the role of government, and the regulatory environment are understood and known. However, this is not the situation facing today's space businesses.

Today's space businesses face significant uncertainty with government and regulatory agencies. Now that private enterprise is operating in space, many important legal issues need to be addressed. The United Nations treaties that most space faring nations have accepted and signed did not contemplate private sector space businesses or individuals claiming space as their own domain for living, working, and playing. Some space entrepreneurs believe that forcing some if not all of the issues arising from these treaties to the courts is the best way to resolve the potential conflicts. Others believe that organizations should be formed to take over the business and legal planning aspects of those wanting to operate in space. The problem is even more local and basic than treaties. In the United States for example, there are potential jurisdictional conflicts involving NASA, the FAA, other agencies within the Department of Transportation, the Department of Commerce, the Department of Defense, the FCC and other federal agencies. These potential jurisdictional conflicts can create problems and run up the costs for businesses considering or operating in space.

Uncertainty and confusion are costly for businesses and our economy. The highest cost results in the avoidance or the unwillingness of the business to pursue operations in the affected field. With avoidance or the refusal to engage in a business, there is an absence of economic

---

[*] P.O. Box 95, Tiburon, California 94920. E-mail: dlivings@davidlivingston.com.

stimulation that would have resulted had the business been pursued. Even with a failed business, money, jobs, ideas, and other contributions are circulated within the economy. Nothing is circulated in the absence of a business undertaking. In addition, a steep price is assigned to the various risk/reward ratios with the businesses that do operate in the uncertain and potentially conflict ridden environment.

An example of this steep price is the regulatory control of private space launches. Largely because of the confusion that came about after the first private space launch of the Conestoga I,[1] President Reagan created the Office of Commercial Space Transportation (OCST) within the Department of Transportation (DOT). By 1986, the OCST was already issuing regulations to govern launches by private companies and after the Challenger accident, the office became a key agency in both space policy and business.[2] The FAA, also in the DOT, was already involved in public and private launches, because all of the launches were and still are subject to FAA regulations in American airspace. Eventually, the OCST received the clear mandate as the agency to supervise private launches. When it was created, the OCST identified at least a dozen federal bureaus, plus states and other districts which could have some jurisdiction in regulating space activities.[3]

Potential conflicts can also come from legislation. The Federal Aviation Act imposes limits on States to make legislation regarding commercial air travel. The Commercial Space Launch Act of 1984 sets up the Federal licensing mechanism, but this Act notes that the "authority of States to regulate space launch activities within their jurisdiction, or that affect their jurisdictions, is unaffected by this Act."[4] It can be costly for businesses to sort out these potential conflicts and it is often a deterrent to even engaging in the business.

Possible solutions to these and other potential conflicts have been discussed. One suggestion has NASA becoming a cabinet level agency, thereby creating a Department of Air and Space. This cabinet model has the newly formed department absorbing the FAA and the National Transportation Safety Board (NTSB), plus other space related activities within the Departments of Commerce and Transportation.[5] Critics are quick to point out that giving NASA additional regulatory and operational duties may adversely affect NASA's obligations in research and development.[6]

Other consolidation plans have been discussed because having a Department of Space as a cabinet-level department would have significant symbolic value, showing that it was important to the U.S. Government.[7] This would send a clear message to the business community and to those interests working with space inside and outside the United States. Still, it is a concept that is not without controversy. Political scientists who have looked into similar reorganizations point out that they are often not successful. Any benefits from reorganizations for symbolic reasons only and without substantive policy changes are almost always temporary.[8]

For the business community, having potential agency conflicts eliminated and elevating the status of commercializing space to the cabinet-level would be positive. It would bring about a new level of awareness for the general business community, helping to open it to the benefits of doing business in space. As the political scientists point out however, if the changes are only symbolic, when the honeymoon period ends, it will be back to business as usual.

Commercializing space extends our terrestrial business environment from earth to several hundred miles above it and beyond. The deciding factors for doing business in space should be similar to those on earth. If there is an advantage to doing business in space, then we should be able to capitalize on that advantage. Getting to this point requires development in many areas, including the ability to obtain long term, low cost financing. Two important potential sources for obtaining this financing are discussed in this paper.

First, the business of commercializing space will require substantial capital. Knowing both the ease and how this capital will be obtained, and the cost for obtaining and using it is important for the space commercialization progress. One idea that addresses this important issue is the concept of a Space Development Bank.

A United States Space Development Bank could help overcome the lack of financing for large commercial space ventures. Dr. Thomas Matula in his paper on this subject, points out that space commercialization and developing nations share the problem of a lack of long-term, low cost financing.[9] Development banks are fairly successful with their assistance to developing nations. The same principals can be applied to developing commercial space industries. Financed by the sale of specific space bonds issued by the federal government, a low cost source of long-term funding would be available to businesses interested in doing business in space. The capital markets would assign a low risk to the bonds since they would be backed by the U.S. Government. Variations on this might include state funded and organized Space Development Banks to support space businesses in specific states. Similar development banks could be formed on an international basis and in specific countries for their own space business development.

Second, venture capital is another often talked about source of funding for commercial space ventures. Exploring the relationship of the venture capital industry to the potential commercial space industry is important. Therefore, as part of my dissertation research, I conducted a random survey of over 600 venture capital firms across the United States to ascertain their interest, awareness, understanding and experience in doing business in space. The amount of funding this industry can provide commercial space entrepreneurs and businesses is significant. Also, this industry represents a primary source of seed, early growth, and development financing.

Table 1 below summarizes the number of venture capital companies surveyed and the number responding. Companies were randomly surveyed from 46 states. Most of the companies contacted were in Massachusetts, California, New York, Texas, Connecticut, and Illinois.

Table 1
STATISTICAL RESPONSE TO STUDY

1. Total of all companies comprising study pool:	1,082
2. Total number of companies surveyed:	640
3. Percent of companies surveyed:	59.14%
4. Total number of responses to the survey:	64
5. Survey Response rate:	10%

Among the many survey findings, common understandings and views held by the responding venture capital companies were revealed. When asked what they thought the most exciting and potentially lucrative commercial space businesses were, the companies reported back with the following statements:

"Space transportation to orbit for non-satellites; More advanced communications; Remote sensing; Scientific applications; Ventures with a large, untapped market with a realistic business plan; Space travel and tourism; Lower cost, higher efficiency launch vehicles; Space businesses that use existing infrastructure or easily deployed infrastructure to deliver high-volume, high-value-added products & services to large, commercial markets; Owning and operating satellites; Growth and technology transfer; Manufacturing or technology in a space vacuum environment; Growth of crystals, semi-conductors, and all types of technology exploitation."

The financial and business risks were deemed to be high and in some cases excessively high to the point of negating any interest in space businesses. Some of the venture capital company responses included these statements:

"Pure play - would be too high for most; Take whatever risk reward you would associate with/similar normal business, multiply by 5X; Government competition/regulation; The risks are high, but so too are the rewards for those that are first and successful; No revenues for several years; Probably higher risk due to large capital requirements and lack of management with long track records; Very low investment potential with astronomic associated risk demanding commensurate rates of return on commercial capital, financial; Relatively significant because of the high costs often involved; Business risks about average, high risk all around, huge amounts of capital required, long time to pay-off, no success models to compare; Possibility of profits nil; It will be a long time before anyone but NASA contractors make any money, if ever; Lack of commercial development; Requires substantial R&D; Uncertainties as to cost and revenues, Breaking new ground; Time horizon outside investment window; Risks totally outside the control of an investor."

All of the venture capital companies wanted very high returns on investments when considering these ventures. Some of their comments included the following:

"Ten to one returns as a minimum; Returns ranging from >30% to >100% IRR; Greater than 30% IRR; Time period of 3-6 years needed; Hundreds of times the return of a normal business."

The companies expressed serious reservations and concerns over other issues including the following:

"International rights/treaties over the use of space; The commercial suppliers to the industry; The minimization of the political part of the commercial equation; Taxation and tariffs; Understanding of distinct market, i.e. exploration vs. science; Interesting legal questions like can I provide phone or pager service in Ghana?; Large capital requirement; Entrenched large company competition; Public financing; Government dependency; What attractions to institutional equity capital are feasible?; Safety, liability exposure; Launch insurance."

As a result of researching this relationship, it is possible to see what the venture capitalist thinks about business in space. The businessman sees what is needed and can then tap into these financial sources. The venture capitalist is concerned with business and management experience, risks, and the usefulness of the product or service. The businessman is concerned that the venture capitalist be open minded. He wants to know that there exists a possibility of obtaining capital, and that quantitative and qualitative evaluation methodology is understood and applicable to the space based business being considered for investment. The businessman wants the venture capitalist to have an interest in commercial space and be supportive of these markets and investments. It will be interesting to note the changes that will have taken place in this relationship when this survey is updated in approximately three years when I revise this work.

New Space Industries (NSI) refers to a combination of private sector space based businesses that have potential commercial value and that are not currently in existence or in their very early stages of development. While many such businesses are discussed, a few seem to have the most potential, especially in the near to intermediate term. A partial list of these businesses includes: a) space manufacturing; b) space resources for both space and earth; c) space business parks; d)satellite and space transfer services; e) travel and entertainment (space tourism); f) R & D services in space; g) space transportation; h)space tethers; i) space infrastructure; j) space utilities; and k) space solar power. Rather than discuss each of these potential space businesses in detail, I would like to identify the needs that these New Space Industries have in common. Next, I will focus on one specific industry that has the potential to be highly successful and lead the others in their development.

Most, if not all, of these industries share key fundamental needs. They all require infrastructure for facilities and operations for their businesses. New and different technology and transportation vehicles are needed as well as financing. Low cost and dependable transportation is also essential. Safety, emergency, traffic control systems, and rescue methods are required. In-space transfer opportunities are necessary, large markets need to be identified and must be attainable. Fuel and propellants need to be available in orbit. Space systems for servicing, repair, construction, maintenance, assembly and in-space transportation need to be created, along with standard interfaces. Policies on governance, zoning, administration and security need to be agreed upon and implemented.

Consequently, of the New Space Industries mentioned, space tourism and entertainment has the best near term potential. Space tourists are like terrestrial tourists, except that they want to go to space to meet their travel needs. They want to experience micro-gravity, see the views, and do many of the things that they would do in the terrestrial environment, plus those that can only be done in the micro-gravity environment. Looking at this potential NSI from the business perspective, one cannot help but wonder if a market exists and what size that market might be.

In 1993 in Japan, and in 1995 in North America, a marketing study was carried out by the National Aerospace Laboratory (NAL) in Japan, under the guidance of Patrick Collins, R. Stockman and M. Maita. This study first examined the Japanese consumer market, then the North American consumer market which is the largest such market in the world. The focus was on the willingness to travel to space for a vacation. These are the most comprehensive studies to date and they tell us why space tourism is a compelling industry that is now being paid attention to by the tourism industry and others.

The study shows that all age groups have a strong desire to go to space with the interest tapering off as people get older. Most important, when the price per person approaches about $10,000 for a five day stay at a space hotel in Low Earth Orbit, including the ride up and back, the market will consist of approximately one million passengers per year. At $10,000 per passenger, this market will generate approximately $10 billion per year![10]

In case there is doubt that such a large space tourism market can exist, think about the already existing terrestrial space market. This is a market that functions daily around the world, especially in the United States, but also in other major countries. This market consists of those who visit space centers, go to launches, space museums, meetings and conferences, and buy space related items from toys to clothing. This market also includes the space related entertainment business, from Hollywood movies to video games and space related television programming. This market is estimated to annually represent well over ten million people spending in excess of one billion dollars per year![11]

There are many challenges and difficulties that must be overcome and resolved for space tourism to flourish. Perhaps the major one is getting the launch cost down to a reasonable level so the $10,000 per person ticket price can be realized. To do so, a new fleet of launch vehicles is needed. In all probability, the launch vehicle that will initially do the job is the Reusable Launch Vehicle (RLV). This may also be a Single Stage to Orbit (SSTO) vehicle or variation of the SSTO model. What are RLV's and why is space tourism so crucial to the RLV program?

Table 2 reflects the probable size of the launch industry for the next decade without space tourism, as estimated by Arianespace. To handle this capacity, Expendable Launch Vehicles (ELV's) can probably do the job and this presents an inherent conflict between the ELV and RLV manufacturers and markets. Note that the total launch industry is not expected to grow substantially from its current level of about $3 billion per year. For perspective, according to Fortune Magazine's 500 Largest U.S. Corporations for the year 1997 as reported in the April 27, 1998 issue, there were 472 U.S. corporations with annual revenues larger than the entire launch industry.[12] Long Island Lighting in Hicksville, NY has revenues approximating the annual revenues of the launch industry. General Motors, the largest company on the list, has annual revenues almost 57 times those of the launch industry.

Table 2
LAUNCH DEMAND OVER NEXT 10 YEARS, IN US$ BILLIONS (10).
Courtesy of Patrick Collins

Geostationary satellites	14-16
LEO	< 6
Space station	8
Earth observation	3
Science & technology	1
Total launch market	< 34

In looking at Table 3, we see a comparison of the two launch vehicles, the ELV and the RLV. In this particular table, the time period is only five years. If 50 launches a year are necessary to handle the capacity estimated in Table 2, 50 ELV's are required to do the job. When we look at RLV's, assuming a quick turn around time and acceptable reliability, one RLV can handle the 50 launches on its own! It is absurd to think that the R&D investment into RLV's will be made for the deployment of only one vehicle. No rational investor, financier, or business person would embark on that course. So for the RLV to become a reality, it needs a market that grows, that is very large, and that is sustainable. The only commercial space industry that has this market potential is space tourism.

Table 3
A COMPARISON OF EXPENDABLE & REUSABLE LAUNCH VEHICLES.
Courtesy of Patrick Collins.

Year	1	2	3	4	5
2 (a) Expendable Launch Vehicles					
Vehicles made	50	50	50	50	50
Launches	50	50	50	50	50
2 (b) Reusable Launch Vehicle: 50 flights/yr.					
Vehicles made	1	0	0	0	0
Flights	50	50	50	50	50
2 (c) Reusable Launch Vehicles: 50 flights/yr.					
Vehicles made	50	50	50	50	50
Vehicles operating	50	100	150	200	250
Flights	2500	5000	7500	10000	12500

The industry is waking up to this fact and to the potential that this huge market represents. Many companies are talking about and designing space hotels for the space tourists. Recently Hilton International disclosed a plan for a hotel on the moon with a beach in an April 19, 1998 article in the London Sunday Times. In addition, over a dozen companies are working on RLV designs, including a team from Stanford University. To help spark the development of a successful RLV, the X-Prize Foundation is sponsoring a $10 million competition for the winning RLV design.

The Adventure Travel industry already knows this. Three adventure travel companies are now selling tickets for sub-orbital flights, ranging from $3500 per person to approximately $100,000 per person. None of the three companies have a vehicle to use, but all think that their vehicle will be available within five years. Of the three companies, only The Civilian Astronaut Corps (CAC) plans on using a traditional rocket, The Mayflower II, which is being designed by Advent Launch Company. Space Adventures, a specialized adventure travel company, plans on using multiple launchers, possibly the Pioneer Rocketplane and the Kelly Space Transportation vehicle. Zegrahm Space Voyages, also part of a large and experienced adventure travel company, has worked on developing its own craft which is similar to the X-15 and is called the Sky Lifter by Vela Technology Development, Inc. The CAC program is the least expensive while the other two companies are selling complete space tourism training packages with the ride for an estimated price of $100,000.

For the present, space tourism is in the domain of adventure travel. This will change however, as the size and power of this market comes to fruition. As the future RLV's make it both affordable and common place to go to space for a few days or weeks, the space tourism market will be identified less with adventure travel and more with the routine, high end tourism market. As the price for getting into space is reduced and a fleet of passenger carrying vehicles is realized, the commercialization of space will be well underway.

It is no longer in doubt that we will soon find ourselves with real opportunities for living, working, and playing in space. The groundwork is already being established. As public sector funding for space exploration and activities takes on new characteristics and becomes more scarce, the private sector is seen more and more as the savior for space activities. With the increasing involvement of the private sector, the development of New Space Industries, and passenger space planes, we will be able to extend our civilization out to the moon, Mars, and beyond. With this comes responsibilities. These responsibilities include the concern for the nature and quality of the business we export to space. It is important to note that not everything about business deserves to be exported from this planet, or to be used in seeding the future generations that will be working and operating in space.

These concerns are hardly new, but in the past they were more specifically directed towards social and political conditions instead of business practices and operations. In 1977, Paul L. Csonka, Director of The Institute of Theoretical Sciences at The University of Oregon, wrote about the social and political concerns of colonizing space in a paper titled "Space Colonization - Yes, But Not Now". Later that year, he had a similar article published in the October 1977 issue of *The Futurist* and then his work was referenced in Hearings Before The Committee on Science and Technology of the U.S. House of Representatives, Jan. 24-26, 1978. His concerns stemmed from the G.K. O'Neill models for space colonization that were popular at the time. Now, twenty years later, we have the technology to live, work, and play off this planet. We have a private sector that can make the financial investment, thus doing more than just filling in the gaps from the once dominant public sector. It is relevant to ask what the character of our business environment in space will be and are our concerns warranted? Do we just go into space doing business as usual and what is meant by "business as usual" in 1998?

As a conservative businessman with twenty-five plus years experience, I've seen and experienced changes in our business environment which even I believe are cause for concern and alarm. The greed and the lack of respect for customers, employees, and investors has become all too real and common in the business community of today.

To cite an example of a business that fuels my concern, I would like to use my son's medical insurance company. Since my son's birth 14 years ago and his Cystic Fibrosis diagnosis, my family has lived a nightmare with the medical insurance industry and the company that insures him. Delayed payments, rejected reimbursements, deferred reimbursements, rejected prescriptions, refused authorizations, intimidated health care professionals, and more. I've watched this company and industry evolve during the past 14 years and sometimes I'm certain that these business executives and industry leaders studied a different American history and different business school concepts than what I studied. Even with numerous complaint letters to the company CEO, the difficulties have been so numerous, so frustrating, and at times so painful that I can only wonder if we live on the same planet.

Seeing this company change classifications from non-profit to profit and reviewing the annual reports says it all. Growing executive salaries and benefit packages, statements showing increasing sales and profits, shareholder reports glorifying the success of the business and statements attempting to demonstrate the company's high regard and concern for the medical needs of its policy holders, are all there to make me believe in the course chartered by company management. I am a policy holder, however, and I deal with their policies and impact on a regular basis. I know the truth behind their success, their profits, and how they can continue to increase salaries and benefits as they do. I also know about their public relations efforts to portray the company as being genuinely concerned about its customers, the policy holders. Good, solid financial and business performance is important, but I question it when it is made at the expense of others, especially those in need.

Medical insurance companies are not the only companies in America that pursue greed and put success above the needs of their customers, employees, and investors as in the above example. I'm confident that virtually everyone if asked could provide their own horror story regarding an experience with a specific company or industry. Today's abuses are not just limited to companies. Our politicians, and our political, legal, economic and social systems all show signs of moving towards the dark side of human nature.

When we talk about commercializing space, about making profits and returning those profits to investors here on earth and maybe someday to investors living in space, are we talking about the medical insurance model of corporate concern and behavior? Or are we talking about a different model, one perhaps we have not even thought or talked about yet.

There are different business models. Each of us can also tell stories about our positive experiences with individual businesses and industries. Our selection of model businesses and industries is not a black and white selection. Variations and shades of gray exist and may largely be subjective. Because of this, we will see the choices from many different perspectives.

Why even bring this issue up, especially since there are no answers at this time, only questions? I bring it up because I would honestly hate to think that our future space settlements on the moon and Mars are driven by the self-centered decision makers, and that their decisions will stem from the same source as those associated with the medical insurance company approach to business. I would hope that the foundation we build in these important and formative years of commercializing space does not mirror the greed, absence of morality and ethical behavior that is now so common in our business community and society.

Questions as to how to guard against this need to be asked, studied, and answered. In the coming years, we don't want to find that healthcare on the moon and Mars settlements is simply an extension of today's HMO's here on earth, do we? Are we to form respected international panels to screen the businesses that operate in space for their behavioral history and track record, along with their employees? Are there other methods to guard against exporting our darker side to space? In the recent hit movie Contact, there was an esteemed international panel which was to select the "right" person to travel in the machine to Vega. In the Contact model, a belief in God was the key for traveling to Vega. What will be our key for taking our businesses to space?

One thing is for sure. Our future generations will be working, living, and playing in space, on the moon, Mars, and beyond. The initial space residents and pioneers will be from earth, but as future generations are born in space and expand outward, their own identity will evolve over time. What springs forth from the seeds that we plant is something that I believe we should all be concerned with today. By addressing these concerns now, we can make sure that when we get going in space, we get going in a way that represents the best we business men and women have to offer.

REFERENCE NOTES

1. Goldman, Nathan C., *American Space Law: International and Domestic*, 2nd.edition, p.193, Univelt, Inc., San Diego, CA. 1996.
2. *Ibid.* 194.
3. *Ibid.*, p. 195.
4. Reynolds, Glen H. & Merges, Robert P., *Outer Space: Problems of Law and Policy*, 2nd. edition, p.280, Westview Press, A division of Harper/Collins Publishers, Boulder, CO., 1997.
5. Kay, W. D., *Can Democracies Fly in Space?*, p.172, Praeger Publishers, Westport, CT., 1995.
6. Goldman, Nathan C., *American Space Law: International and Domestic*, 2nd edition, p. 209, Univelt, Inc., San Diego, CA 1996.
7. Kay, W. D., *Can Democracies Fly in Space?*, p. 172, Praeger Publishers, Westport, CT., 1995.
8. *Ibid.*, p. 172.
9. Matula, Thomas, *The Potential Role Of A Space Development Bank In Accelerating The Commercial Development of Space*, The Space Studies Institute, Inc., 1997.
10. Collins, Patrick, Stockmans, R., Maita, M., Demand for Space Tourism in America and Japan, and its Implications for Future Space Activities, a market research project sponsored by the National Aerospace Laboratory, Japan, 1995.
11. Rogers, Tom, Director, Space Transportation Association, quote from remarks made in Albuquerque, NM, April 1998.
12. *Fortune Magazine*, "The Fortune 500 For 1998," p.211, April 27,. 1998, published by Time-Life Publishers.

MAR 98-035

FUNDING THE FIRST HUMAN EXPEDITION TO MARS

George Osorio[*]

INTRODUCTION

Much has been written about the technology, the science, and the socioeconomic factors involved in safely sending humans to the planet Mars and returning them to Earth. The literature is replete with mission scenarios, propulsion concepts, and Mars colonies. One need only look at the various papers presented at the different "Case for Mars" conferences to realize just how much coverage the subject of sending humans to Mars has received. For more than thirty years, even before humans first walked on the Moon, missions to Mars have been proposed but never enacted. Various reasons have been offered to account for this vacillation on the part of humanity to delay the first human expedition to the Red Planet, which I will refer to as Mars Human Mission One, or MHM1. Chief among them is the seemingly exorbitant cost of MHM1. With estimates ranging from as low as $30 billion dollars to as much as $450 billion and more for a journey that is expected to last at least two or more Earth years, it is easy to understand why, in today's economic climate, MHM1 has yet to materialize. Another reason given for the delay is that there is not enough public resolve; that, somehow, the general public needs to "jump start" the MHM1 effort. From all this two things have become clear: 1.) Although we already possess sufficient technological know-how to physically propel and return a human (or several humans) to Mars and back; there is a sense among the decision makers (those upon whom we have come to depend for initiating a MHM1 effort) that the effort is too risky and costly; and 2.) Despite all the innovative proposals for sending humans to Mars, no one, including these same decision makers, has seriously taken up the challenge of following through with any one plan, or even to simply agree on one mission scenario as the best for initiating MHM1.

I believe we've beaten to death the discussions on technology, missions, and scientific return. What has eluded us is how to pay for that first mission so that we might move forward and execute. When it comes to paying for MHM1, all missions proposed to-date are contained within the "government funding" paradigm, including those which considered international cooperation. That is to say, implicit in their cost estimates was the underlying assumption that the entire effort would be directed by political leaders using government funds (i.e., through NASA). Recall president Kennedy's call to put a man on the Moon, Reagan's commitment to build the Space Station, and Bush's Space Exploration Initiative. Kennedy's mandate was executed rapidly, but at a high price. Reagan's legacy, the Space Station, is being executed in a painfully inefficient and costly manner, while Bush's SEI has yet to materialize in any form. As a result, given these past government funded space exploration efforts, and given the volatile nature of politics, the estimated cost for even the cheapest MHM1 effort has been, until now, considered out of reach. Today, however, a paradigm shift has occurred in our thinking. There is now a widely held belief that using capital obtained from private sources, MHM1 can

---

[*] Aerospace Engineer, Irvine, California.

be executed much more efficiently than if done relying solely on politicians, using government funds.

In this paper I propose to show that a privately funded, properly planned MHM1 can be accomplished quickly such that some representative of the human race will set foot on the surface of the Red Planet by the year 2010. Furthermore, I maintain that, based on the history of government funded space projects to-date, no single government or international effort will accomplish this job; only by bringing together the global human spirit, while taking advantage of market forces, will MHM1 ever be accomplished. I also feel strongly that the cheapest MHM1 effort, going directly to Mars without first developing the admittedly expensive infrastructure needed to establish a continuous route, is the best approach. This is the Zubrin model detailed in his book, *The Case for Mars*. I will say more about this in the next section, but, for a justification of using this approach over others, I think Zubrin makes very good arguments and so I refer the reader to Zubrin's book. Two areas in which I disagree with Zubrin, and on which I will expand in the next section, are the objectives of MHM1 and its crew composition.

Finally, on a more philosophical note, it may be argued that human progress is dependent on the maturity of the human race such that, in order for human colonization of another planet to take place, it is necessary for humans to have achieved a single global culture that permits utilizing the entire world for this effort. If such a requirement exists, then we are, indeed, far from being capable of sending humans to Mars. However, I believe that what I propose in this paper, if properly executed, can have the effect of galvanizing all humans on Earth to, effectively, create that global culture. In any case, I'm convinced that going directly to Mars, and initiating the effort now, is the only way to see this happen in our lifetime. I hope this paper will more than stimulate those gathered at this Founding Convention of the Mars Society so that those who hear what I have to say, and who agree with my views, will assist me in starting the process. By starting the MHM1 effort now we can, within a decade, witness the first expedition to the Red Planet knowing fully well that it was through our dogged determination that it happened when it did. Let's get on with it!

THE FIRST HUMAN EXPEDITION

The debate over how and when to send humans to Mars has raged since Neil Armstrong first set foot on the Moon three decades ago. Since then, scientists, engineers and analysts have all conceptualized countless missions, technologies, cultures and scenarios associated with sending humans to Mars. However, during these same three decades, no one mission or scenario has emerged as the choice for a first human expedition to the Red Planet. Although a thorough evaluation of the different mission proposals is beyond the scope of this paper, in developing a cost figure for MHM1, I have assumed that MHM1 will consist of a straight shot to Mars, similar to the Apollo project, and following very closely along the lines of Zubrin's "Mars Direct" concept. Why this approach? Because, as Zubrin points out, this is the lowest cost effort and one that will put a human on the surface of Mars in the shortest time. It also bypasses, initially, the need to develop the expensive infrastructure needed to support other, more complicated concepts. Zubrin's estimate of the cost for Mars Direct was in the neighborhood of $50 billion dollars. For the purpose of developing my arguments in this paper, and to support certain assumptions that deviate from Mars Direct, I have placed the cost of MHM1 at $100 billion. Using this figure, a very conservative estimate for a Mars Direct-type mission, there can be no doubt as to whether MHM1 can be funded from entirely private sources, assuming we start executing now and unless the analysis proves otherwise.

Let's now consider in a little more detail what MHM1 might be and how it differs from Mars Direct. All space exploration concepts that I have read about have had as their primary mission to increase scientific knowledge. With the exception of satellites and the now popular but still nascent "space tourism" industries, space exploration projects have traditionally been executed by scientists for the sake of science. I propose that MHM1 have as its primary objective to simply "prove" that we can safely send humans to Mars and return them to Earth. There is no other objective that needs to be fulfilled by this first mission. Any science we derive above and beyond this objective will be "icing on the cake". We can always send instruments and robot probes to gather data and return samples as we've been doing since the first Viking lander. What we would be looking for from a first human expedition is the human perspective and nothing else. There is no need to evaluate the scientific merits of the expedition, or the technological breakthroughs promised by the effort. However, given that even Mars Direct would include a protracted stay on the surface of Mars, it would be prudent to assign the first crew the task of "proving" certain human capabilities while on Mars. From testing out the atmosphere to constructing raw materials out of the Martian surface, to looking for water, the first humans can not only keep themselves quite busy, thereby reducing possible strains on crew cohesiveness, but could potentially greatly facilitate a trip by a second crew to the Red Planet. So, without seeking information on geologic formations or Martian paleontology, we can have a successful and productive mission, filled with information worthy of scientific analysis back here on Earth.

Having established that we are going to Mars "because it's there for humans to explore", let's consider the crew composition. That is, how many humans are we sending there and who are they? Zubrin established a crew of four for Mars Direct. Since I've already determined that MHM1 is to only prove our ability to send humans to Mars and back, and nothing else, I suggest that the first crew consist of five individuals. From psycho-social considerations, an odd-numbered crew is preferable to an even-numbered crew. Zubrin had established early in his book that the minimum number of individuals for this first crew might be three, and I agree that three constitutes a minimum.. However, considering the possibility of the loss of one life, we may find ourselves with a crew of two, or even one, sometime during the crew's long stay on Mars. Therefore, I believe that, if our primary objective is to send humans to Mars and return them, then we need to establish redundancy in the human cargo, and assure the return of our space vehicle, by sending a minimum of five individuals.

Next, I would like to consider who these individuals must be that make up our first crew. Again, citing Zubrin's concept, Mars Direct called for four individuals. As Zubrin put it, two "Scottys" (referring to the television series Star Trek characters) and two "Spocks". Or, in the vernacular, two "mechanics" and two "field scientists." However, Zubrin based this on the assumption that the mission of MHM1 is to gather scientific data. In addition, Zubrin defined a mission failure in terms of the failure of one of the technical systems (as in "the spaceship"), rather than considering the failure of individuals to gather the data, or simply considering the death of one or several of the crew members. Assuming that MHM1 is simply to prove that we can send humans to Mars and back, and having established a minimum redundancy in the human cargo, I suggest that the individuals whom we send on this first voyage should all be "mechanics". No scientists are needed. Why? Because, if we are trying to insure that we get humans to Mars and back safely, the only way we can do that is if all the crew members understand the technical systems equally well. Anyone who is not versed in the technical systems used to deliver them to Mars and back, stands a good chance of being doomed should he/she be the only one left alive. Furthermore, having several "field scientists" with no hope of fixing their return launch vehicle is equivalent to not having them there at all, resulting in mission failure. On the other hand, if repairs are needed to successfully effect a launch for the return trip, having several "mechanics" will accomplish the job much more efficiently. The only caveat I would add to this is that, during crew selection, crew members be selected who have broad backgrounds founded in science and the humanities, so that Earth-bound observers can obtain a more "complete" human perspective of Mars. It will also help when we consider obtaining funds for this mission, as we will see in the next section.

To summarize, MHM1 will consist of a crew of five "mechanics", with broad backgrounds in science and the humanities, who will go to Mars with the single purpose of "proving" that humanity is ready, willing, and able to go to Mars and back. Any scientific knowledge gained from this first human mission (and I assure you there will be much despite the lack of "scientists" on board) will be "icing on the cake." As for the cost of this mission, my assumption of $100 billion remains a conservative estimate, even after accounting for the additional crew member, when compared with Mars Direct. For comparison, Table 1 shows the cost associated with past space projects.

Table 1
COST OF SELECTED SPACE PROJECTS

PROJECT	COST (REAL YEAR $)
Gemini	1.28 billion
Apollo	24 billion
Shuttle	30 billion and counting (est.)
Space Station	30 billion and counting
MHM1	100 billion (est.)

Source: Historical data obtained from NASA web page (http://www.hq.nasa.gov/

FUNDING OPTIONS

Now that we've established a proposed mission for MHM1and a conservative estimate of the cost, the next question to be answered is, how will we pay for this effort? As discussed earlier to some extent, the answer, after three decades of debate, appears to be pointing towards private funding. This is evident from the many papers on privately funded concepts being presented at this Founding Convention of the Mars Society. What sort of private funding can we obtain to pay for MHM1? I propose that, despite the seemingly large amount needed to pay for MHM1, $100 billion dollars, we can do it, starting now, through a well planned and well coordinated business plan and marketing effort that takes into account each phase of the mission from systems design and fabrication to crew selection and training, to mission execution, and, even, to post-mission activities. We can pay for MHM1 from the same private funding that is generated daily across the globe in support of charitable causes, the same funding that supports our entertainment industries, and the same funding that supports our capitalist socioeconomic complex.

There are several sources of money that pour daily into the economies of the world and which can easily be tapped to pay for MHM1. These include individuals, corporations and governments. In addition, there are several methods by which the Mars Society can go about obtaining these funds. Several of these methods are presented here and, in some cases, are discussed separately in presentations made by other individuals at this Founding Convention. However, I have not had an opportunity to share any of my ideas with the other presenters.

Therefore, where any of the material contained in this paper also appears in separate papers by other authors, it is purely coincidental. The fact that many Mars activists are starting to sing similar tunes on how we pay for a MHM1 mission portends the paradigm shift taking place as we move away from soliciting government money and, instead, realize the greater potential inherent in using more efficient, and effective, capitalist methods for getting the job done. What follows is a list of different methods of obtaining funds to pay for MHM1 and brief discussions of how the Mars Society might apply these methods to the MHM1 effort.

CONTRIBUTIONS - INDIVIDUAL, CORPORATE, GOVERNMENT

If we examine how much disposable income is in the hands of every individual in this world (recall that, for MHM1, we want to get the entire world involved, not just America - more on this later), we might get an idea of whether we can generate the necessary capital to execute a MHM1. According to the *World Development Report*, published by the World Bank, there were the equivalent of an estimated $4 trillion dollars in savings worldwide in 1988. Let's assume that, ten years later, in 1998, taking into account population and inflationary trends, the amount of money held in savings worldwide is only slightly larger than that. Estimating how much of this amount can be considered discretionary disposable money is a difficult task. However, if we look at the relationship between what is needed for MHM1 and this amount, we see that MHM1 can be paid for by inducing the world to contribute just 2.5% of its savings. How many of us would be willing to take 2.5% out of our savings to contribute to MHM1?

Now let's see how we might go about inducing the world to contribute this sum. Looking first at individuals around the world, we would be interested in people in countries who have the highest level of disposable income; this includes most developed nations and some developing countries. Table 2 shows the countries with the highest per capita income as reported by the World Bank for the year 1988 (I have used this report only for illustrative purposes; more up-to-date information can be obtained from other sources and the same process applied). As shown in the table, per capita income for the top ten countries in 1988 exceeded $15,000.00 dollars. In addition, the same report showed that global savings amounted to a total of more than $4 trillion dollars. These figures are important because they provide a basis for estimating the amount of individual disposable income available based on per capita income, population, and rate of savings. In addition, one can look at how much money is spent on non-essential things like entertainment to discern total discretionary spending and then develop a per capita discretionary disposable income figure. Establishing these figures at this time is beyond the scope of this paper, but is a valuable exercise when developing a marketing plan for MHM1.

Table 2
PER CAPITA INCOME OF SELECTED COUNTRIES

COUNTRY	Per Capita Income	COUNTRY	Per Capita Income
Switzerland	$29,880	United States	$20,900
Japan	$23,800	Denmark	$20,450
Norway	$22,290	Germany	$20,440
Finland	$22,120	Canada	$19,030
Sweden	$21,570	France	$17,820

Source: *World Development Report*, 1991, World Bank

If we now consider the wealthiest individuals in the world, who might represent our first target group for soliciting contributions, we notice that there is a great deal of wealth ($700+ billion dollars) held by the one hundred (100) wealthiest people, as reported in *Forbes* magazine. Just one seventh of their net worth would pay for MHM1. As shown in Table 3, even just the ten wealthiest individuals in the world possess enough wealth to support MHM1. Naturally, this would require that they contribute a considerable percentage of their net worth. So, it would seem that a successful campaign to obtain contributions from the wealthiest people may generate a considerable portion of the amount needed, at least for the initial phase, for MHM1.

Table 3
WEALTHIEST INDIVIDUALS AS REPORTED IN *FORBES* MAGAZINE

NAME	COUNTRY	NET WORTH	BUSINESS
William H. Gates III	USA	$51 billion	Microsoft Corp
The Walton Family	USA	$48 billion	Wal-Mart Stores
Sultan Hassanal Bolkiah	Brunei	$36 billion	oil, gas
Warren E. Buffett	USA	$33 billion	Berkshire Hathaway
King Faud Bin Abdul Aziz Alsaud	Saudi Arabia	$25 billion	oil, real estate
Paul G. Allen	USA	$21 billion	Microsoft Corp
Sheikh Zayed Bin Sultan Al Nahyan	UAE	$15 billion	oil, investments
Sheik Jaber Al-ahmed Al-jaber Al-sabah	Kuwait	$15 billion	oil, investments
Kenneth Thomson	Canada	$14.4 billion	Thomson Corp
Jay and Robert Pritzker	USA	$13.5 billion	Finance

Source: *Forbes* Magazine on the internet (http://www.forbes.com/)

Another source of money in the form of contributions can come from corporations around the world. Corporations tend to be worth more than individual human beings alone, but, in some cases, may not have the flexibility to contribute as much as a single, wealthy individual. However, many corporations do contribute a small percentage of their profits to worthy causes or charities. Estimating the net profits of all corporations and looking at contributions made to charitable organizations, and assigning a percentage that might be contributed to MHM1 is an exercise that is beyond the scope of the present discussion. However, as with discretionary disposable income, this type of calculation is useful in developing a plan for obtaining such funds. In addition, as explained in a later section, corporations may be induced to contribute as sponsors of MHM1 in some capacity. For reference, Table 4 shows net profits of selected American corporations.

Finally, there are governments. Considered as individual entities, governments, like corporations, hold greater wealth than can be obtained from any one human being alone. The danger in obtaining funds from governments lies in the possibility of falling into the trap of allowing the same governments partial or total control over the MHM1 mission. Therefore, a carefully drawn plan for soliciting funds from government entities must include careful scrutiny of the laws and regulations affecting the disbursement of the funds and any strings that may be attached. As with corporations, governments can contribute as sponsors by providing, for example, launch services or facilities.

Table 4
NET PROFIT FIGURES FOR SELECTED AMERICAN CORPORATIONS, FOR FY 97

COMPANY	NET PROFITS FOR FY 1997
General Electric	$8.2 billion
Exxon	$8.4 billion
General Motors	$6.7 billion
IBM	$6.1 billion
Ford Motor Co.	$5.9 billion
Travelers Group	$4.6 billion
Citicorp	$3.6 billion

Source: *Forbes* Magazine on the internet (http://www.forbes.com/)

In summary, one method of obtaining private funds is to solicit contributions from individuals, corporations and governments around the world. In the United States alone, in 1996, over $120 billion was given away to charitable organizations. Part of this figure represents some of the discretionary income we are concerned with as we try to induce individuals to contribute to MHM1. If Americans have more than $100 billion that they can "give away" in one year, there is more than enough money available to fully support MHM1 and a well orchestrated and successful effort at soliciting money from all these investment sources might be sufficient to pay for the entire mission. However, realizing that it's difficult to make everyone part with his or her money, we must consider other methods to "skin the cat".

MEDIA BLITZ

Most of us are very familiar with the manner in which advertising and the media induce us to direct our disposable income towards certain specific products, entertainment, and services. A well orchestrated "media blitz" campaign used to generate the funds needed to support MHM1 can generate a tremendous response from individuals, corporations, and governments all around the globe. This is evident from the response to such profitable efforts as sending human DNA samples to Saturn, spreading ashes of the deceased in orbit, and even selling extraterrestrial real estate.

One part of any successful media blitz campaign includes advertising. As everyone in private business knows, this can be a very expensive method of obtaining customers. Using the movie industry as an example, we typically get news stories about an upcoming motion picture a few months before the actual release of the movie. In the case of summer hits, some movies are advertised as much as a year before their release. This can be done for MHM1 to begin developing a psychological mind-set in the general public about the importance of this mission, and to support any merchandising effort, as discussed in the next section. By continuing to hammer at the public with ads, the support for the project will grow during the two or three years leading to launch date. For example, during the crew selection phase of the project, the right advertising can have an electrifying effect on the general public as they follow (and are intimately involved with) the crew selection process and the different crews.

Another aspect of a well executed media blitz is use of television, motion picture, books and the Internet computer media. While this aspect has fuzzy overlaps with the advertising discussed in the previous paragraph, here I point to coordinating with these industries to support the effort by including some aspect of the project in their programs. For example, web sites can be created for the project (this is a given), a TV show can include the project in its setting, and even motion pictures can have some part of the project (or the entire project) addressed in their content. Obviously, this type of publicity needs to be planned and coordinated at the earliest stages of MHM1 planning to derive its greatest value.

Finally, there are the other types of publicity accomplished through such media as merchandising, sponsorships, and special events. These are discussed separately below. Although a media blitz may cost a substantial amount of money without the possibility of a measurable return, it has a value that is difficult to quantify. Nevertheless, this value needs to be addressed somehow during the early planning stages of MHM1 to determine if the cost of a media blitz is justified.

MERCHANDISING

I'm giving this topic a separate heading because, unlike pure advertising, merchandising has a measurable return on the money invested in creating the merchandise. Most space activists are familiar with selling T-shirts or mugs to promote a cause. This, on a very basic level, is the idea behind merchandising. We invest some capital in specialized products which are then sold at a profit. Again, there is some fuzzy overlap with other methods of raising funds discussed below, but, as an example, let's discuss the frenzy that takes place during some special sporting event such as the World Series or Super Bowl. For some limited time during these events, every item one can think of is sold at highly inflated prices (thereby greatly increasing profits) and, in most cases, they are sold quickly. Since these events are of short duration, the peak selling period is usually very short; on the order of days. Because of this, any miscalculation on the type of product being sold can quickly turn what appears to be an obvious profitable concept into a nightmarish loss.

A good example of the profit potential that can be derived from merchandising a space project is the phenomenal success of *Sojourner* replica toys sold during and after the Pathfinder mission to Mars. This is a good omen for MHM1. Since the Pathfinder mission was carried out under the auspices of a government agency, NASA did not realized much in the way of royalties from the sales of these toys. However, an MHM1 mission that is fully funded by private investors will need to develop copyrights, patents and other protection that will impact profitability. These need to be closely monitored during any discussions with the media (as outlined in the previous section) as well as during any part of a merchandising campaign.

A good start for a MHM1 merchandising campaign might be to simply sell the rights to (or collect royalties from) toy manufacturers for toy versions of the different systems used to execute MHM1. As with the media blitz, a merchandising campaign needs to be carefully integrated into the entire MHM1 project to derive maximum profits. I believe merchandising should play a role in generating funds for MHM1, and be included in its Business Plan, if only to help pay off the initial investment.

OLYMPIC GAMES/WORLD CUP SOCCER

We are all familiar with the quadrennial events known as the Olympic Games and the World Cup (soccer). These events, symbolizing the human spirit of cooperation over competition, are carried out by several nations and paid for by different entities within each country. Since patriotic pride plays a large role in these games, each country (through government action and public support) pays for its group of athletes to participate. Looking at the American method of supporting our athletes in these events (other countries are similar, though some have more government backing), we see that a great deal of support is in the form of sponsorship. From individual contributions to help individual athletes, to corporate sponsors of specific athletes or teams, to corporate sponsorship for the entire American contingent of athletes, enough money and other support are obtained to pay for American participation in these events.

MHM1 is an event that lends itself well to obtaining sponsors in different forms. It is a unique event in human history taking place over a period of approximately four years from launch. As such, it shares many common traits with the Olympic Games or the World Cup. A suggested strategy here is to obtain sponsors for different phases or aspects of MHM1, thereby increasing sponsorship participation. For example, sponsors (individual, corporate and government) can support specific hardware, a specific portion of the project, individuals working on the project, or the entire project itself. As with other methods discussed above, there is some fuzzy overlap between this method of obtaining funds and the solicitation for contributions that was discussed above. The major difference here is that, in some cases, the sponsor solicitation might involve goods or services, rather than money.

For an example of a space project that relies on global sponsorship, let's look at the Space Station program. The space station is "sponsored" by several countries, each with its own share of responsibility for the project. Each country contributes according to the maturity of its aerospace industrial base. Unfortunately, because the sponsoring entities are governments, rather than private investors, this project is experiencing the traditional cost overruns and schedule slips associated with previous government projects.

For MHM1, I envision the international community doing the same and contributing as sponsors wherever and however they can. However, to maximize investment through sponsor participation, we need to consider creative strategies that permit full participation by all. As an example, at one time it was suggested to NASA that they seek additional funding for the Space Station by selling advertising space on the outside of their modules to corporate businesses such as soft drink manufacturers. The idea being that, for example, the Coca Cola company might pay quite a bit of money to have one of the space station modules painted in the Coca Cola logo to look very much like an oversized can of the popular soft drink hanging from the space station. Unfortunately, to the bureaucrats running the program this did not seem like a good idea and the space station is still, today, experiencing costly overruns and schedule delays.

To summarize, MHM1 can obtain funding from sponsors in the form of individuals, corporations and governments. As with solicitations for contributions, the potential for obtaining enough funds to cover the entire mission exists. A carefully planned sponsorship program will permit participation by everyone across the globe, thereby maximizing the amount of funds generated.

LOTTERIES/RAFFLES/AUCTIONS

The last source of funding that I will consider here is holding lotteries, raffles, or auctions. There are other sources that I will not discuss at this time, though they merit discussion. Among these are contributions from foundations and other charitable institutions. The reader is left to determine how these sources might contribute to an overall funding program to support MHM1. Lotteries, raffles and auctions share similar characteristics in that a single prize is bid upon by several individuals who spend, or invest, an amount much less that the value of the prize to be won. In the case of lotteries, a ticket is purchased containing a possibly winning combination of numbers such that the dollar amount to be won is many times greater than the purchase price of the ticket. For raffles, there is usually an item to be won, rather than a dollar amount, such that the value of the item is much greater than the purchase price of the ticket. Similarly, in an auction, different individuals bid a certain amount of money for an item with the expectation that what they pay for the item is less than that at which the item is valued.

Looking at the major lotteries across the country, one quickly gets the impression that discretionary disposable income in the US is quite high. Even lotteries across the globe seem to generate large amounts of money as individuals seem to pour what they can into the highly improbable dream of obtaining a winning ticket. In California, the state LOTTO game generates enough money to award more than $4 million dollars to the winner each week. As with lotteries and other games of chance everywhere, as the winning amount increases (due to, in the case of the LOTTO, no winning numbers having been drawn), there is more participation, driving the winning amount even higher. Can such methods of obtaining funds work for MHM1?

I believe a carefully planned and executed MHM1 effort should consider establishing lotteries, or raffles or auctions to generate funds. However, given the complexity of this project, and the overhead associated with coordinating and maintaining any one of these fund raising activities, the return to the MHM1 project may wind up being negligible. There is also the danger in these type of activities of raising unrealistic expectations such that, if no winners are identified (because no one selected the winning combination, for example), participation will drop off to the point of making the activity a money losing proposition.

A NEW CHALLENGE

Up to this point, we have discussed the first expedition to Mars in context of the mission and how much it might cost. We also examined possible sources of private funds that could be used to pay for this mission. And, we've even considered methods used for obtaining these private funds. The question before us now is, who will act to make this happen? The answer, I believe, is here, at this Founding Convention of the Mars Society. We, the members of the Mars Society, are the only ones who believe in this mission enough to get the process started.

Today, I propose that we form a planning committee immediately to lay down the groundwork for the MHM1 effort. This committee will map out a schedule to launch, and will outline a business plan that we can then sell to potential investors. This committee will also be responsible for developing marketing strategies that will lead to making MHM1 a success. I have set a personal goal to develop a schedule that can put a human on Mars by the year 2010. I have considered some of the necessary steps that are involved in making this launch date, and I have even looked at the sequence of events required to make everything come together at the appropriate time. However, a thorough discussion of these concepts is beyond the scope of this paper.

One individual alone cannot work this complex project and make it come to fruition. Indeed, one nation, alone, cannot carry it out. This project, by its very nature, requires the backing and participation of the entire world. But we cannot wait for the world to clamor for MHM1. Instead, we must come together at this Founding Convention to begin the process and bring our ideas to the world. I look forward to working with Mars Society members to bring the greatest adventure in human history to life.

SUMMARY

Although I am not a well known activist in the space exploration or scientific communities, I have been following the space program with a great deal of interest from the early days of Apollo. Like many Mars exploration advocates, I, too, believe that we humans should expend a good proportion of our collective energy in proving that we now are capable of sending humans to another planet and back. I realize that The Mars Society has developed a cautious approach, incrementally proceeding from small, hitchhiker payloads on government funded missions to stand-alone private robot missions and, eventually, human exploration. Unfortunately, for me as well as for many other Mars exploration activists, this approach is unnecessarily slow and may actually delay human exploration of Mars even more. My reason for saying this is that the complexity involved in executing a well planned mission, as well as the public, governmental and economic factors influencing this type of activity could, at any time and under the right circumstances, derail the best effort to send humans to Mars.

In this paper I have presented what I consider clear evidence that a MHM1 mission can be accomplished starting today, using privately obtained funding. The seemingly exorbitant cost of MHM1 is no longer an obstacle to carrying out this mission; we only need the will of a "critical mass" of individuals to begin the process. I believe that if we begin now, we can see a human landing on the surface of Mars by the year 2010. Therefore, the time to begin the human exploration of Mars is now.

PROMOTING PRIVATELY FUNDED SETTLEMENT OF MARS

Alan B. Wasser

Why is there no really significant effort being made to develop affordable human transport to Mars?

Everyone knows "the answer" to that question: because the government and the taxpayers are unwilling to pay for it without an incentive like a space race with the communists.

But is taxpayers' money really the only way the habitat of humanity can ever be expanded beyond the Earth?

Anyone who has heard Robert Zubrin, or read his book, knows that private enterprise could raise the kind of money needed, and build Mars ships cheaper, better, and faster than any government — *if* there were a sufficient profit in it. It is private enterprise, not government, that has quietly raised the multi billion dollar cost of filling the sky with competing constellations of communications satellites. Several small firms have recently raised large amounts of capital to begin developing privately funded satellite launch vehicles, some reusable.

Unfortunately, private enterprise invests big money in risky ventures only if there is a prospect of a really big profit, and there is currently no product we could bring back that could possibly produce enough profit to justify the cost of developing the ability to send people to Mars in the first place.

But there is a way we could "create" such a "product".

Throughout history, the value of newly claimed land has often been the justification for the cost of human expansion, and settlers' rights have been the basis for making such land claims. Those same rights could yet be the economic basis for humanity's settlement of Mars. What it would take is the passage of a rather simple law that is currently being debated in key Congressional offices and NASA headquarters. Officially it is called an "Act for the Promotion of Privately Funded Space Settlement" but it is more commonly known as the land grant law.

The biggest hurdle for such a law to overcome is the 1967 "Outer Space Treaty" which prohibits national appropriation or claims of national sovereignty on Mars or other celestial bodies. That treaty was designed to, as an only recently declassified State Department document put it, "defuse the space race" so that money could be diverted to the cost of the escalating Vietnam war. U.S. government funding for space went up every year until that treaty was ratified, but has gone down every year since then.

Fortunately the U.S. and most spacefaring nations refused to ratify a subsequent treaty, usually called the "Moon Treaty", which would have gone on to ban private property. Therefore, while nations cannot claim land on Mars, private entities can, if they base the claim on something other than national sovereignty, such as settlers' rights.

While the U.S. cannot grant land ownership on Mars, it could choose to recognize a claim, made by those who have actually established a privately funded Martian settlement, of private land ownership around their base.

The proposed law would disavow any claim of U.S. sovereignty but direct all U.S. courts and agencies to immediately grant full legal recognition to a land claim of up to a specified size made by any private entity which has established a genuine permanent human space settlement that meets the specified conditions. Of course, to maintain a permanently inhabited settlement would require at least one ship going back and forth between the Earth and the settlement. The most important condition in the law would be that the settlement, and passage on that ship, must be open to any peaceful person who is willing to pay for it.

With U.S. recognition of their land ownership, the investors who paid to establish a settlement on Mars (most likely to be a consortium of multi-national companies) could start recovering their investment by selling sections of their land, back on Earth, just as soon as the settlement was established. If the land grant is made large enough, that could represent a very big incentive, even if the value of each acre of Martian land was not great.

The proposed law calls for recognition of a Martian settlement's claim of up to 3,600,000 square miles, approximately the size of the United States, about 6.5% of Mars's surface. At even a very conservative ten dollars per acre, that would be worth 23 billion dollars. If that proves insufficient to promote the development of a privately funded settlement, there is plenty of room to enlarge the grants later.

The same law would allow recognition of a Lunar settlement's claim up to 600,000 square miles, approximately the size of Alaska, about 4% of the Moon's surface. At ten dollars per acre, that would be worth 4 billion dollars.

Of course, once a true settlement is established, with regular transport open to any paying passenger, Martian land will be worth much more than it would be now, when there is no way to reach it.

A privately funded settlement could produce income from such things as selling transport and services to scientists, explorers and tourists and exporting raw materials and manufactured products. None of those could justify the cost of developing affordable human access to Mars in the first place, but once that is done to win the prize of the land grant, they will pay the settlement's operating costs and eventually make a profit. A dozen teams are competing for the ten million dollar X Prize. How many will try for a prize worth at least twenty three billion dollars?

The recent reports from the Clementine and Lunar Prospector missions finally put to rest one of the most common arguments against land grants; that there is no such thing as "valuable property" on celestial bodies. Think of private ownership, officially recognized by the U.S. government, of a claim the size of Alaska, centered on the Lunar south pole's crater of permanently frozen water and the mountain on its shore with the almost permanently sunlit top, (which Ben Bova, in his wonderful book *Moonrise* was kind enough to call "Mt. Wasser"). That would be worth a fortune even now, with no way to get there. How many times more than that would such a claim be worth, once there really is a permanent settlement on the mountain.

The consortium that wins such a grant can immediately start producing significant income by selling off parcels of a few acres each, down in the crater with water mining rights, or on the mountain top near the tower that gathers full time solar power. If the buyers are the kind who want to visit or use their land, they become paying passengers on the consortium's space line. If whatever they do with it produces freight, either direction, or brings in customers or tourists, even better.

The law allows a space line to set appropriate standards of behavior and safety for passengers and cargo and the use of its facilities, but not to act in an anti-competitive manner. It may not unreasonably deny landing rights, and the right to transport passengers and cargo, to any other safe and peaceful vehicle willing to pay a reasonable fee for landing rights.

Of course, most of the early buyers of land on Mars or the Moon will be speculators and investors just looking to make a profit re-selling the land when the price rises, sooner or later. That's good for the consortium. The primary sales bring in money quickly, and the resale market increases the value of the land the consortium still owns. Land is one thing people buy, hold and sell even when there is no current way for them to "use it" because they can make a tremendous profit buying and holding it either until a use arises, or a "greater fool" is willing to pay even more for it.

Clearly, an internationally recognized private property regime is urgently needed as soon as possible, but it will be much easier if the U.S. initiates and administers the process until an international body is formed to do it, rather than trying to get a new international agreement first.

The legislation urges other countries to adopt similar laws and guarantees U.S. recognition of claims by citizens of all countries which agree to reciprocity. It instructs the State Department to try to negotiate new treaties making the same rules international law. It automatically defers to any such international agreements as soon as they are ratified by the U.S. It pledges to defend extra-terrestrial properties by imposing sanctions against aggressors. If need be to secure international agreement, the State Department is authorized to agree to treaties which require that all claimants must be consortia of companies or citizens from several different countries. It could even be required that at least one of the partners in each consortium be from a developing country.

Land grants attracted private funding for the building of the trans-continental railroads in the last century, thus minimizing the cost to taxpayers. In that case the grants were given in advance, in return for promises to build the railroads, which led to graft, favoritism and expensive bailouts. On Mars, nothing need be awarded until an actual settlement has been established. That will lead to a competitive race to design and build affordable human transport as soon as possible. Those interested will fear that, if they don't rush to establish a settlement soon, someone else (perhaps from another country) will get there first, cutting them out.

When I first began promoting the idea of land grants a decade ago, the main problem was convincing skeptics that there could be land ownership in space and that real estate on the Moon and Mars might someday be valuable. Since then, most space activists and even key people in NASA and Congress have begun to accept that once-radical idea. Now, a new problem has arisen: the urge to squander that value for the quickest possible gratification, by awarding it for easy missions like robotic surveys, instead of saving it to pay for true privately funded space settlements.

Several people have proposed claim registries, mining patents and other mini-awards that aren't real ownership but would, in effect, hold claimants' places in line. But why would we want to give someone a land grant for some small step and allow them do nothing more for the next twenty years except stop anyone else who is ready to settle and develop the land? The existence of a permanently inhabited settlement is the economic point of no return for development. Only then is it easier to justify going forward rather than delaying expenditures.

Under most plans, the mini-claims based on robotic surveys would not confer enough rights to make them saleable. They would not bring even enough to repay the cost of the survey. Therefore, they would do the recipient little good, and reinforce the idea that the land is basically worthless. Worse, they would detract from the psychological value of a real claim; the ego-boost that investors could get by being able to look up and say "I own a piece of that" which might tip the scales and get them to risk investing in a settlement effort.

Some people object to the idea of anyone "owning" land beyond the Earth because they want it all to be "the common heritage of mankind". This feeling was much stronger in the days before socialism was proven to achieve only uniform poverty. In space, too, what no one owns, no one cares for or develops. Clearly, mankind as a whole would benefit greatly if private enterprise developed cheap human access to Mars and offered it to any peaceful person willing to pay a fair price for it, regardless of nationality. It is well worth making ownership of a mere six and a half percent of Mars' surface a prize for doing that.

For the Mars Society, supporting the passage of the "Act for the Promotion of Privately Funded Space Settlement" involves little or no cost, and can be done while also pursuing all other options. For the individual members, it may well be their only hope of seeing a Martian settlement established in their lifetimes.

Chapter 8
ROBOTIC EXPLORATION

NASA Ames Robotics Researcher, Mike Sims, explains what can be done soon to Mars scientist Peter Smith. To the left of Sims is visionary scientist Geoff Landis.

JPL Mars Program Chief Engineer Rob Manning lays out the plans for upcoming missions.

MAR 98-037

NAVIGATION AND MOBILITY SYSTEMS ARCHITECTURE FOR PLANETARY ROVERS

Pablo Flores[*]

The design of the mobility system is associated with the hardware and software used by the navigation system. A rover moving on the surface of Mars is directly predicated by the terrain characteristics and environment where the rover could be deployed. These will predicate the navigation systems, hardware and software to be supplied. To obtain better results, it is necessary to get better information about the geometry terrain, taking into account rock distribution, soil compositions and mechanical properties, abrasive qualities of Martian soil, etc.

Actually any models are only approximated and they will be optimized when more data will be forthcoming from current and future Martian missions.

A standard six-wheel rover called Marsokhod is presented in this paper, supporting an innovated chassis design, moving at approximately 0.15 meters per second, adapting to the terrain to be crossed.

It will carry multiple instruments on its three platforms and could be assisted by an orbiter. The finished mission and Marsokhod design is not determined yet. For that reason this work is limited to the use of conventional navigation systems and landing site.

When the surface model is determined, taking into account the objectives assigned to the mission, we can begin a selection of the more reliable and inexpensive development of the structure and mobility system, led by the information captured by the navigation systems. The computer on board will be the interface between both systems. An architecture model of the mechanisms and movement of the Marsokhod on the surface of Mars is given in this paper.

INTRODUCTION

The use of rover vehicles appears to be a key tool for planetary exploration. Certainly, they can carry out dynamic studies searching targets of scientific interest. Providing a cheaper intelligent resource of mobility and navigation compared to a static-standard lander is the way to change near-future planetary missions, making them more profitable from the point of view of the balance between mission cost and information obtained.

A preliminary study of geometrical characteristics of the terrain to be selected as the landing site is the first essential step to begin an optimal rover design for any planetary exploration.

An approach, with a high grade of accuracy, could be made in simulated Martian landscapes where the rover is tested and evaluated [1,2,3].

---

[*] Student of aerospace engineering at the Moscow Aviation and Technology University (MATI) Konstantin Tsiolkovsky. Moscow. Russia (under a scholarship from the Ministry of Culture and Education of Argentina). Affiliations: Member AIAA; Member BIS. E-mail: flores@rosmail.ru. Mailing Address:
Мосвскюя ОблЋст, Щелковский рюйон, Пос. БиокомбинЋт, Д 41, КВ 46. (141142), Russia.

Based on the information provided by the two Vikings and the Mars Pathfinder and its rover, Sojourner, the Martian surface shows irregular sand dunes with highly dense rock distribution per square meter. On the other hand, the rocks show sharp angles and erosion from the wind, dust particles and the action of thermal changes [5].

This paper is centered on the Russian Marsokhod [4] (now international project) design. At the mini-landscape test facilities, following the scheme proposed in this paper, the rover showed excellent performance traveling on sand dunes and geometrically irregular terrain, and in dust storms it is not more effective than other rovers, like the Sojourner class, in crossing planitias type terrain, by statistical calculus based on the lower distance chassis-surface, where the Marsokhod wheels were getting more hits from the various sizes of Martian rocks, transferring more kinetic energy. But the compact design of the Marsokhod provides reliability and good performance to survive the planned time of one year traveling on the Martian surface.

This work is part of a first-stage investigation looking for the best performance in autonomous navigation systems of planetary rovers. The second-stage, planned to be carried out by the author, is the development of a rover navigation system based on a virtual reality approach.

MARSOKHOD

The vehicle is composed of three sections or platforms based on a six-wheel device.

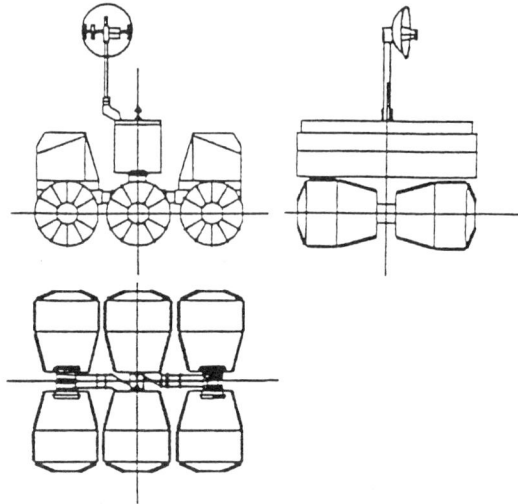

Figure 1 Marsokhod Preliminary Views.

The weight is symmetrically distributed in such a form that the central platform contains 20% more than the other two. The 93.8 Kg mass summary of the Marsrover is given in Table 1. Each titanium alloy wheel has 350 mm of diameter containing autonomous engines powered by a Nicad battery supplying 8.3 watts for each one.

Table 1

Systems, elements	Mass, Kg
Chassis	25.0
Power System	13.8
Radio System	11.0
Control System	6.0
Scientific Instruments	10.0
Structure (trusses, beams, arms)	12.0
Thermal Control System	9.0
Cable Net	7.0

TEST FACILITIES

A scale model of 1:10 was built to observe its performance in an 0.90 by 0.30 meters Martian landscape.

In the same way, 1/10 of the weight of the real model was distributed among the three platforms and reduced to 39.8 percent by gravity difference.

The velocity of the center of mass, placed symmetrically at the center platform, was variable, from 0.012 to 0.018 m/s (0.15 m/s for the real model).

The observations were limited to geometrical and kinematic studies, given the difficulties of obtaining titanium wheels, and mechanisms were replaced by lighter and inexpensive material. The Marsokhod moving was filmed using an 8 mm videocamera, digitizing images with an Image Motion Capture Card and software, assisted by C++ programs to calculus and graphic mechanics and statistics of the move, working in a PC Pentium 133.

A parallel method was established, designing a 3D Marsokhod on Autocad 13, and afterward putting the graphic in a 3D Studio Software, where it did a virtual travel on a terrain similar to the region reproduced at the Martian landscape test facilities.

Figure 2   Marsokhod view at the mini-landscape.

Both results were compared in order to select an appropriate mobility and navigation system for the Marsokhod, and with a new approach to virtual reality configuration.

OBSTACLE PERCEPTION AND NAVIGATION SYSTEM

Usually the acquisition of 3D pictures around the rover is desirable. At this time a combination of stereo cameras and laser are the main devices of the Obstacle Perception System. Taking into account the capacity to develop relatively high speeds: 0.15-0.25 m/s, the Marsokhod image acquisition system must have the ability of look a long distance ahead to maneuver. However, since a laser finder to detect obstacles is not included in the current Marsokhod Project, the use of a combination of laser scanning with a stereo camera to obtain an optimal surrounded image acquisition is recommended.

The use of inclinometers gives to the computer on board information of the surface inclination under the vehicle, allowing it to stop if such inclinations exceed 35°.

Based on the information received with the perception system (stereo cameras-laser scanning), an elevation map is elaborated (Figure 3). Then the operator, using a previous virtual travel of the terrain, analyzes the path variants. If the path presents a safety level assigned, the Marsokhod can roll ahead for the next distance, at first calculated in 20 meters. Using the proposed architecture, with high reliability, each stop distance would be increased to 30-40 meters (more autonomy obtained).

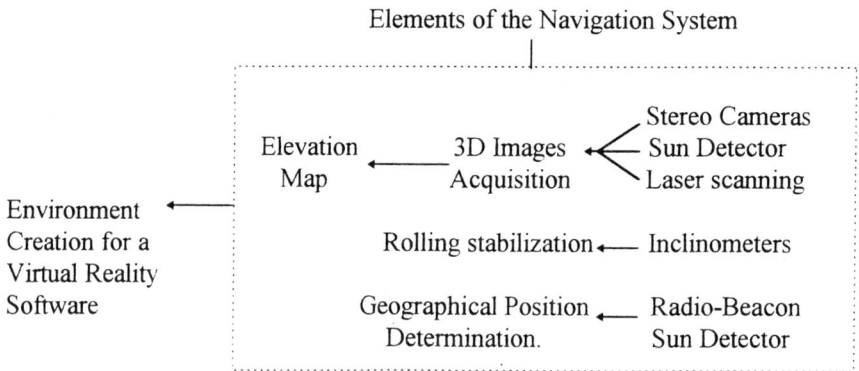

Figure 3 Ways to obtain a virtual reality environment.

The current Marsokhod Program contemplates rolling during the Martian day. At night it stands without moving, measuring some atmospheric parameters and checking out the systems. The use of beacons—not included in the actual program—can help to improve Marsokhod positioning, making communication with the lander and/or the orbiter more effective [6].

Taking into consideration a related landing place (sand dunes with a little rock distribution per quadratic meter), the horizon detection appears as the following target to be acquired for positioning [7].

Also recommended is a semi-autonomous navigation system, using an on-board autonomous path planning and selection to ensure that it could operate if anything goes wrong by system failure or operator errors. This process offers a way to find the algorithm needed.

Its travel would be assisted by a lander-orbiter scheme to provide what is necessary to increase the rover power supply.

Probably the Marsokhod will collect samples as part of a Mars Sample Return Mission. The Marsokhod has the capability to carry out such a mission. So, new instruments and a robot arm could be attached, influencing the total mass and changing all performances of the Marsokhod. This was demonstrated by distributing additional weight to the three Marsokhod platform. The observations indicate that the ideal weight for a Marsokhod class rover to overcome obstacles and gain mobility, without meaningful loss of performance and kinetic energy, was between 40-55 kg.

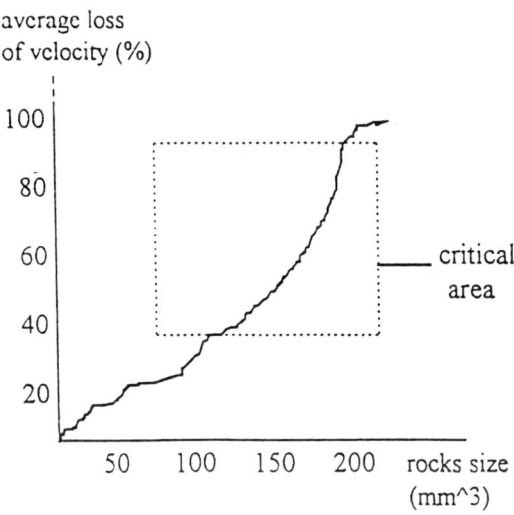

Figure 4 Dependence between loss velocity of the center of mass and rock size.

MOBILITY SYSTEM

The special configuration of the Marsokhod wheels, and their low height, allows a less complex structure for the suspension design, mechanisms and actuators, facilitating also its hermetization and protection from the Martian dust. However, since the lightweight aluminum structures were selected, there are many discussions about replacing the axles for composite materials.

The two lateral axles can be displaced from the central one in a one wheel size diameter, given better climbing capability at hard angles.

The Marsokhod can turn in place using steering actuators, maneuvering the front and back axles.

The Marsokhod design only allows a relative "go ahead" movement, compared to the Sojouner capacity to turn around at almost 360 degrees from the staying point.

There are no equations that would allow depicting the axles and wheels moving on the irregular surface. Instead, the use of a virtual reality approach to understanding the future move (steps) the rover could do is strongly recommended.

DISCUSSION

Despite the fact that the test was carried out without helpful practice on a real model, and under economic restrictions of obtaining needed hardware and software, a simple and effective approach was made in researching geometry interaction of surface-chassis, and choosing the appropriate architecture.

The use of a virtual travel to test the performance before a future real-step appears to be a most delicate operation. A *virtual travel system* operation would be the key to research in autonomous systems for planetary robots.

ACKNOWLEDGMENT

The author would like to thank to Dr. Boris N. Martinov of the Lavoschkin Association for his advice, the Ministry of Culture and Education of Argentina for the annual scholarship and support for the research activities; The Center for Mars Exploration for providing The Mars Educational Multimedia CD-ROM, and to Dr. Erik P. Krotkov of the Robotics Institute, Carnegie Mellon University for providing helpful information.

REFERENCES

1. Eric Krotkov, Reid Simmons, Fabio Cozman and Sven Koening, "Safeguarded Teleoperation for Lunar Rover," SAE Technical Paper 961585. California (1996).
2. Eric Krotkov, Martial Hebert and Reid Simmons, "Stereo Perception and Dead Reckoning for a Prototype Lunar Rover," *Autonomous Robots*, 2, 313-331.
3. NASA Space Telerobotics Program, World Wide Web Site, 1996.
4. Lavochkin Association, Mars Mission "Molniya" Version, Technical Note, 1995.
5. The Center for Mars Exploration, The return to Mars Educational Multimedia CD-ROM, 1996.
6. M. Kazharinov, P Grishin, V. Ipatov, A Kolomenski, D. Ulianitski, Radio-Technical Systems, Editorial High School Moscow, 1990.
7. Introduction to the Spacecraft Navigation, V. Shevshaievich, Editorial Radio-Soviets, Moscow, 1970.

MAR 98-038

ON THE DEVELOPMENT OF AIRBORNE SCIENCE PLATFORMS FOR MARTIAN EXPLORATION*

David W. Hall and Robert W. Parks

The presence of an atmosphere on Mars provides planetary scientists with the opportunity to extend their coverage through airborne robotic exploration. This paper discusses basic considerations for the design and development of atmospheric science platforms for Martian missions including similarities and differences with terrestrial applications. Included will be discussion of aerodynamic and propulsion requirements as well as an overview of the Mars Flyer development program at NASA/Ames Research Center.

INTRODUCTION

The Pathfinder/Sojourner mission to Mars in the summer of 1997 excited public imagination and piqued widespread interest in exploration of our nearest planetary neighbor. Data return exceeded the expectation of mission planners and set the stage for new missions of discovery to the red planet. Sojourner provided data not available from space-based telescopes in earth orbit or satellites in orbit around Mars by actually touching and sampling rocks in the vicinity of the landing area, a circle roughly 30 meters (100 feet) in diameter. As soon as pictures of the landing site showed up on our television sets, the human desire to know what's on the other side of nearby ridge lines reared up and mission planners immediately began discussing ways to find out in future missions.

A year previously, however, staff at NASA/Ames Research Center, some of whom would play key roles in Pathfinder/Sojourner science, began looking at ways to extend our scientific knowledge beyond the immediate landing area. During the autumn of 1996 and winter of 1997, a small team of NASA staff and private contractors designed, built and flew two scaled prototypes of the first of several Airplanes for Mars Exploration (AME) created to enhance a variety of challenging science missions. The first small remotely piloted vehicle (RPV) provided qualitative and some quantitative assessment of aerodynamic performance at Reynolds Numbers similar to those expected on Mars. The second small RPV provided proof-of-concept testing of the deployment scheme and involved design and construction of a third RPV to carry it. The goal of these flight tests was to prove the viability of an atmospherically deployable Mars science platform which they could propose to NASA/Headquarters during the December 1996 Discovery opportunity.

Types of Missions

Since then, we have identified several candidate missions for fixed wing airplanes on Mars. The 1996 Discovery proposal represents the high end "the airplane IS the mission" approach, where the goal is to fit the largest possible airplane into the available launch volume and mass. These airplanes tend to be about 9 to 12 meters in wingspan and 150 to 200 kilo-

---

* Presented by David W. Hall and Robert W. Parks, David Hall Consulting, in conjunction with the NASA Ames Research Center, Space Advanced Concepts Branch (Code SFS), to the Founding Convention of the Mars Society, held at Boulder, Colorado, August 13-16, 1998.

grams initial mass. The result is an airplane that can carry tens of kilograms of scientific instruments and is capable of flying thousands of kilometers. On a different scale, we are currently examining a much smaller version of this type of mission, one that can take advantage of a very low-cost ASAP launch on an Ariane 5 mission.

A third concept being investigated is an "airborne scout" aircraft that would launch from a Surveyer class soft lander. Its purpose would be to provide very high resolution mapping of an area in the immediate vicinity the landing site. This type of mission is quite attractive when combined with one of the "mobile lander" concepts, as data from the aircraft would enable mission planners to pick the most interesting sites for a close investigation by a lander, as well as allowing them to determine the best path to get to those sites.

Finally, there is an intermediate class of mission, where we have an austere lander, with only minimal retrorocket and communications ability. This would allow a ground deployment of the airplane, launch during favorable weather conditions, as well as data storage and relay. The penalty is that the weight of the lander results in a significant reduction in airplane size and capability, but there may be an increase in overall mission reliability.

PACKAGING CONSIDERATIONS

The first set of constraints to be applied before flight can commence at a Martian site concerns how we get there. In other words, how can we package a useful airborne science platform in the limited volume and mass available for interplanetary launches? Stowage schemes will be highly dependent upon the type of launch platform available, the volume of the aeroshell and its back covering, and payload mass allowance.

Airframe stowage within a 2.65 m (8.69 ft) diameter aeroshell will affect aircraft complexity (number of folds), wingspan, wing aspect ratio, fuselage length, tail boom length, horizontal tail span, and vertical tail span. For the aeroshell to be used in the AME mission, total folded span must be less than 2.4 m (7.87 ft), as shown in Figure 1. Aeroshell height creates a constraint as well, as does the taper half angle of the aeroshell back cover.

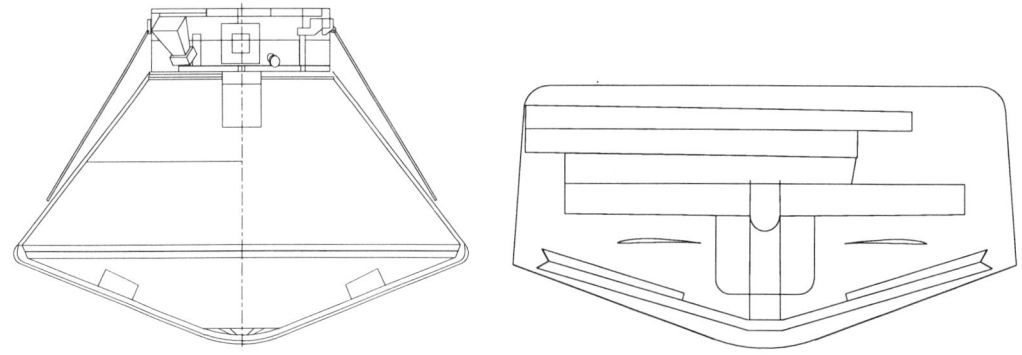

Figure 1 A Stock Aeroshell (left) Constrains Overall Vehicle Geometry and Must be Modified by Increasing the Back Shell Half Angle as Much as Possible (right).

SCIENCE CONSIDERATIONS

Of the approximately 600 kg which can be lifted toward Mars, only 330 kg will arrive there and, of that, only 210 kg is available for an RPV. By the time designers account for airframe, deployment mechanisms, propulsion, navigation and control, approximately 10 to 20 kg will be available for science instruments. Table 1 presents the AME payload suite.

Table 1
A SMALL PAYLOAD CAN YIELD MUCH USEFUL DATA

ITEM	MASS kilograms	LENGTH meters	WIDTH meters	HEIGHT meters	POWER watts
Scan Platform (fuselage)					
Assembly	0.924				
Four Video Cameras	2.000	0.400	0.400	0.250	6.0
Near Infrared Mapper	1.250	0.303	0.150	0.100	11.0
Mini-TES	2.450	0.203	0.152	0.076	5.0
Robotic Arm (port wing)					
Assembly	1.814				
Camera	0.900	0.080	0.060	0.060	10.0
Raman Spectrometer	1.900	0.100	0.040	0.040	2.0
APX Spectrometer	0.500	0.080	0.070	0.065	0.3
Atmospheric Science (fuselage)	0.500	0.120	0.120	0.120	0.5
TOTAL	12.238				34.8

THE MARTIAN FLIGHT ENVIRONMENT

There's good news and there's bad news, and enough of each to make Martian flight possible, but quite challenging. First, the good news: Gravity on Mars is 3/8 that on earth, so airplanes weigh less than on this planet. That adds range and endurance to mission performance.

The bad news, however, can potentially offset this. The Martian standard atmosphere is presented in Figure 2.[*] The mission proposed two years ago flew at one kilometer (3,280 ft) above Martian datum, or an altitude roughly equivalent in density to 33 km (108,000 ft) on earth. Standard ambient temperature one km above Martian datum is about 206 Kelvin (-85 Fahrenheit) and the freestream speed of sound is about 227 meters per second (mps), or 745 feet per second (fps). It is this combination of low temperature, low density, and low speed of sound that defines the flight domain of Mars science aircraft.

The low density means that airplanes will have to fly relatively fast to obtain enough lift, while the low speed of sound places an upper limit on the speed, particularly when propeller tip speed limits are considered. Further complicating this already gloomy picture of the Martian atmosphere is a 20% seasonal variation in atmospheric density as carbon dioxide sublimates into and out of the atmosphere.

INTERPLANETARY GEOMETRY

Mars is on the order of one astronomical unit from earth even at its closest approach. This translates to an eight to ten minute light time between terrestrial mission control and aircraft flying over Mars, so real-time control from earth is not a near-term possibility. Therefore, the aircraft will have to be designed to be fully autonomous with robust enough navigation, guidance and control systems to determine where it is, which direction it's headed, and where it wants to go.

---

[*] In keeping with the spirit of fun so prevalent in an otherwise serious and focused planetary science community, these data are available on the Martian Advisory Committee on Aeronautics (M.A.C.A.) web page, which is at the Red Peace website (www.redpeace.org). It's our intention to use this web page as a forum for dissemination of aeronautically related information related to Mars.

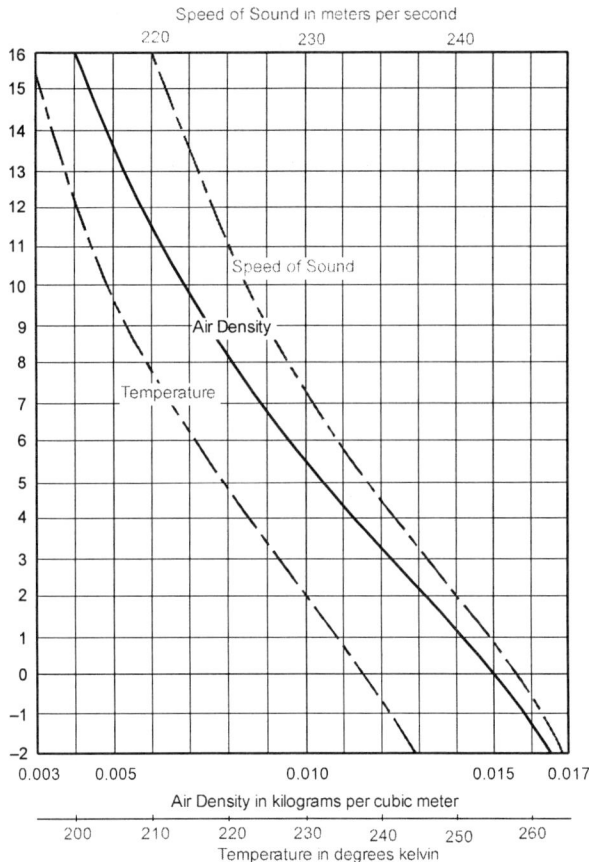

Figure 2 The Martian Standard Atmosphere Behaves Predictably.

If multiple-day flights are required, then the aircraft must have sufficient onboard intelligence to choose a safe landing site, set down, establish communications with earth and/or a satellite, and then take off again. Complicating this scenario is the amount of time each day when earth is above the Martian horizon and/or in view of a satellite in orbit around the planet which could collect mission data and re-transmit it to earth.

Transfer to earth of flight data will determine the minimum ground time interval before AME attempts a second flight. Requirements for data transfer will be sun position for ground power generation, earth position in the Martian sky, Mars Surveyor satellite position, and data transfer rate. The resultant ground time will be approximately ten days to two weeks depending upon local weather conditions.

CANDIDATE PROPULSION SYSTEMS

Complicating proposed long-range extraterrestrial mission scenarios is the lack of oxygen in the Martian atmosphere, which limits near-term propulsion systems to either electric motors or monopropellant engines. In fact, the Martian atmosphere contains only trace amounts of oxygen and is almost entirely (95.5%) carbon dioxide. This translates to a near-term requirement to carry both fuel and oxidizer onboard or to use some form of stored electric power.

Lithium/Hydrogen Peroxide Propulsion System Configuration

The Discovery mission which Arizona State University and NASA/Ames Research Center proposed in late 1996 was based on using electric propulsion in the form of a lithium/hydrogen peroxide fuel cell being developed by Pinnacle Research Institute of Los Gatos, California.

The lithium/hydrogen peroxide propulsion system is divided into three principal parts: a cell stack, an oxidizer reservoir, and a reactor/storage system. The overall propulsion system includes electronic controls to monitor stack output voltage. These controls either add hydrogen peroxide solution when voltage begins to decline or withhold it when voltage rises too far. Thus, the controls create a constant-voltage powerplant even if load power changes, and as the internal concentration of chemical constituents varies. Figure 3 presents a schematic of the AME fuel cell propulsion system.

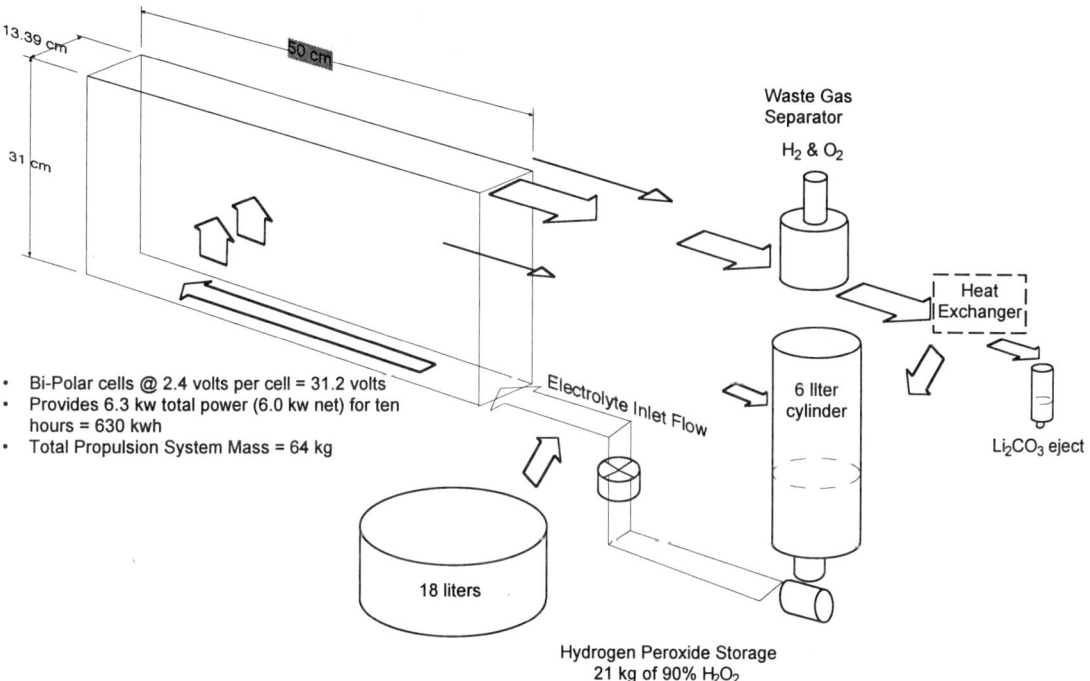

Figure 3 The Lithium-Hydrogen Peroxide Fuel Cell Appears Promising with Development.

The selected system voltage (or bus bar voltage in powerplant terminology) is nominally 31.2 volts for a stack of thirteen, 2.4 volt cells. An approximate 30 volt propulsion system appears to be the best compromise with motor designs currently available for this mission.

The minimum energy requirement is to provide 6 kw of power to the motor for the duration of two flights of roughly five hours each. There will be another 0.3 kilowatts of power required for operation of the circulation pump, auxiliary subsystems and science instruments. Total energy required from the fuel cell, then, will be 63 kilowatt-hours (kwh). To generate this amount of energy, the fuel cell will require 6.6 kilograms of lithium and 20.1 kg. of hydrogen peroxide.

Hydrazine Engine

In the 1970s, Dale Reed and Jim Akkerman built and flight demonstrated a hydrazine engine at NASA/Dryden Flight Research Center on an RPV. Figure 4 presents a cutaway of their hydrazine engine design. Note its integral heat exchanger buried in the fuel tank.

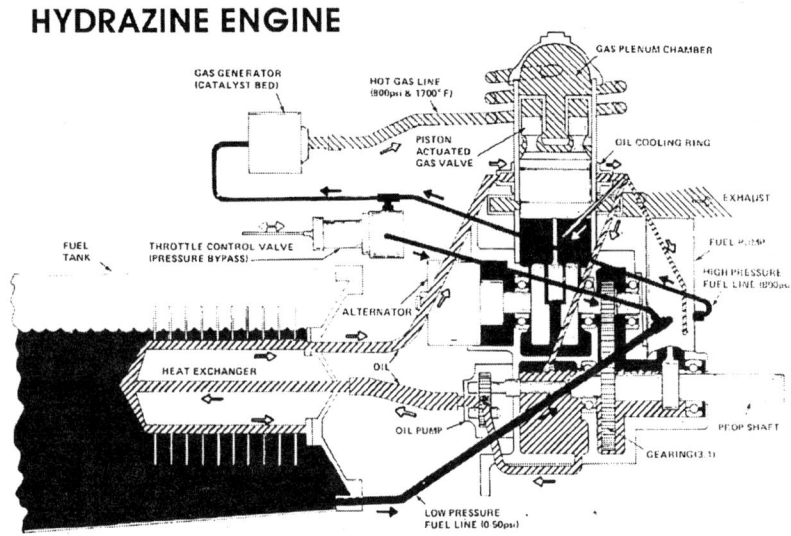

Figure 4 The Mini-Sniffer Hydrazine Engine was Extensively Tested in Both Altitude Chambers and Free-Flight in the 1970s.

While the lithium/hydrogen peroxide fuel cell promises very high energy density, it is an unproven system, and significant development must be accomplished before it is ready to fly on Mars. The hydrazine engine has less than half the net energy density, but hydrazine technology is well proven in space flight. There would be some development work needed for the engine itself, but it is a much lower risk. The hydrazine engine is also very easily scaled to smaller sizes, and thus may be the system of choice for any of the smaller aircraft.

MISSION REQUIREMENTS

Because of the need for reliable deployment, the precariousness of the Martian atmosphere, and the varying elevations of Martian terrain, science mission choice will have a strong impact on platform design. To illustrate this, consider four potential mission scenarios:

- Deploy over Crater Gusev (185° West, 15° South) and reconnoiter geophysical features over approximately 4,000 km (2,480 mi.).*

- Deploy over Hebes Chasm and fly back and forth across Valles Marineris for a distance of 3,000 km (1,860 mi.).†

- In conjunction with a surface rover, scout likely sites at which life may now exist, or may have existed at one time over a multiple kilometer radius from the landing site.

---

* This was the mission which Arizona State University, in conjunction with NASA/Ames Research Center, proposed for the Airplane for Mars Exploration (AME), a Discovery Mission, submitted in December 1996.

† This is the subject of the current Discovery proposal from Malin Space Science Systems, Orbital Sciences, Naval Research Laboratory, and NASA/Ames Research Center and was submitted in June 1998.

- Provide a piggy-back package for a non-Mars-bound launch which will be fired from earth orbit toward Mars. On arrival, at Mars, it will conduct airborne science for at least one hour.

Crater Gusev Mission

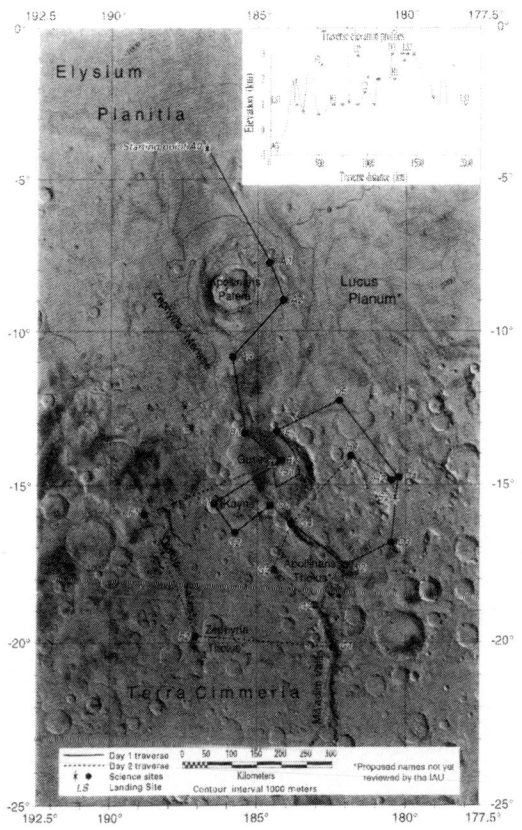

Figure 5   AME will Methodically Meander Across the Martian Surface.

During NASA's early AME proposal effort, scientists knew where they wanted to fly, but hadn't yet determined exact ground tracks and altitudes. Hence, the airplane design team iterated with mission scientists until firm mission requirements could be established. The design mission is in the vicinity of the Gusev Crater just south of the Martian Equator from Latitudes 5 South to 15 South and from Longitudes 187 West to 182 West. Figure 5 presents mission ground track and profile. Total flight path length is 3,200 kilometers (1,988 statute miles) over a desired two days of operation.

This apparently meandering flight track allows close examination of a wide variety of geological features over a large geographic area. Onboard visual and infrared sensors will collect data during the first day of flight, which will be transmitted to earth after landing. Mission and payload controllers will have an opportunity either to analyze the first day's data and reprogram AME to revisit particularly promising sites for more aerial data and/or ground data, or to continue the original mission flight plan.

Note in Figure 5 the dramatic change in terrain elevation over the mission track. Flight actually starts over terrain whose elevation is below nominal Martian sea level, at minus one km (-3,280 ft). The first day's flight will progress over increasingly higher terrain until AME traverses the Gusev Crater edge, which is 3 km (9,840 ft) above sea level. The second day's flight will start at a surface elevation of 0.5 km (1,640 ft) and end in a high elevation landing at just over 3 km (9,840 ft).

This mission requires two flights. The first flight begins with successful deployment from a Viking-type atmospheric entry aeroshell and lasts approximately five to seven hours. Flight altitudes will vary from 2 km to 4 km. AME must then be able to transmit data to earth via an intermediate Mars Surveyor satellite link. Instruments onboard will continue making ground-based observations which also must be transmitted to earth. At the end of the data transmission period, AME must take off using hydrazine thrusters and resume flight for another three to five hours, then land and transmit its newly collected data to earth.

As might be expected, the kinematic requirements for takeoff and landing, and total ground time will affect design and operation of the propulsion system. These considerations have a driving impact on overall propulsion type, size and mass. Hydrazine, battery and fuel cell power trains were examined as well as tailed and tailless designs. For a nine hour duration flight over a total distance approaching 3,500 km, only the fuel cell propulsion system proved viable, particularly when ground dormancy and restart were taken into consideration. Surprisingly, neither the tailed nor tailless designs proved appreciably different in performance at this level of detail, as shown in Figure 6.

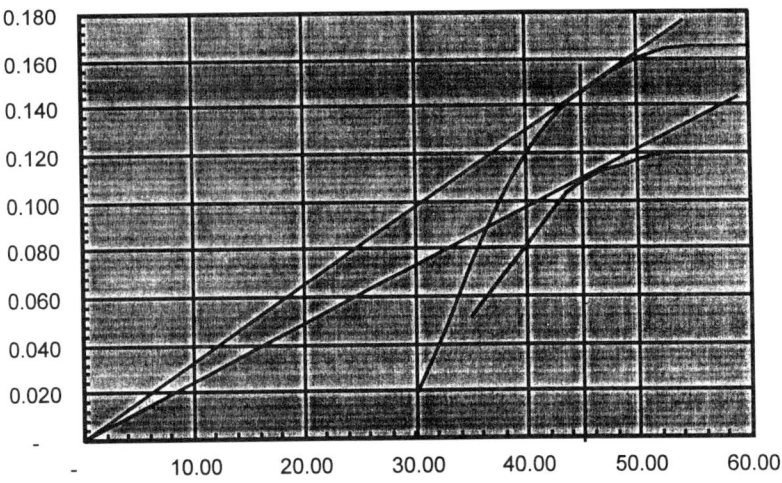

Figure 6 The Highest Payload Fraction Corresponds to a 42.5 foot Wingspan for Both the Aft Tail and Tailless Configurations.

The airplane is shown in Figure 7 in both its stowed and deployed configurations, and characteristics of the baseline AME are:

Wing area	12.25	sq.m.
Wing span	12.45	m
Cruise Lift Coefficient	1.00	
Duration	8.8	hrs
Range	3500.	km
Aspect Ratio	12.65	
Cruise Lift-to-Drag Ratio	16.39	
Cruise Reynolds Number	100,000.	
Climb Motor Power	11.75	Kw
Cruise Mach Number	0.474	
Cruise True Airspeed	111.	mps
Structural Mass	36.0	Kg
Payload Mass	25.0	Kg
Communications & Navigation Mass	46.0	Kg
Fuel Cell System Mass	59.4	Kg
Hydrazine Rockets & Fuel Mass	34.7	Kg
Motor Mass	2.7	Kg
Total Aircraft Mass	203.8	kg

Figure 7 The Current AME Configuration Meets All Mission Requirements.

Valles Marineris Mission

The mission chosen by the science team for their 1998 Discovery mission proposal (Mars Airplane for Geophysical Exploration, or MAGE) involves three traverses of the great rift on Mars at its widest spot, and flight in the vicinity of Hebes Chasm. Total traverse distance is about 1,900 km (1,025 n.mi.) in terrestrial units.

The configuration which the Naval Research Laboratory (NRL) conceived for this challenging mission is shown in Figure 8. The configuration is a zero sweep flying wing which deploys from an aeroshell on atmospheric entry, the entry sequence being similar to that used with the Airborne Mars Explorer (AME) vehicle.

The MAGE aeroshell represents a more radical departure from the Pathfinder aeroshell configuration than the AME aeroshell. It provides much needed volume to house a low density airplane, and also allows a much simpler and more reliable deployment scheme. While this configuration is very different than existing planetary exploration aeroshells, it is very similar to the successful shapes used for Mercury and Gemini spacecraft.

Figure 8   The MAGE Configuration is a Flying Wing.

Figure 9 presents the stowage scheme for this aircraft and Figures 10 present views of parachute deployment. Table 2 presents descriptive data for the MAGE configuration.

Pole Cat Mission

There is benefit to providing an airborne scout type of vehicle to operate in conjunction with a rover. Available volume and mass would constrain the RPV to a fairly small size, but it could be designed to extend mission planners' views of the landing site to uncover promising directions and safe paths for the rover to transit. The RPV would be stowed in a box during flight to Mars and arrival at the landing site. It would then be deployed at the top of a launch pole and rail, weathervane into the wind, and launch using hydrazine thrusters. The propeller would begin turning and the RPV's main propulsion system would take over. Flight would be within line-of-site of the lander and rover and only long enough in duration to map a roughly one kilometer circle around the lander.

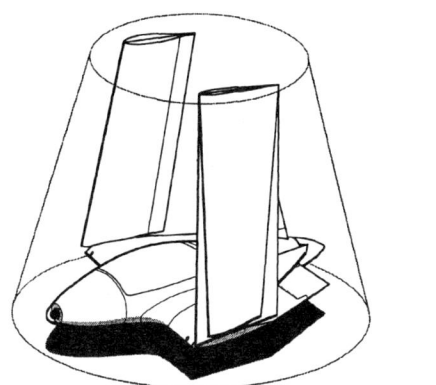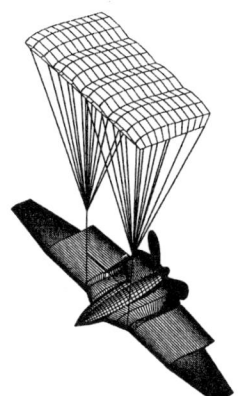

Figure 9 (left)   In-Transit Stowage Differs from AME in Using More Volume in a Back Shell.
Figure 10a (center)   MAGE Deploys Under a Parasail.
Figure 10b (right)   Note Its Relative Angle-of-Attack.

Table 2
THE MAGE CONFIGURATION IS SIMPLE AND EFFICIENT

ITEM	VALUE (SI)		VALUE (English)	
Geometry				
Wingspan	9.73	m	31.91	ft
Reference Wing Area	9.25	sq.m.	100.00	sq.ft.
Aspect Ratio	10.25		10.25	
Aerodynamics				
Cruise Reynolds Number	120,000		120,000	
Cruise Lift Coefficient	0.60		0.60	
Maximum Lift-to-Drag Ratio (L/D)	20.72		20.72	
Masses (Earth reference)				
Structure	27.10	kg	60.00	lbf
Propulsion Group	15.00	kg	33.00	lbf
Avionics, Surface Controls, & Fixed Equipment	24.00	kg	53.00	lbf
Empty	66.10	kg	146.00	lbf
Payload	20.00	kg	44.11	lbf
Fuel	55.90	kg	123.28	lbf
Gross	142.00	kg	313.17	lbf
Propulsion				
Cruise Power Required	10.00	kw	13.5	HP
Hydrazine Engine Specific Fuel Consumption	2.43	kg/kw/hr	4.00	lb/HP/hr
Propeller Diameter	2.20	m	7.22	ft
Propeller Efficiency Factor	0.60		0.60	
Performance				
Range	1,800	km	971	n. mi.
Endurance	3.33	hrs	3.33	hrs
Cruise Mach Number	0.66	M	0.66	M
Cruise Altitude above Martian Datum	8.5	km	27,880	ft
Power-to-Weight Ratio	0.0704	kw/kg	0.043	HP/lbf
Initial Wing Loading	15.35	kg/sq.m.	3.13	psf

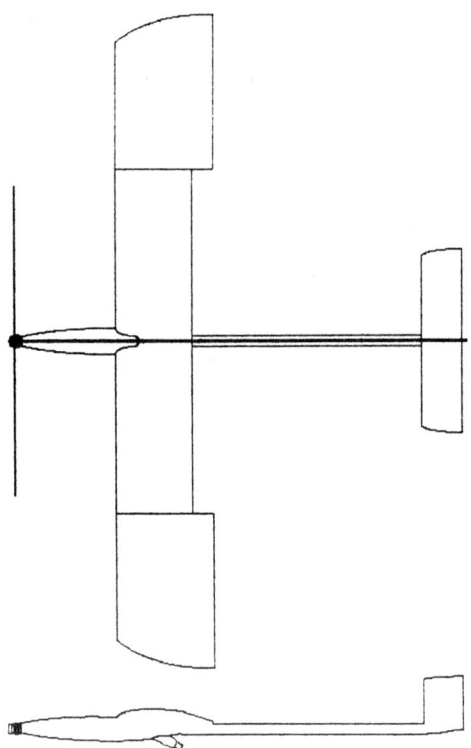

Figure 11 The Pole Cat Launches from a Mast on a Mars Surveyor Lander.

Piggy-Back Mission

The final type of mission to be discussed here would start with space being made available on an Ariane 5 launch into earth geosynchronous orbit for a self-contained payload and propulsion system to go to Mars. The As Small As Possible (ASAP) system would weigh approximately 50 kg and contain an RPV weighing no more than 20 kg. Because of the need to fit the container inside the existing Ariane 5 nose, a different wing deployment scheme may be necessary. Figure 12 presents one concept for an ASAP mission which includes a small RPV with externally-braced rollout wings.

CURRENT EFFORTS TO DEVELOP TECHNOLOGIES

Subsonic flight at terrestrial altitudes equivalent to mission altitudes on Mars varies from 30 to 35 km (99 to 115 kft). Two current NASA research and development programs, (APEX) and Environmental Research Aircraft and Sensor Technology (ERAST), are charged with developing vehicles to fly at these extreme altitudes. APEX will launch a modified Schweizer sailplane from roughly 36.6 km (120 kft) from a large balloon. The sailplane will stabilize its descent at about 31.4 km (103 kft) and glide power off to ground level, collecting aerodynamic test section data and atmospheric data during its flight.

The ERAST/Pathfinder, Pathfinder Plus, and Centurion are in development with components of Solar Centurion flying during summer 1998 on Pathfinder Plus out of Barbers Point Hawaii. Before ceasing operations, Pathfinder Plus attained over 24.4 km (80,000 ft). Both research and development programs demonstrate the existence of a small but growing database of experience at altitudes and airspeeds equivalent to those required for flight on Mars.

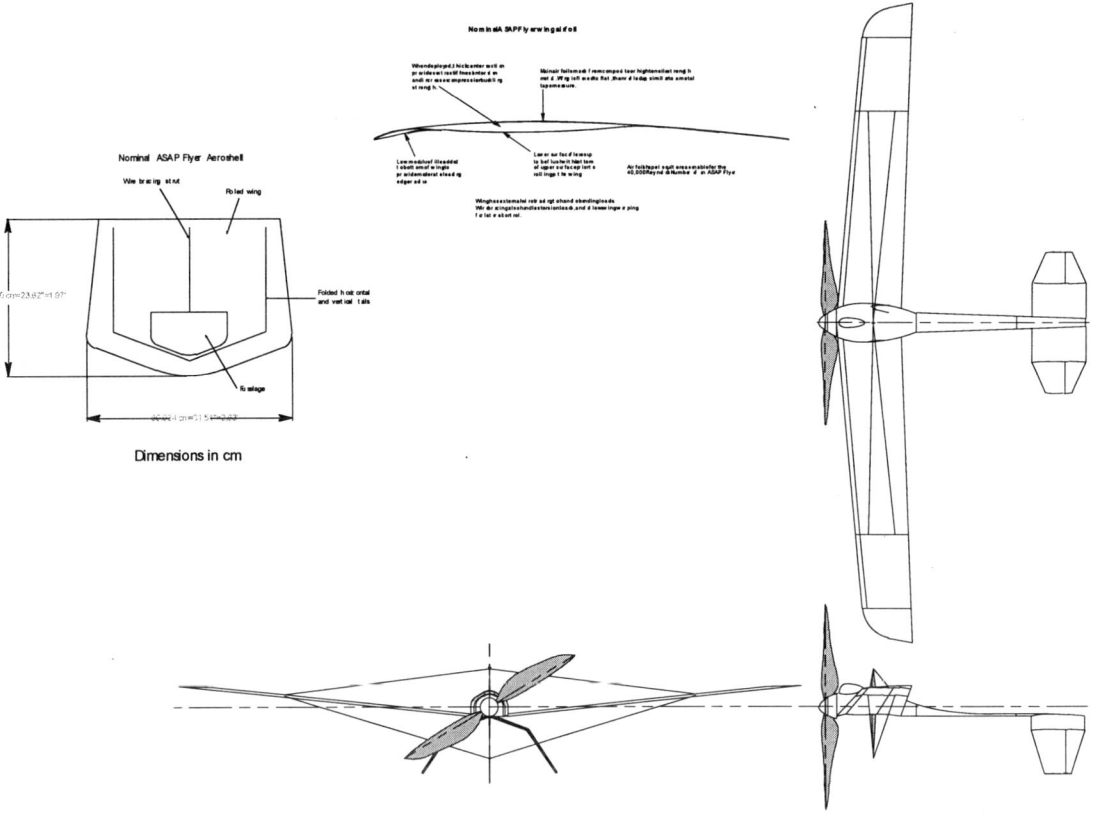

Figure 12 The Trick is Finding a Way to Reliably Unroll the Wing.

SUMMARY

This paper has presented four potential airplane concepts for a variety of Mars exploration missions. The intent has been to show that airborne flight at Mars enhances mission science while adding as little risk and budgetary increment as necessary. At least one of these four approaches should be suitable for inclusion in existing and planned Mars missions.

REFERENCES

1. Reed, R. Dale, "High-Flying Mini-Sniffer RPV: Mars Bound?", *Astronautics & Aeronautics*, June 1978, pp. 26-39.
2. Hall, D. W., Parks, R. W. & Morris, S., *Airplane for Mars Exploration (AME): Conceptual Design of the Full-Scale Vehicle; Design, Construction, and Test of Performance and Deployment Models*, David Hall Consulting, May 27, 1997. May be downloaded in pdf format from DHC's www.redpeace.org website.
3. Hall, D. W., Parks, R. W., Tsai, K. C., & Galbraith, D., *Airplane for Mars Exploration (AME): Conceptual Design of the Full-Scale Vehicle Propulsion System*, David Hall Consulting, June 30, 1997. May be downloaded in pdf format from DHC's www.redpeace.org website.

MAR 98-039

REDUCING RISK AND COMPLEXITY OF ROVER AND ROBOTIC OPERATIONS ON MARS

Russell R. Mellon* and Thomas R. Meyer†

The Mars surface exploration activities planned for the next decade involving Landers and Rover vehicles depend upon the successful interplay of a number of time critical operations. In particular, mission planning for the '03 and '05 Landers assumes that a very small positioning error will be achieved for placement of the '05 payload in close proximity to the '03 lander, which will contain cached samples intended for return to Earth. The Rover vehicle contained in the '05 payload will be used to obtain these samples and will deposit them in the ascent stage. A number of scenarios can be envisioned in which this plan could fail.

If the proposed typical 100-300 m ground targeting error for the '05 Lander with respect to the '03 Lander location is exceeded, perhaps increasing to a value of 1 km or more, navigation of the Rover between the two destinations becomes much more hazardous and time consuming if no a priori knowledge of the intervening surface obstacles exists. Moreover, one or both Landers might be sited on local butte formations whose steep sides could preclude access by the Rover. In this case, a terrain map generated during the final descent phase of the '05 Lander mission would alert controllers of the necessity for acquiring contingency samples from the immediate area surrounding the Lander. Even if no impassable barriers exist between widely separated ground locations, the time consuming nature of blind navigation could exceed the time window established for sample collection and return.

We propose to increase the chances for sample return mission success through the acquisition of a detailed terrain data base obtained via an atmospheric reconnaissance vehicle deployed by the '05 Lander on its final descent phase. We envision this vehicle, which we call the Mars Raptor Reconnaissance System, to be capable of generating a detailed surface description through deployment of an optical imaging sensor package on an unpowered glider platform. The glider would be released at approximately 5 km altitude and would make a number of orbits of about 1 km radius during its descent to the surface. Depending upon the exact configuration chosen for the glider, its helical descent trajectory could take between 10 and 60 minutes to complete. The glider payload ("Bird") would acquire ground images, store and data compress them for RF transmission to a Raptor control element ("Nest") on the Lander and provide for 3-axis guidance and control of the glider trajectory.

In addition to the mapping function, a Mars Raptor could also be used to perform valuable ancillary missions, such as obtaining large scale images of wide areas surrounding the Lander as well as high resolution views of select areas within the total reconnaissance area, permitting both the large and meso scale terrain relationships of the Lander region to be placed

---

* Director of Research and Development, Equinox Interscience, Inc., 6420 S. Quebec St., Ste. E, Englewood, Colorado 80111. E-mail: optics@eisci.com.

† Boulder Center for Science and Policy, P.O. Box 4877, Boulder, Colorado 80306.

in proper geological context with the small area directly observed through Lander imaging and by Rover visitation. Furthermore, the development of Raptor technology provides the means for deployment of multiple reconnaissance platforms on future missions. Raptor technology, either glider based or exploiting reaction powered platforms, could provide a remote sensing capability over large areas of the planetary surface when deployed from a central Lander location. Such wide area image reconnaissance surveys (mapped at much higher ground resolution than is possible from an Orbiter) could also include payloads delivered to the surface for conducting surface and subsurface resource detection. Examples of these surface delivered payloads could include seismometers and impulse generators for use in prospecting for the postulated Mars subsurface water table.

A number of investigators have explored the design of aircraft for operation in the Mars atmosphere. The development project which we propose here assumes that a suitable design database is already in place allowing the required glider platform to be built. The objective of our work is to demonstrate that the necessary Bird and Nest elements can be built to meet the performance, mass, volume and electrical power consumption requirements that would make the Mars Raptor Reconnaissance System feasible.

Bird Element. The Mars Raptor Reconnaissance System's data acquisition payload is contained within the Bird element of the deployed glider platform. The Bird payload element contains an optical imaging system consisting of 2 fixed focus refractive lens assemblies which image onto a pair of 1024 X 1024 pixel focal plane detector arrays. Our analysis indicates that two separate miniature imagers offer weight savings over more complex systems with selective focal length or zoom capability. Stereoscopic image pairs can be taken with each lens using the translational motion of the glider platform to produce the required parallax offset between data frames obtained at closely separated time intervals. A variable image scale (zoom focal length adjustment) can be obtained electronically by combining focal plane pixels into multiple binned units.

We anticipate that an electronic zoom range of 8:1 is feasible for each lens assembly, yielding ultimate ground object lineal resolutions in the range of 9-12 cm. The conflicting requirements of covering a wide angular field of view and resolving objects of the small dimensions that could pose navigational difficulties for the Rover requires that 2 lenses and focal planes be used. A summary of the angular coverage and lineal resolution of the Bird imaging systems appears in Table 1.

Table 1
BIRD IMAGER CHARACTERISTICS

Lens EFL	Total FOV (°)	Ground Res. @ 5 km	Ground Res. @ 1 km
100 mm	5.28 X 5.28	45 cm	9 cm
10 mm	49.5 X 49.5	4.5 m	90 cm

The Bird electronics consist of the following subsystems: Battery and Power subsystem, Antenna, Transmitter, Flight Actuators, Sensors, and the Electronics Package, as shown in the Block Diagram, Figure 1.

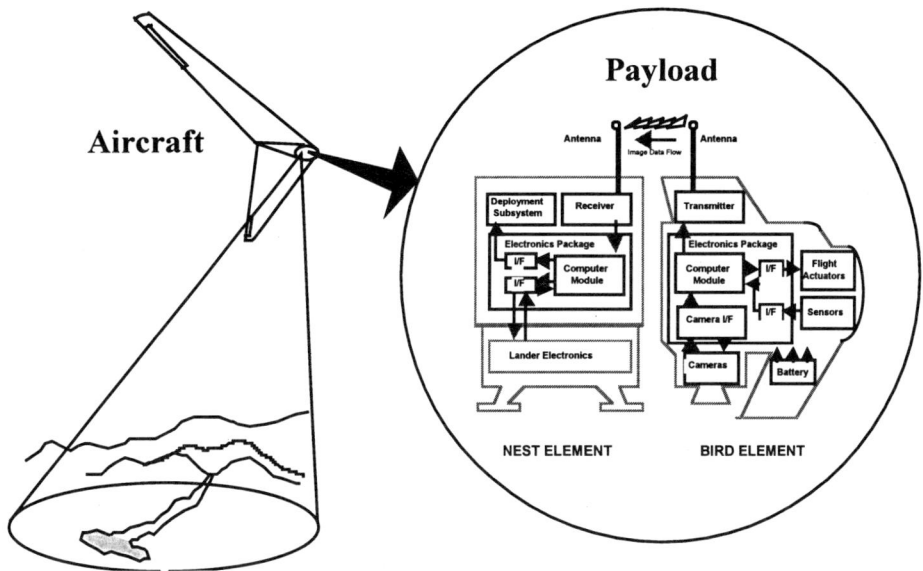

Figure 1 Mars Raptor Reconnaissance System Block Diagram.

The Battery / Power subsystem consists of high density Lithium technology cells which power all electronics directly, and a switch to enable power prior to Bird deployment. The battery operates continuously from initiation of the Raptor system launch sequence to the end of reconnaissance (10 - 60 minutes) without need for recharging during the mission.

The Antenna and Transmitter combination operate at L Band frequency, chosen because of the wide availability of portable devices from the cellular telephone industry. Data is transmitted at a bandwidth of 4 MHz. The Flight Actuator Subsystem provides lightweight servo control of the wing, elevator, and possibly the tail air control surfaces. It operates from control signals output by the Bird Electronics. The Sensor Subsystem provides analog input data from three axis accelerometers and an airspeed sensor. These signals are conditioned then Analog-to-Digital converted in the Electronics Package to be processed by the Computer Module.

The Electronics Package contains the Computer Module and electronic interfaces between this module and all other subsystems, including the Camera Interface. The Camera Interface outputs timing signals to drive the Camera's CCD focal plane arrays, and inputs pixel data which is converted to digital values and written to the Computer's memory at high speed. Since images are taken from only one camera at a given time, both CCD's can use common electronics.

The Computer Module is the mastermind of the Bird, affecting the flight control, navigation, engineering data acquisition, camera image acquisition, and data transfer functions. This high performance, power saving general purpose computer is packaged in an extremely low weight module, and (except for removable submodules) is identical between the Bird and Nest elements in order to reduce development time and construction costs. Such a computer under development at Phillips Laboratories is the Advanced Instrument Controller.

Software executing in the Computer Module performs self test, camera data transfer, navigation, and flight control algorithms. Camera data is read by software from memory, where it was previously placed by the computer's Direct Memory Access hardware, then sent in packets to the transmitter for relay to the Nest Element. Navigation software processes airspeed and three axis accelerometer data to determine actual course, which is then compared with the predetermined spiral flight path, providing feedback to the flight control software. This software operates at with a 100 millisecond update rate.

Engineering data will also be acquired by the Computer Module and transmitted to the Nest with a time tag so it can be correlated with the time tagged images for processing on Earth. This data, consisting of temperature, airspeed, and other navigation data, will be used to provide additional context for the reconnaissance images.

Navigation and flight control software will operate continuously, with an image transfer occurring on an average every 10 seconds and at a peak rate of one every 2 seconds. Each image consists of 1024 x 1024 one byte pixels, occupying 1 MByte or 8 Mbits (note: 8 Mbits / 2 secs = 4 MHz bandwidth). If the computer in the Bird stores and transfers one image while it is acquiring the next one, it needs 2 MByte of memory for data storage.

Using advanced packaging technology (discussed in the next section), it is anticipated the entire Electronics Package will have a mass of 25 - 50 grams, and the Battery subsystem will have a mass of 30 grams. Depending on component selection, it may be necessary to add a mass of 20 grams for local radiation shielding.

Nest Element. The primary function of the Nest Element is act as an interface between the Bird and the Mars 2005 Lander. The Raptor Nest Element deploys the Bird, receives both image and engineering data from the Bird, compresses this data, then saves it and transfers it to the Lander (for relay to Earth via a Mars orbiter). It acts as a digital data buffer between the high acquisition rate of the Bird and the low orbiter transfer rate.

The Nest subsystems are mechanically mounted on the Lander platform in such a way that the Deployment Subsystem has unobstructed access to the atmosphere in a lateral axis. The omni-directional whip antenna can be stowed in either a folded or extended position within the aeroshell, but is extended during Bird deployment, and must have an unobstructed view of the Martian sky during descent and after immediate landing.

The Nest's power is enabled before the mission's descent phase, which allows the Lander to communicate with the Computer Module. Software is downloaded before descent, and the Raptor system commanded to perform Self - Test. At the appropriate time in the mission timeline, the Raptor deploy command is issued by the Deployment Interface, which ejects the Bird into the Martian atmosphere at its initial altitude. The Computer immediately enables its Receiver to input image data from the Bird. This compressed data is transferred directly to computer memory where it is stored, compressed, then sent at a low rate via the Lander Interface for transmission to Earth.

The Nest consists of the following subsystems: Antenna, Receiver, Deployment Subsystem, and an Electronics Package, as shown in Figure 1. The Antenna, Receiver pair are tuned to the Bird's L Band transmitter frequency. The Receiver demodulates the signal, detects it as digital data, and transfers this data into the Computer memory via Direct Memory Access.

The Electronics Package consists of the Computer Module, the Lander interface, and the Deployment Subsystem interface. The relatively simple Deployment Subsystem interface consists of the high reliability discretes to command and detect the actuation of required mechanical or pyrotechnic devices. The Lander interface connects to the Lander electronics using the latter's standard power and bus protocol. Only 5 Watts of peak power will be required by the

Nest during the Bird deployment phase (60 minutes maximum), and 0.5 Watts immediately before deployment.

The Computer Module, composed of standard computing components, is shown in Figure 2. This architecture is the same between the Nest and the Bird computers. It consists of a Central Processing Unit (CPU), Read Only Memory (ROM), Random Access Memory (RAM), Direct Memory Access (DMA), and Input / Output (I/O). ROM is used for storing essential software instructions since it is nonvolatile. RAM, which faster and larger, is used to store data and downloaded software, but it must be powered to retain its data. The I/O uses parallel ports to directly transfer data between the CPU and external modules. The CPU itself will consist of a CMOS microprocessor with the capability of slowing its clock cycles during non-processor intensive periods to reduce power consumption.

The Nest Computer requires more memory than the Bird computer. These data storage requirements are dominated by image data. Each image is 1 MByte in size before data compression, and there are 200 images acquired during the mission. Modern data compression algorithms with a 5:1 compression ratios (University of Arizona IMP Martian imager) can be applied to the data. The total memory storage requirements for images is 40 MBytes (= 1 MByte / image * 200 images / 5). Engineering data storage require only 300 KBytes total assuming 100 bytes of data are received every second for the entire Bird deployment phase (3000 seconds). An additional 3 MBytes will be required by the software instructions and data.

A common Computer Module design between the Bird and Nest was chosen to meet interplanetary mission requirements for very low mass and minimal power consumption. A common design will afford the additional benefits of reducing cost, development risk, and software complexity.

One packaging approach to be considered is High Density Interconnect (HDI) which could implement the Computer Module in a package with a mass of less than 25 grams. By reducing the hardware elements to bare dies which are mounted to a thin ceramic substrate, then interconnected by micro-vias deposited on Kapton layers above the substrate, HDI eliminates the additional mass of IC packages, circuit boards, and extra connectors. Support hardware for thermal and vibration support is reduced accordingly. Packaging the substrates as submodules that mount on a larger substrate has advantages for testing and integration, making the computer architecture shown compatible with an HDI approach.

HDI also has benefits for radiation considerations. Because the area to be shielded is so small due to proximity of the dies, local shielding for radiation becomes feasible. This, coupled with the relatively low radiation exposure within the aeroshell and the very short deployment life outside of it, allows some non-hardened components to be used, increasing capability while reducing cost. Through state of the art packaging techniques and use of powerful commercial electronic and software technology, we will be able to produce a high performance, highly reliable instrument suite at a low cost.

Expected Results. The Bird and Nest elements described above can be built and tested in order to demonstrate that the necessary degree of miniaturization, mass savings and performance factors have been achieved which will permit a flight model of the Mars Raptor Reconnaissance System to be built and flown on the Mars '05 mission. We intend to demonstrate a Bird payload element which has a total mass of 230 g or less and a Nest element of 200 g or less. Electrical power dissipation goals for the demonstration breadboards are: ≤ 5 W for both the Bird and Nest elements.

We plan to test the breadboard demonstration system by means of a captive cable deployment approach. The Bird payload element will be mounted in a cable car gondola permitting motion with respect to the ground, where a scale mockup of Mars surface terrain features can be "overflown" as the gondola vehicle descends on a downward sloping trajectory while attached to a support cable. During the cable tests, image data will be transmitted to the Nest element for storage and subsequent retrieval and evaluation. Another test station will exercise the Bird's on-board inertial reference and 3-axis glider active control systems through captive testing of the Bird Element on a rate table. The Bird will be mounted on the edge of a rotating platform, where pitch, yaw and roll disturbances can be introduced in order to change the attitude of the Bird with respect to the rate table. These disturbances will simulate the effects of wind gusts that could be encountered by a real Mars Raptor. The rate table test will then assess the degree to which the induced attitude disturbances are sensed and compensated for by commands to the appropriate glider pitch, yaw and roll control surfaces.

SOIL SAMPLING ON MARS*

John L. Paterson[†]

EXECUTIVE SUMMARY

In response to the need to develop a device for collecting and isolating soil samples on a landed Mars mission, a preliminary study was conducted. The need arises from a new class of experiments in exobiology dealing with the search for organic material and the nature of soil oxidants. This work was originally intended as a follow-on to the MOX '94 experiment onboard the ill-fated *Mars '96* Russian lander.

It is thought that organics may be present at some depths and locations on Mars. To analyze organics it is necessary to decompose a sample by heating it within a chamber and testing the evolved gasses present. The basic instrument to perform this analysis can be configured many ways, *but all require a known quantity of soil from a known depth* to be captured in a sealed container and humidified similar to what will be done in the Thermal and Evolved Gas Analyzer (TEGA) experiment scheduled for the Mars Polar Lander. The soil required may come from just under the surface, perhaps only a few inches, or from depths of several meters.

The ability to capture soil in either case *as simply as possible with very limited mass and power* is the context for the following study. This paper examines how near surface soil conditions are measured and replicated, and discusses test results of three candidate designs of soil capturing devices. The study is presented Chronologically.

REPLICATING PHYSICAL SOIL CHARACTERISTICS BASED ON THE VIKING LANDER RESULTS

The Necessity of Simulating Mechanical Soil Characteristics:

In the initial design phase of both near-surface and sub-surface sampling mechanisms (or any device that interacts with soil) it was decided to duplicate the conditions found near the surface according to Viking. While these conditions do not represent the entire range of soil characteristics found on Mars, it served as a good starting point (particularly for the near-surface samplers) and was useful for two reasons:

1. A knowledge of *in situ* working conditions is of great aid in designing a device. and
2. Good fidelity simulated conditions are required to validate a device with any degree of confidence.

---

*   This work was originally performed for C. McKay and L. Lemke at NASA Ames Research Center in 1994.

†   Systems Engineer, Lockheed Martin Missiles & Space. E-mail: john.l.paterson@lmco.com.

Mars Soil Simulation:

For our purposes it was decided that the first six inches of the Martian surface would be replicated according to the results obtained by Viking [1]. In order to create a simulant, knowledge of grain size distribution, cohesion and density were of interest. The most important parameter in the design of soil capturing devices (particularly penetrators) is *cohesion*. This is the ability of individual particles to 'stick' together, and dictates the level of difficulty of penetrating material. The cohesions of Martian soil-like materials, (albeit at only two locations at depths to 10") were indirectly measured by Viking 1 & 2 during the 'Backhoe Touchdown Tests'.

These tests employed the use of Viking's motor-driven arm and Backhoe mechanism. As the Backhoe dug through soils of known depths, ammeter readings were taken at the extend drivemotor and converted to a known force exerted by the collector head. Terrestrial touchdown tests were later performed by the Science Test Lander using Lunar soil simulants of varying known cohesions. The forces exerted on the Test Lander's collector head were correlated to the known value of cohesion allowing the cohesion value of the Martian *in-situ* to be backed out. This was the method employed to gather the cohesion parameter.

Once the cohesion of the soil is established, the cohesion of the representative sample can be measured easily using a device known as a *Shear Vane Torquemeter*. This same device was used on the Science Test Lander's simulant to determine its cohesion property. A mathematical constant is employed which coverts torque (as seen by the Shear-Vane when the soil fails) to cohesion. A Shear-Vane device was fabricated according to the personal notes of H. Moore, and was used to calibrate the cohesion of our Mars analogue soil.

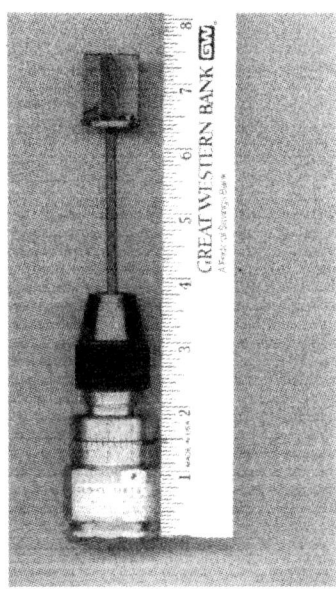

Photo of Soil Cohesion Calibration Tool or Shear Vane Torquemeter.

Grain Size Distribution:

The grain size distribution of the soils found by Viking were not well established, and it was decided to use material similar in grain size distribution to the Lunar analogue used in the Viking Science Test Lander Backhoe touchdown tests.

Simulant Sample Preparation:

The material used for our simulant was crushed basalt, which is the same material typically used for Lunar soil analogues. Soil was obtained from a local quarry, sieved and re-combined to a known grain-size distribution [2]. It was then compacted in a press until a known cohesion was reached. For our purposes it was decided that a medium cohesion of 2 KPA would represent the drift material found by Viking and would be suitable for near-surface penetration tests. The soils cohesion was measured using the shear-vane torquemeter. The sample container was 4" in diameter and 6" deep.

Definition of Terms Relating to Samplers:

1. **Near-Surface** means intended for the first 1-2 inches of loose drift material (little or no soil cohesion).
2. **Sub-surface** means intended 2 to 6 inches in depth of more blocky material (little-medium soil cohesion).
3. **Deep Sampler** means intended for samples 6 inch to 3 feet in depth (strong soil cohesion).

DESIGN OF NEAR-SURFACE SAMPLER

The requirements of the near-surface sampling instrument include the ability to:

1. penetrate beneath the first inch of soil
2. capture approximately 1/2 cc. of soil
3. retrieve the soil inside a gas tight container capable of maintaining 15 psi
4. humidify the sample
5. provide O_2 and pH sensors.

An initial working model was concepted and fabricated out of clear Plexiglas™ for visibility. The goal being to test a hand held device and crudely measure the force required to penetrate soil and to study soil capturing capability. The ability to humidify/pressurize and instrument the device were only concepted into the samplers general configuration. The sampler was tested by pushing it approximately 2" into the simulated drift material by hand.

While penetrating, material falls through window slats near the tip, where it is deposited on a platform. After the platform is covered, the 1 cc of material is raised via a plunger and brought to the top of the container where it would be trapped inside a sealed chamber for analysis.

The design is simple, requiring only to movements, lends itself to being fully automatable and performs the job of acquiring a known volume of material from a known depth.

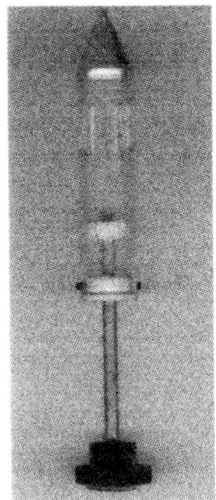 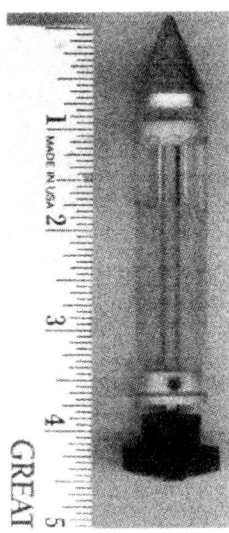

Photos of Near Surface Sampler in the Deploy (left) and Retrieve (right) Positions.

Several techniques were discussed and critiqued regarding how the sampler could be pushed further into the soil or should blocker stronger material be encountered. These included:

1. brute force (i.e. a large mass on top of the sampler) (not very aerospacey)
2. an explosive device that drives the sampler in (scientists aren't very keen on the possibility of sample contamination by combustion products)
3. a pneumatic percussion mechanism (requires compressed gas, not good from a systems standpoint)
4. vibration via an electrically driven solenoid.

The vibration method was chosen for further study because it lends itself to being the most mass/energy efficient system of inducing work on the soil [3].

After several iterations, a new penetrator was fabricated out of stainless tubing. The spiked shaped tip was designed with a 1 to 4 taper. The body size chosen was 3/8", (much smaller than the original 1" diameter). This was done to reduce frontal cross-sectional area which reduces the downward pressure required to penetrate soil (tool pressure). A linear actuator was attached to the penetrator, and driven by an signal generator/amplifier. A soil testbed was fabricated which allowed the penetrator to be suspended and counterweighted above the soil, to vary the tip force.

Testing of Near-surface Sampler:

Trials were performed with the device vibrating at frequencies from 10 to 50 Hz. in calibrated Mars simulant with a cohesion of approximately 2-4 Kpa. After permeating the soil 1-1/2" past the tip, (but not far enough for soil to fall into the slats), the device stopped and would not push any deeper into the soil. This was probably due to vibratory compaction of soil ahead of the tip as it worked its way into the sample. The mass of the sampler (load on the tip) was changed from 1/8 lb. to 1/2 lb. without any effect other than to dampen the vibration.

Conclusions:

This method of near surface sampling shows promise as a mass/energy efficient system for sampling blocky material to a depth of 2-3 inches. A follow-on study would probably include contacting the investigator of previous related work [3] to get details of their experiment. Reducing the diameter and shape of the tip may reduce tool pressure. Further experimentation is warranted after more key issues have been identified.

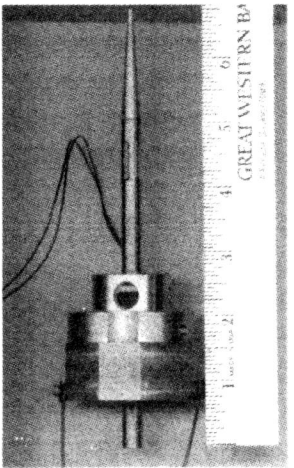

Photo of Vibratory Sampler.

DESIGN OF A SUB-SURFACE SAMPLER

A novel device to sample soil between 2" and 6" in depth was investigated following the non-conclusive results of the near-surface vibratory sampler. The inspiration for this design came from the *in-situ* Lunar sampling devices used during Apollo (the Rotary Impact Hollow Core Drill) and the Russian sample return mission which returned a 2 m deep column of soil using a helix drill. In both cases the entire column of soil was captured for later analysis. While this is useful for studying stratigraphy of the soil, only about 1 cc of soil is required from the bottom of the hole for the purposes of exobiology and related studies on Mars. Both devices were also designed for one time use only, whereas for the purposes of exobiology, several samples from different locales using the same device repeatedly is warranted. The location of each subsequent sample may be dependent on the results of the previous test. This assumes that that the sampling device is mounted on a rover or other mobile platform.

It was decided that the first candidate concept would be a derivative of a core drill. However, instead of being a continuous hollow core tube, is was configured from a very short 1" OD core drill, only 1" in length. This was attached to a 3/16 shaft via a secondary chopping blade at the top. A hole saw was used as the core drill body, with the secondary blade and shaft welded on.

The theory of operation is that as the drill spins and is pushed into the soil, the soil *flows through* the hollow core until stopped at the desired depth. When it stops the core is full of slightly compacted soil, held inside the core and retrievable if pulled up out of the hole. Operationally, the drill would be placed over a container and the sample pushed out by separate forks from the top. It is worth noting that when drilling through material, *every cross-sectional part of the tool that comes in contact with the soil must disturb the soil ahead of it*. A drill will not readily penetrate soil under brute downward force such as in the case of the spiked tip used in the near-surface sampler.

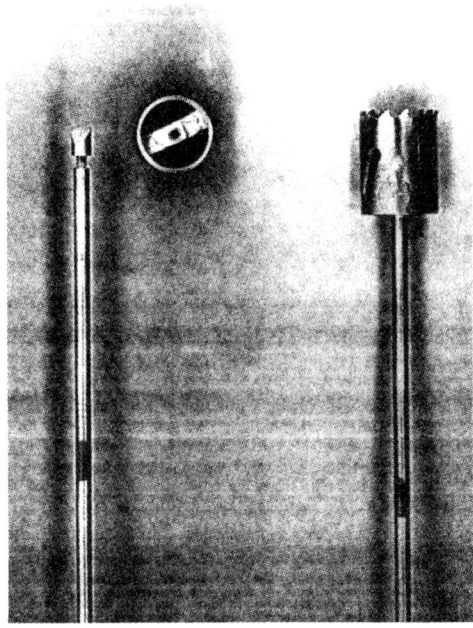

End and Side View Photos of Core Drill for Deep Sampling.

Soil Testbed Preparation and Testing of Sub-Surface sampler:

A first soil testbed was made by compacting (to 2-4 Kpa initially representing Crusty/ Cloddy material) crushed basalt into a 4" OD steel tube container 6" in depth with 1/2" thick brightly colored chalk at the bottom. The theory being that if the core-drill stops at the chalk level, the core will be full of only that material represented at that depth.

Spinning at 300 RPM, with a downward force of 5 lbs., the drill took 2 minutes to reach the full depth of 6". Once there, it was lifted straight up (while still spinning clockwise), halted, and soil pushed out of the core by hand from above. This was performed successfully several times with soil of varying cohesions. It was found that if the soil's cohesion is too low, material would fall out of the core immediately after extraction, and if too high, then the drill would stall, delivering the torque to the container.

Conclusions:

This candidate concept shows promise as a means for sampling soil to a depth of 6". These initial tests made use of a hole saw designed for wood, not soil. Although success was attained to this depth, it became apparent that the system relies heavily on known soil cohesions and grain size distribution. For example, it would not work if small shards (1/8" pea-gravel) type of material were encountered. Further success will require experimentation on tooth design, spacing, shaft interface etc. if this configuration is pursued.

DESIGN OF A DEEP SAMPLER

In response to the desire of obtaining soil samples from a depth of as much as 1 m, a unique device was concepted. The requirements included the ability to obtain a known volume of soil from a known depth, and the ability to obtain samples from different locations. While an end-to-end system was concepted, the 'front end' of sample collection at depth was demonstrated. The task was split into two parts: drilling to depth and soil collection/retrieval. The drilling to depth is performed by a helix type tube drill. The inspiration for this design stemmed from an ice auger (a stainless tube with spiral flytes) used in mountain climbing. The sample collection approach is unique and uses a deployable drill tip with a cylindrical chamber to capture material. The drill was intended to handle drift, crusty/cloddy and blocky type material.

Method of Operation:

The system works by drilling into the soil to a desired depth and deploying the tip of the drill via a solenoid at the top of the shaft. The deploying of the tip away from the body of the drill allows material into the cylindrical chamber. Once material has entered the chamber, the tip is retracted, capturing the soil. The drill is then reversed and backed out of the hole and swung to a position above a sample collection container. The tip is deployed again, allowing the captured material to fall into a sample collection container where it is analyzed. Disposable cups would be used in the analyzer (ejected out the bottom after use) so that the whole process can be repeated.

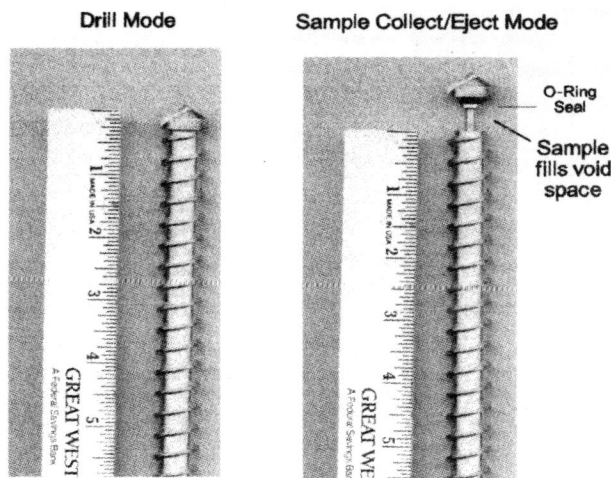

Photo of Deep Sampler.

Testing the Deep sampler:

A soil testbed was constructed consisting of a 3-ft. deep thickwall PVC pipe, 14" in diameter attached to an aluminum plate at the bottom. The pipe was filled with unconsolidated crushed basalt with three 1" wide bands of colored chalk placed at 6", 18" and 3-ft. depths. The drill was powered with a 120 V. AC electric drill turning at approximately 300 rpm. A stand was built with vertical shafts and cylindrical roller bearings which held the electric drill upright and steady to produce a vertical hole. The mass on top of the drill was approximately 3 lbs. The diameter of the helix on the sampler was 1".

Three trials were performed in the testbed by drilling down to the first, second, and third bands of colored chalk. At each banded layer, the drill was halted, the tip deployed (by pushing a plunger by hand) material captured and brought to the surface where it was placed into a small clear container. A total of nine samplings revealed the attainment of approximately 90% colored chalk and 10% gray basalt. This led to a high degree of confidence that only the soil at that specific depth was sampled. The samples were not weighed but judged visually to be of similar quantity in each case (approximately 1/2 cc).

Conclusions:

The results of testing the deep sampler show good promise as a candidate for sampling the soils of Mars at any depth, given that the physical characteristics such as cohesion, grain size distribution and grain shape will allow it. The tip, helix and size of the drill warrant further study, as does the interfacing with a rover and the sample analysis container it would interact with.

INTEGRATING A SAMPLE ACQUISITION SYSTEM ON A ROVER

After the success of the initial deep drill testing, it was decided that the next step would be to layout the rest of the system, with emphasis on rover interfaces, mechanical components, controls, kinematics, stowage, and how the system could be integrated with the sample analysis instrument(s). This also included concepting a system that was flight-like. The deep drill/sampler was originally designed to be fit on a rover of the same class as the Russian *Marsokhod*, a six wheeled, 100 kg, 1 m long very capable platform. A preliminary systems design study revealed the following:

Stowage: The 1 m. drill would be stowed horizontally on a rover bed for stowage, and would unstow to a vertical position when needed operationally. It would not need to be restowed unless found to upset the center of gravity of the rover or interfere with other instruments (although it might be advantageous to have a wind measuring instrument or camera on top of the drill, as it would provide an existing platform 4 feet from ground level).

Deployment: when the rover stops at a site to be sampled the drill would swing into a vertical position, rotating about the bottom of the drill. This one movement brings the drill vertical and positions it out over the edge of the rover in the front or rear. The drill would require a reversible motor with a variable torque clutch head (or equivalent) for motor protection in a stall situation. This can be accomplished with a motor/gearhead the size of a D battery, as is done on small cordless electric screwdrivers. A rack and pinion or leadscrew linear slide would accommodate vertical movement (similar to an automated drill press). Once the sample is acquired the drill would be backed out of the hole and be rotated to a position over the sample cup located in the instrument chamber where the sample would be deposited. The drill tip chamber would then be purged with pressurized CO_2 from the atmosphere for cleaning between uses to avoid sample contamination between samplings.

SIZING A SYSTEM FOR A *SOJOURNER* CLASS ROVER
(1998 update)

Current plans include the deployment of a *Sojourner*-like vehicle (*Athena*) on the surface of Mars in 2001. The drill sized for the larger rover was 1" in diameter, this being the largest diameter practical, given a projected rover mass which drives available tool pressure in the soil. For a smaller platform, tool/sampler size could be reduced to about 1/4" to 3/8" to provide the same tool pressure on the soil given the reduced mass of the rover. The deep drill would necessarily be only about as long as the rover, although it could perhaps be longer than the rover bed if stowage room is available. The *TEGA* soil analysis instrument might be placed on top of the non-rotating portion of the drill and samples inserted from an elevator within the drill (similar to near-surface sampler) from beneath, to eliminate the need for precious deck space on the rover.

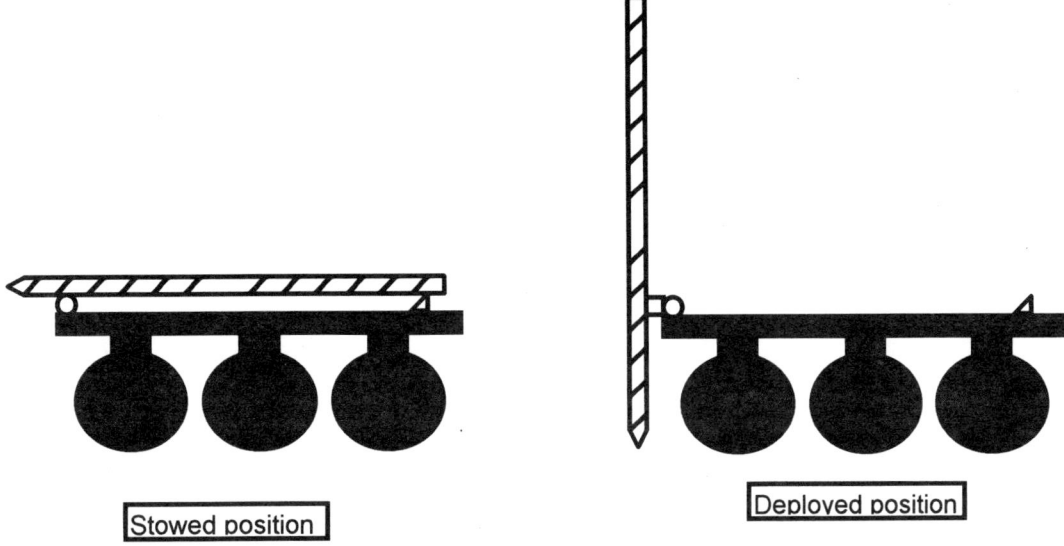

If the downward force required to drill is projected to exceed what the rover can provide without lifting the rover off the ground, then some type of anchor may be required [4].

ACKNOWLEDGEMENTS

The following contributions to this work are greatly appreciated:

Henry Moore: Help with understanding soil dynamics/Viking history and use of personal notes to construct a shear-vane torquemeter

Peggy Kacsmar: Editing and photo-imaging

Langly Hill Quarry: Gratis crushed Basalt

REFERENCES

1. Moore *et al.* (1987). "Physical Properties of the Surface Materials at the Viking Landing Sites on Mars," USGS Professional Paper 1389 USGPO, Wash. DC p.76.
   Moore, H. J. (1991) "Estimates of some Physical/Mechanical Properties of Martian Rocks and Soillike Materials," Open-File Report 91-568 USGS, Menlo Park CA.
2. Bernold. L. (1988). "Earthmoving in the Lunar Environment," Proc. of Space '88. Engineering, Construction, and Operations in Space (S. W. Johnson and J. P. Wetzel, eds.) ASCE, New York, NY pp. 205-207.
3. Nathan *et al.* (1992). "Mass and Energy Tradeoffs of Axial Penetration Devices on Lunar Soil Simulant," Proc. of Space III, Engineering, Construction, and Operations in Space (W. Sadeh, S. Sture and R. Miller eds.) ASCE, New York, NY pp. 441-457.
4. Paterson, J.(1992) "Mobile Continuous Lunar Excavation," Proc. of Space III, Engineering, Construction and Operations in Space (W. Sadeh, S. Sture and R. Miller eds.) ASCE, New York, NY pp. 1076-1077.

MAR 98-041

MARSPLANE - FLYING ON MARS WITH EXISTING AIRCRAFT

Fabrizio Pirondini*

NASA's ERAST program is developing unmanned aerial vehicles to fly high altitude, long endurance missions. This paper studies the performances of these aircraft in low Martian atmosphere, an environment very similar to Earth's stratosphere. Since no oxygen is available in the Martian atmosphere, a solar powered aircraft is analyzed. Such an aircraft could help to deploy planet-wide weather station and microrover networks, greatly enhancing their flexibility. Results show that with current technology it is possible to develop an UAV capable of continuous flight in low Martian atmosphere. Pathfinder and Centurion aircraft, developed by AeroVironment Inc., could meet that goal. Pathfinder was found to be the best suited to the Martian environment, being able to fly continuously at latitudes between 30°N and 50°N with a 20 Kg payload for up to 160 sols. Both Pathfinder and Centurion are able to fly all year long during daylight hours with a 100 Kg payload near the equator, and in spring and summer months in the ± 50° latitude range. Flight latitude and time of year play significant roles in determining payload mass for both aircraft. Improvements in fuel cell and solar cell efficiency could greatly enhance the performances of this kind of aircraft.

1. INTRODUCTION

In the last decade a renewed interest has developed towards studying advanced unmanned aerial vehicles (UAV), which have the potential to complement existing platforms for Earth and atmospheric sciences (without mentioning obvious military purposes). There was a need for aircraft with the combination of subsonic capability, very high ceiling, long range and endurance. In response to this need NASA has started in 1994 the ERAST Program, aimed at developing and flight testing a new generation of UAV, designed for very high altitude (up to 30 Km), long duration (up to 100 hours) flights.

In the meantime, after a decade of almost complete paralysis in Solar System exploration, NASA is starting a new golden era in planetary studies. The American commitment to pursue a launch attempt to Mars every launch window during the next decade, and maybe further on, will hopefully start the long-term colonization of the Red Planet.

Accidentally the aerothermodynamic environment that the advanced UAV are going to face in upper Earth atmosphere is very similar to the one they would find in low Martian atmosphere.

The purpose of this paper is to investigate if, with existing or soon-to-be-developed technologies, it would be possible to build a UAV able to fly in the Martian atmosphere for very long missions (days to weeks). After establishing a model for solar irradiation on Mars surface that takes into account latitude, seasonal variations and dust storms, the performances of ERAST UAVs in Martian atmosphere will be analyzed. Specifically we will try to point out in which condition it will be possible to have overnight and daytime flights.

---

* Fabrizio Pirondini, Aerospace Engineer, Via Lolli 28, I-42100 Reggio Emilia, Italy. E-mail: pirro@tin.it.

2. THE ERAST PROGRAM

In response to the growing scientific need for in-situ measurements at altitudes and with duration greater than the current fleet of science platforms, the NASA Office of Aeronautics has sponsored the Environmental Research Aircraft and Sensor Technology (ERAST) Program (Ref.[1]). Started in FY94, the program is scheduled for seven years with flight demonstrations of vehicle technology and payload capability.

The two primary science mission requirements for the advanced remoted piloted aircraft (RPAs) are:

1. very high altitude (up to 30 Km), long endurance (up to 50 hours) with science payloads of 100-1400 Kg.
2. high altitude (24 Km), long endurance (96 hours) with 50-1600 Kg payload capability.

Both these missions require aircraft that would fly at subsonic speeds at this altitudes in order to meet key science measurement requirements. The critical technologies that would be developed and demonstrated under the ERAST program include high subsonic Mach number low Reynolds number aerodynamics, propulsion, lightweight structures, autonomous flight.

The main projects within the ERAST program due to fly higher than 75,000 ft, and their characteristics, are summarized in the following table.

Table 1
ERAST Program aircraft (with max. altitude over 75,000 ft) specifics
(From Ref.[2],[3],[4],[5],[6],[7])

Name	Manufacturer	Overall Length [m]	Wing Span [m]	Aspect Ratio	Wing Area [m^2]	Body Diam. [m]	Launch Weight [Kg]	Cruise Speed [m/s]	Endurance [hours]	Maximum Altitude [ft]	Power plant
Pathfinder	AER	3.66	30.50	12.3	75.6	N.A.	218	7	-	75000	a
Perseus A	AUR	8.11	18.30	-	-	0.79	795	36	5	75000	b
Perseus B	AUR	8.11	18.30	-	-	0.79	1135	36	24	75000	c
Centurion (*)	AER	2.44	62.80	25.7	153.2	N.A.	500	9	-	100000	d
Theseus (**)	AUR	14.95	42.09	-	-	0.92	2770	23	20	82000	e

Code	Powerplant
a	6 x AeroVironment electric
b	1 x Arion 1 closed cycle
c	1 x Arion 2
d	14 x AeroVironment electric
e	2 x Arion 3

Code	
AER	AeroVironment Inc., Simi Valley, CA, USA
AUR	Aurora Flight Sciences Corp., Manassas, VA, USA
(*)	Not yet flown (only Proof of Concept has flown)
(**)	Not within ERAST program

3. THE MARTIAN ENVIRONMENT

The surface pressure on Mars is less than one per cent of that on Earth, and as on Earth pressure decreases with increasing altitude. At the lowest point on Mars, in the Hellas basin, the pressure is 8.9 millibars (mb); at the summit of the large volcanoes in the Tharsis region the pressure is as low as 2 mb. There is no convenient "sea level" on Mars, so in this analysis an arbitrary datum level is chosen at 560 Pa pressure (Ref.[8], pag.31). Temperature on Mars are far lower than on Earth: a typical temperature at the equator is about -50°C, though in some equatorial regions the temperature may climb above the freezing point of water and as high as 30°C by noon. At the poles a typical winter temperature is around -150°C. Maximum wind speeds were around 9 m/s (32 Km/h).

The following tables summarize models of Earth and Mars atmospheres used in this paper.

Table 2
Temperature, pressure, density and sound speed in Earth atmosphere.
(From Ref.[9])

h [m]	h [ft]	T [K]	p [Pa]	ρ [Kg/m³]	a [m/s]	a [km/h]
0	0	288.16	101325	1.2250	340.3	1225
20000	61000	216.66	5529	0.0889	295.0	1062
30500	100000	232.81	1102	0.0165	305.8	1101

Table 3
Temperature, pressure, density and sound speed in low Martian atmosphere.
(From Ref.[10])

h [m]	h [ft]	T [K]	p [Pa]	ρ [Kg/m³]	a [m/s]	a [km/h]
0	0	218.0	560	0.0170	235.9	849
1000	3050	215.7	540	0.0167	234.7	845
2000	6100	213.4	520	0.0164	233.4	840
3000	9150	211.1	500	0.0161	232.2	836

The sound speed, a, is derived from the well-known perfect gas formula:

$$a = \sqrt{\gamma(R/m)T} \qquad (1)$$

where γ is the ratio of specific heats (c_p/c_v), R is the perfect gas constant (= 8314 J K^{-1} kmol^{-1}), m is the mean molecular weight ([g]), T is the air temperature ([K]) and a is the sound speed ([m/s]).

The Martian atmosphere is mostly composed of carbon dioxide, with minute traces of nitrogen, argon, water, carbon monoxide and inert gases like krypton and xenon. For different gas compositions, such as Earth and Mars atmospheres, there are changes in values of γ and m. Earth atmosphere is mainly a diatomic gas ($N_2 + O_2$), while Martian atmosphere is triatomic (CO_2). This translates to different γ, while different molecular weights prompt changes in m.

Table 4
Earth and Mars atmosphere chemical composition and properties
(From Ref.[8], [10])

Planet	Atmosphere chemical composition	γ	m	R/m
Earth	N_2=78% (m=28), O_2=21% (m=32), H_2O =1% (m=18)	1.40	28.97	287
Mars	CO_2=95% (m=44), N_2=3% (m=28), Ar =2% (m=40)	1.33	43.44	191

4. VEHICLE MISSION

Various studies (Ref.[11]) have already stated how a Mars airplane could be a very important complement to Mars exploration of the Red Planet. Its gondolas could be equipped with lots of sensors. Just to give some examples:
- Camera (Surface imaging, mapping, geological studies, meteorology)
- Spectrometers (Surface, subsurface physical properties)
- Neutron Analyzer (Soil composition and water content)
- Ion Beam (Soil composition)
- Radar (Surface imaging, mapping, soil composition, geological studies)
- Magnetometer (Magnetic properties)
- Atmospheric package (Atmospheric composition, meteorology).

Moreover it could be possible to load some small weather stations or microrovers on the aircraft and to release them while on the ground or even to parachute them on any chosen spot on the planet. In this way it should be possible to quickly deploy a planet-wide weather station system and a network of microrovers to explore the most interesting areas. It would likely be a cheaper and more flexible approach than deploying the system from orbit.

In case two different manned outposts are built, the aircraft could be used to transport samples from one lab to another, since it is unlikely (because it is obviously more expensive) that both outposts will have exactly the same scientific instruments.

5. ERAST AIRCRAFT ANALYSIS

While there are many aircraft available that fly slowly at low altitudes, and there are high performance aircraft that can fly over 80,000 ft at supersonic speeds, there are no aircraft presently available that fly higher than 72,000 ft subsonically. Current UAV record is 71,500 ft (21,800 m) reached by Pathfinder in June 1997, even if Pathfinder itself is scheduled to reach 80,000 ft in summer 1998. Within the same project Centurion is due to reach 100,000 ft. The dynamic pressure available to a subsonic aircraft at higher altitude is very limited: such an aircraft must therefore be lightweight with wing loading more like a sailplane than a powered aircraft.

This study will show how ERAST UAVs would perform in low Martian atmosphere. In order to study their performance we shall provide them with a propulsion system designed for the specific Martian environment.

Martian atmosphere is almost completely void of oxygen, therefore it would be a lot easier to use a completely electric aircraft instead of combustion engines. Various alternatives explored (Ref.[11], [12] p.160), like an UAV launched by catapult or rocket and powered by a piston engine driven by steam generated by chemical breakdown of hydrazine, seem much more complex and anyway leak the main characteristic of solar propelled aircraft, i.e. its almost unlimited range.

The great advantages of a solar propelled aircraft (i.e. an UAV with solar panels which move electric motors and charge the batteries or fuel cells needed to fly at night) are:
- Days or weeks endurance
- Very long range
- No need for fuel or oxidizer (no fuel plants or storage facilities).

The efficiency of converting solar flux to propulsive power is determined by the product of the various efficiencies. In this paper we will assume the following efficiencies (Ref.[7],[11], [13]): solar photovoltaic array ~ 19%, power conditioner ~ 92%, motor/gearbox ~ 87%, propeller ~ 86%. The total efficiency, e_{TOT}, is ~ 13%, but should be increased in the next decade to ~ 18-22% (Ref.[11]). For a representative future design we use e_{TOT} =18%.

Colozza study (Ref.[13]) shows that, with all other parameters held constant:

1. Increasing photovoltaic (PV) cell efficiency does not greatly affect aircraft size
2. A reduction in PV cell specific mass is directly proportional to a reduction in aircraft size
3. Increasing fuel cell efficiency gives modest size reductions
4. The size reductions gained by an increase in the specific energy of the fuel cell are substantial.

We will now go on determining aircraft power requirements based on mass and flight altitude. This is done by using the power conservation equation, that gives the total power that must be available to fly:

$$\text{Preq,tot} = \text{Preq} + \text{Ppay} + \text{Pres} \tag{2}$$

The expression for the power required to fly, Preq, is obtained by using the velocity for minimum power or maximum endurance, vmin, given by (Ref.[9]):

$$\text{Preq} = \frac{1}{2}\rho v^3 \cdot Sw \cdot c_{D,0} + \frac{(mtot \cdot g)^2}{\pi \cdot e \cdot AR \cdot Sw \cdot \frac{1}{2}\rho v} \quad \text{where}$$

$$v = v\min = \left[\frac{4(mtot \cdot g)^2}{3\pi \cdot c_{D,0} \cdot e \cdot Sw \cdot (\rho \cdot b)^2} \right]^{0.25} \tag{3}$$

In this expressions ρ is air density, v is cruise speed, S_W is wing area, $c_{D,0}$ is the drag coefficient, mtot is the total mass, g is the gravity acceleration (g=9.81 m/s^2 on Earth, g=3.72 m/s^2 on Mars), e is Oswald efficiency factor, AR is the wing aspect ratio and b is the wingspan. We assume $c_{D,0}$ = 0.04 (Ref.[7]) and e = 0.8 (Ref.[13]).

The payload power, Ppay, is a constant which can be changed depending on the mission being considered.

Reserve power, Pres, is power required for maneuvering or gaining altitude. It is given by:

$$\text{Pres} = mtot \cdot g \cdot RC \tag{4}$$

where RC = v sin α is the rate of climb (or vertical velocity) of the aircraft, when the flight path is inclined to the horizontal at the angle α.

The solar radiation that must be available for flight is given by:

$$\text{Ireq,tot} = \text{Preq,tot} / (S_W \, e_{TOT}) \tag{5}$$

It is now possible to yield an expression for solar intensity required for flight in terms of mtot, Sw and AR, therefore having a parameter to evaluate ERAST aircraft performances in low Martian atmosphere.

Table 5
ERAST aircraft performances in low Martian atmosphere (z = 3000m).

Aircraft	Min. Power -Max. Endurance Speed (vmin) [m/s]	Power Required (Preq) [W]	Solar Intensity Required (Ireq) [W/m^2]
Pathfinder	26.5	1802	132
Theseus	78.9	50525	3509
Perseus A	78.9	21967	3487
Centurion	23.3	2496	91

Since the solar flux on Mars is always smaller than 800 W/m^2, the above results show that Pathfinder and Centurion are the only UAVs capable of ever achieving sustained flight on Mars. The bad performances of Theseus and Perseus-A are surely affected by the assumption made in this tables to substitute their propulsion system (engines, fuel, etc.) with Pathfinder's one (which is surely a lot lighter) without changing any weight. Anyway it is to be said that on Pathfinder or Centurion baseline, that is without making any change on the aircraft, it should be possible to fly on Mars. The following analysis will focus on these two airplanes.

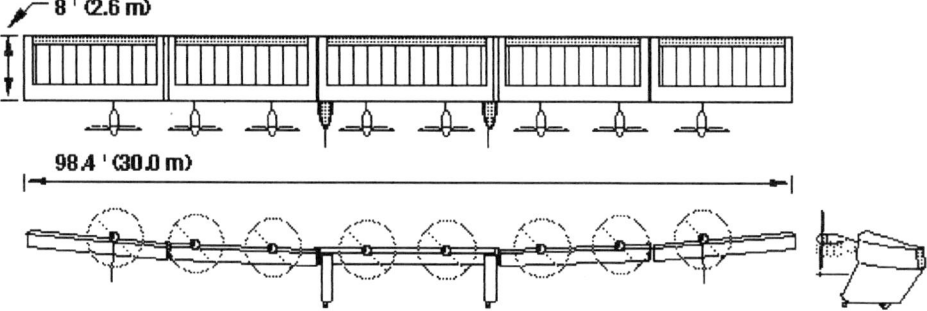

Figure 1   Pathfinder Plus aircraft (from Ref.[3]).

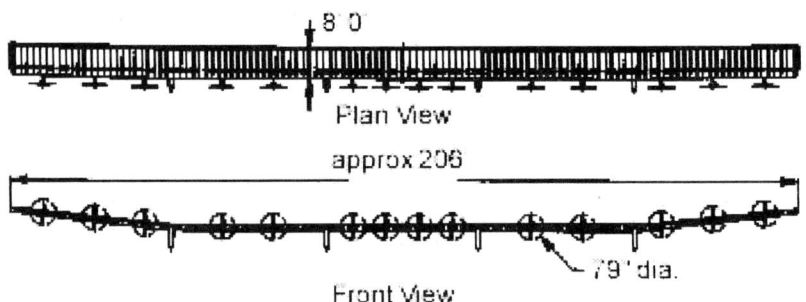

Figure 2   Centurion aircraft (from Ref.[7]).

If we consider different payload masses, keeping payload power requirement at 100 W, and considering reserve power as power necessary for a 5° climb, we have the results in Table 6

Table 6
Solar intensity required (I req.) at "maximum endurance-minimum power required" speed (vmin) for Pathfinder and Centurion with various payload masses. Reserve power is computed with $\alpha=5°$. Flight conditions are those at 300m, as of Table 3.

	PATHFINDER							CENTURION						
mpay (kg)	Wtot (N)	vmin (m/s)	Preq (W)	Pres (W)	Ppay (W)	Ptot (W)	Ireq. (W/m²)	Wtot (N)	vmin (m/s)	Preq (W)	Pres (W)	Ppay (W)	Ptot (W)	Ireq. (W/m²)
0	744	25,3	1562	1638	100	3301	242	1860	23,3	2496	3783	100	6379	469
100	1116	31,0	2870	3010	100	5980	439	2232	25,6	3281	4973	100	8354	614
200	1488	35,7	4419	4634	100	9153	673	2604	27,6	4135	6266	100	10501	772

In Figures 3 and 4 total solar intensity required (Ireq) curves are plotted for different payload masses for both aircraft. These are the levels required to maintain level flight, without reserve power (Pres=0). Note that the most important values are those around minimum, since it is likely that batteries can easily permit take off and acceleration to cruise conditions.

It should be noted that with speed around 25 m/s an aircraft can cover 1080 Km in 12 hours, and circle Mars equator in 240 hours.

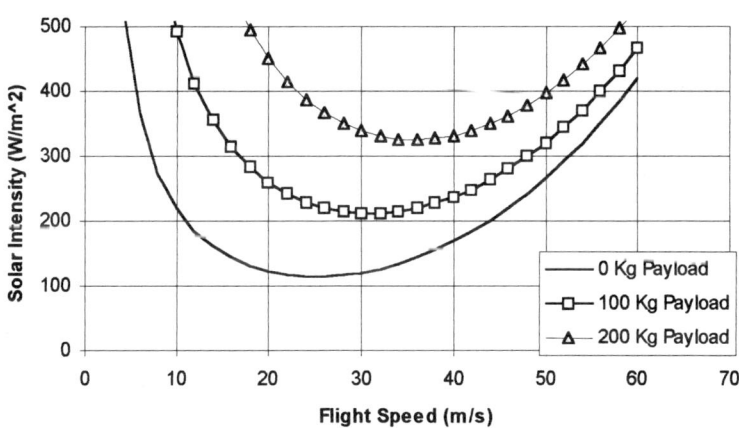

Figure 3 Pathfinder solar intensity required (Ireq) as function of speed for various payload masses. Total mass without payload is assumed to be 220 Kg. Flight conditions are those at 3000 m, as of Table 3.

6. SOLAR POWER AVAILABLE ON MARS

Solar Radiation Intensity on Atmosphere Top

Now that aircraft power requirements have been determined it is time to analyze which is the level of solar intensity available in low Martian atmosphere.

The intensity of solar radiation that reaches the surface of a planet is dependent on the energy incident at the top of the atmosphere, the absorption of radiation by gases and dust, and the scattering of radiation by molecules, dust and clouds. Moreover, the orbital eccentricity, solar declination and rotation rate of the planet also affect solar radiation intensity at the surface.

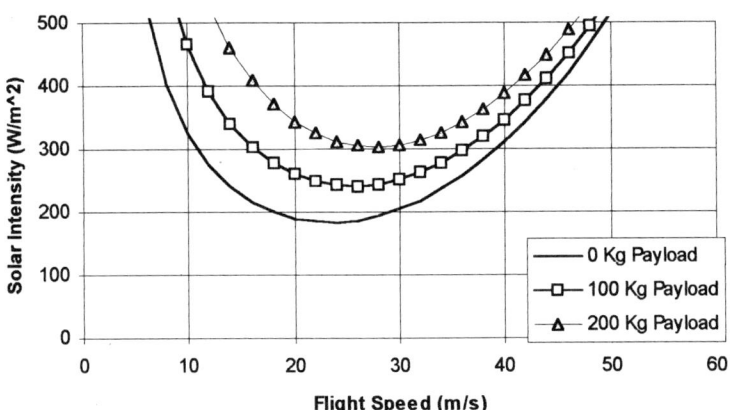

Figure 4 Centurion solar intensity required (I_{req}) as function of speed for various payload masses. Total mass without payload is assumed to be 500 Kg. Flight conditions are those at 3000 m, as of Table 3.

When calculating solar radiation intensity on a planetary surface, it is first necessary to determine the distribution and variability of solar irradiance above the atmosphere. The intensity of solar radiation incident at the top of the Martian atmosphere, I_0, can be expressed as (Ref.[11], [14]):

$$I_0 = [I_{SC} / (r / a_e)^2] \quad (6)$$

where I_{SC} is the solar constant, defined as the average amount of solar radiation in near Earth space (1353 W/m²) and a_e is the semimajor axis of Earth orbit (1 AU ≈ 1.496 10⁸ Km). The instantaneous Sun-Mars distance, r, is given by:

$$r = a(1 - e^2) / (1 + e \cos \theta) \quad (7)$$

where a is Mars' semimajor axis (1.524 AU), e is its orbital eccentricity (0.0934), and θ is the true anomaly, i.e. the angular distance from Martian perihelion, which occurs at an areocentric longitude of 248°. Due to the large eccentricity of Mars' orbit, the solar irradiance varies from 495 W/m² at aphelion to 725 W/m² at perihelion. This seasonal variability is significantly larger than that on Earth.

Throughout our analysis we will always assume horizontal solar cells, since it is highly likely that solar panels will cover the wing surface and that the plane will spend the great majority of its flight time in level flight. For a horizontal collector the irradiance is diminished by the cosine of the solar zenith angle z:

$$I_{hor} = I_0 \cos z \quad (8)$$

The zenith angle of solar incidence, z, can be expressed in terms of various altitude and azimuth angles:

$$\cos z = \sin \delta_s \sin L + \cos \delta_s \cos L \cos h \quad (9)$$

In this expression, L is the planetary latitude, h is the local solar hour angle measured from local noon and δ_s is the solar declination (equal to zero on the date of the equinoxes). The solar declination can be expressed in terms of ε, the Martian obliquity (24.94°), and λ, the areocentric longitude, which is measured westward from the Martian vernal equinox:

$$\sin \delta_s = \sin \varepsilon \sin \lambda \quad (10)$$

It is often useful to write Eq.(9) in terms of the local hour angle at sunset, corresponding to a zenith angle of 90° (found from Eq.(9) with z=90°.):

$$h_{ss} = \cos^{-1}[-\tan L \tan \delta_s] \quad (11)$$

One of the most important factors to consider when designing a solar energy system is the number of hours per day that the Sun is above the horizon. The total daylight hours are the number of hours per day that the Sun is at zenith angles less than 90°, which from Eq. (11) is found to be:

$$T_d = 2(24.623/\pi) h_{ss} \quad (12)$$

where $24.623/\pi$ is to convert from radians to hours. Notice that if $[-\tan L \tan \delta_s]$ is greater than one, the Sun either never sets (if $z_{min} < 90°$) or never rises (if $z_{min} > 90°$). The results of Eq.(12) are plotted in Figure 5.

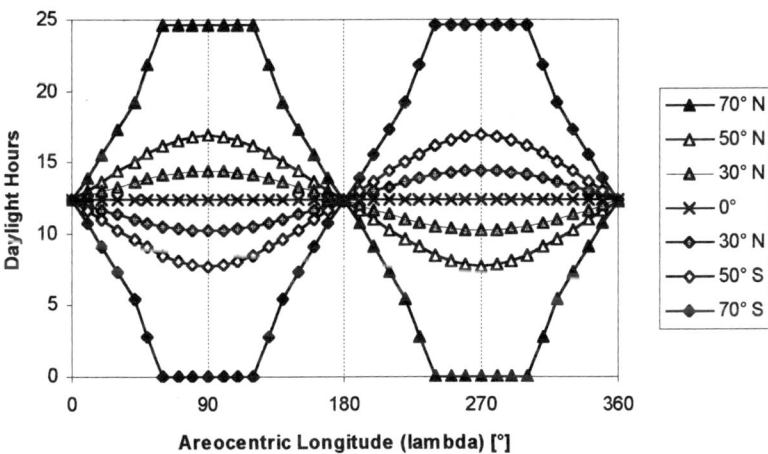

Figure 5 Daylight hours on Mars vs. areocentric longitude.

With these formulae, an expression can be derived for the diurnally-averaged intensity at the top of the Martian atmosphere by integrating Eq. (8) over the entire day. When this integration is performed, the following expression is obtained (Figures 6 and 7 for northern and southern latitudes respectively):

$$I_d = (I_{sc}/\pi)(a_e/r)^2 [\cos \delta_s \cos L \sin h_{ss} + h_{ss} \sin \delta_s \sin L] \quad (13)$$

where h_{ss} is the local hour angle at sunset corresponding to a zenith angle of 90°:

$$h_{ss} = \cos^{-1}[-\tan L \tan \delta_s] \quad (14)$$

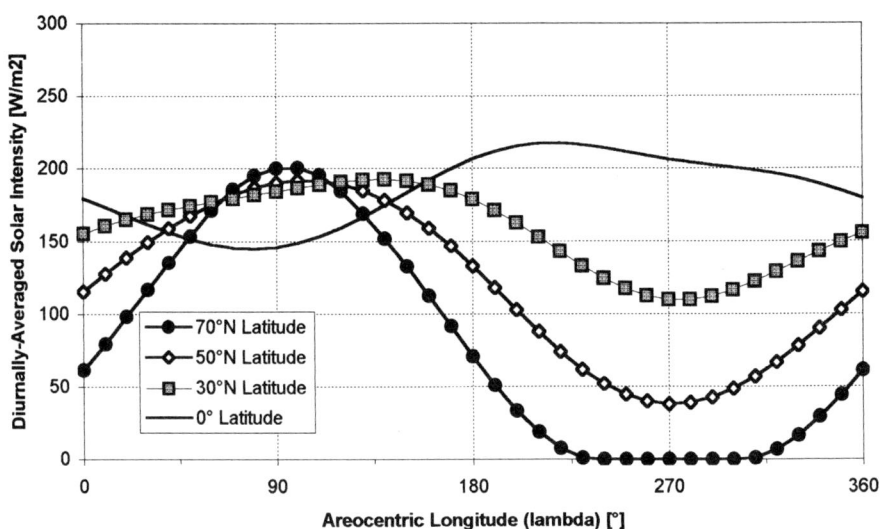

Figure 6 Average solar radiation intensity (without atmospheric attenuation) for northern latitude sites.

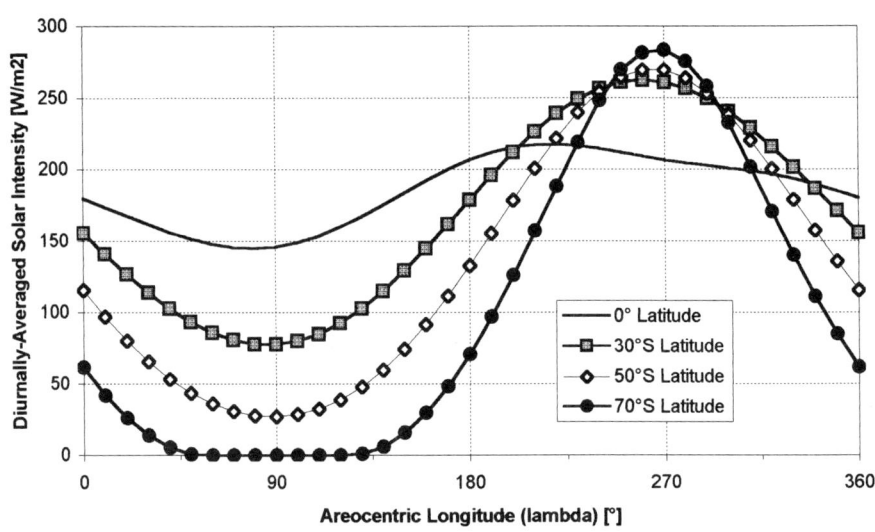

Figure 7 Average Solar radiation intensity (without atmospheric attenuation) for southern latitude sites.

Solar Radiation Intensity on Ground

On Mars an important role in defining atmospheric properties is played by dust storms. Global dust storms that occur mainly at Martian perihelion can increase the optical depth (τ) of the atmosphere significantly, reducing the direct transmission of solar radiation by up to 95%. The effects of dust storms must be therefore accounted for in this study.

Even if a number of different methods have been employed to measure the opacity of the Martian atmosphere, probably the most reliable one is the one in which the opacities are derived from the Viking Lander images of the Sun and Phobos. By comparing images taken at the same time but at different elevation angles, optical depth values were derived with a direct application of Beer's law:

$$I_{surf} = I_0 \exp(-\tau / \cos z) \tag{15}$$

This calculation was performed by Pollack *et al.* (Ref.[14]) and the model used in this paper is a simplified version of the one implemented by Geels, Miller and Clark (Ref.[14]). The model optical depth is shown in Figure 8 and the results of Eq.(15) are in Figure 9.

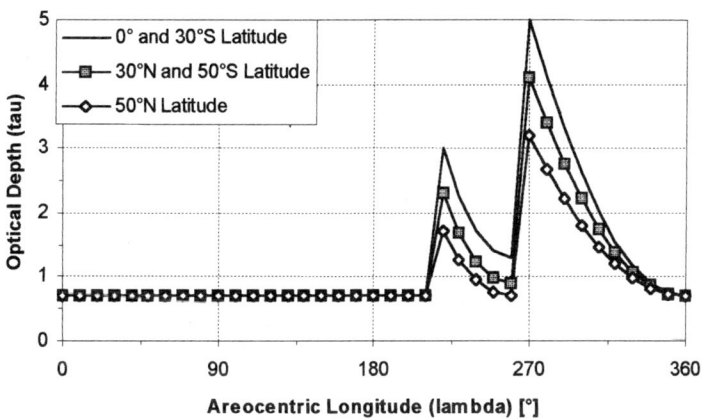

Figure 8   Model optical depth vs. areocentric longitude.

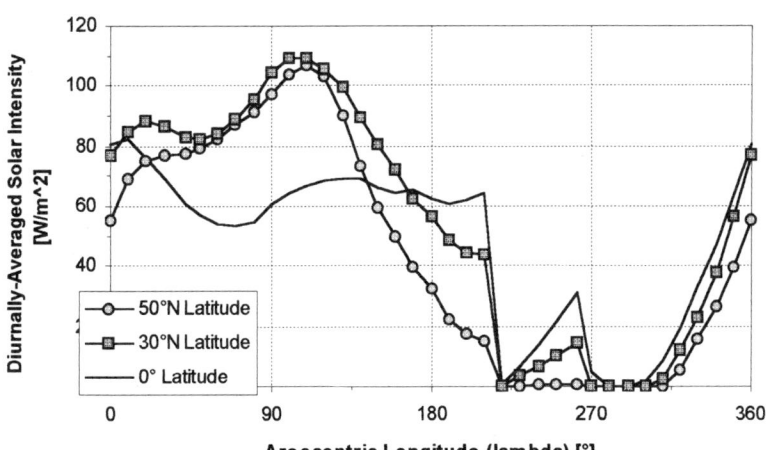

Figure 9   Direct (unscattered) solar radiation intensity, including model dust storm.

A work by Pollack (Ref.[14]) tabulates total radiation on a horizontal surface as a function of τ and cos z. Replacing the term (exp [..]) in Eq.(15) with an interpolation of Pollack data results in Figure 10.

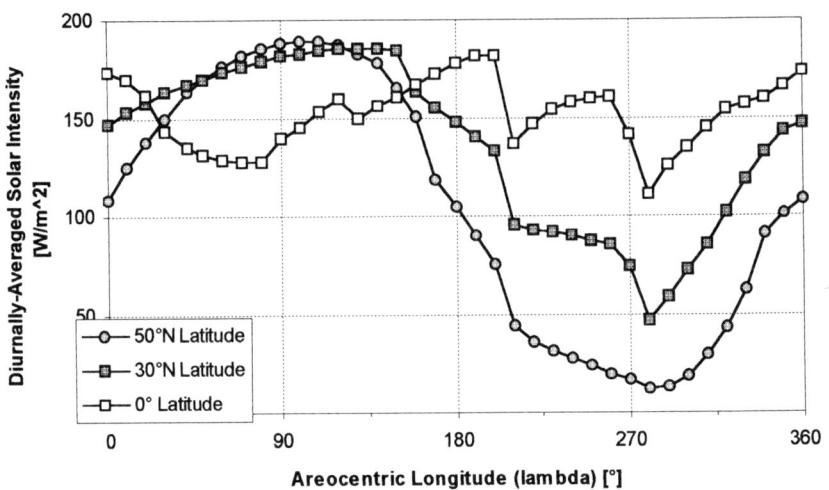

Figure 10 Total (including scattered) solar radiation intensity, including model dust storm.

A comparison of Figures 9 and 10 shows that the estimated total radiation is very much larger than the direct radiation.

The scattered light will be strongly filtered by the dust in the Martian atmosphere. Although this filtered spectrum is not precisely known, the strong reflectance of Martian surface material toward the red and the infrared is compatible with the spectral responses of typical solar cell materials (Ref.[14]).

The assumption is made in this article that the spectral response of the total solar radiation as of Figure 10 and that of the solar cells are the same.

7. RESULTS

The daily-averaged solar intensity level (Figure 10), together with the knowledge of the number of daylight hours (Figure 5) can be used to determine the amount of solar power available for flight in the Martian atmosphere. Two different case studies are analyzed.

Continuous Flight

If a solar powered aircraft is to be capable of continuous flight, enough energy must be collected and stored during the day to both power the aircraft and to enable the aircraft to fly throughout the night. Assuming a fuel cell efficiency (eff_{FC}) of 60% (Ref.[13],[15]) and a first order model of solar radiation variability during the Martian day (solar radiation constant in daylight hours), the amount of solar radiation available for overnight flight (I_{CONT}) is given by:

$$I_{CONT} = I_{AVE}\, \text{eff}_{FC} / (1 - x + x\, \text{eff}_{FC}) \tag{16}$$

where I_{AVE} is the daily-averaged total solar intensity (Figure 10) and x is the fraction of daylight hours (number of daylight hours / 24,36).

A comparison between the resulting level of solar intensity (Figure 11), the levels of Figure 10 that mimic a battery efficiency of 100% and the requirements of Figures 3 and 4 shows that Centurion would be unable of continuous flight on Mars, even with 100% efficiency batteries (at least at Northern latitudes between 0° and 50°). Pathfinder on the other hand is able to fly at 30°-50°N for most of the northern spring and summer, but with a very small payload. Requirements for a 20 Kg - 100 W payload are shown in Figure 11. Improvements in battery efficiency should lead to a significant increase in payload mass.

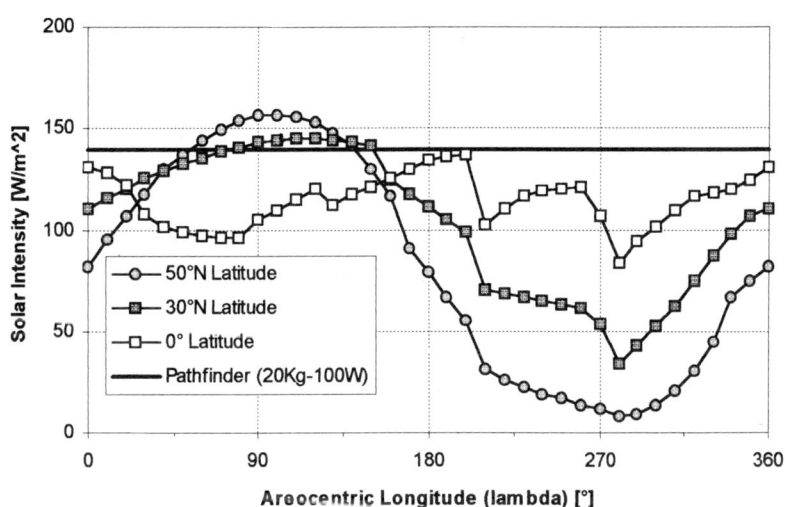

Figure 11   Total solar radiation intensity available for continuous flight (assuming fuel cell efficiency = 60%).

Daylight Flight

If we restrict our analysis to daylight hours flights the available solar intensity are obviously much higher, because the batteries do not have to store energy for flying at night. Assuming again a first order model of solar radiation variability during the Martian day (solar radiation constant throughout the daylight hours), the amount of solar radiation available for daytime flight (I_{DAY}) is given by:

$$I_{DAY} = I_{AVE} / x \qquad (17)$$

A comparison between the resulting level of solar intensity (Figure 12) and the requirements of Figures 3 and 4 shows that both Centurion and Pathfinder should be able to fly with a 100 Kg payload during daylight hours at the equator during the whole Martian year. Pathfinder requirements for a 100 Kg - 100 W payload (which are the same Centurion has for a 50 Kg - 100 W payload) are shown in Figure 11. Moreover, flights could reach higher northern latitudes in the warmer months (areocentric longitude < 180°).

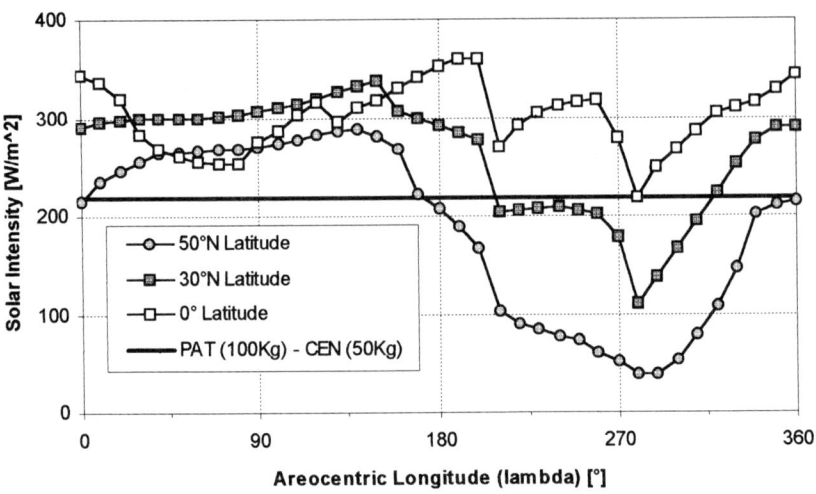

Figure 12 Total solar radiation intensity available for daytime flight.

A more detailed model of solar radiation variability during the day and the relative energy balance diagrams can be developed from methods in Ref.[13],[16].

8. CONCLUSIONS

Results of this study show that with current technology it is possible to develop an UAV capable of continuous flight in low Martian atmosphere. Actually, the Pathfinder and the Centurion aircraft developed by AeroVironment Inc. for NASA's ERAST project could already meet that goal. It was found that Pathfinder could be able to fly continuously at latitudes between 30°N and 50°N with a 20 Kg - 100 W payload for up to 10 months, while on higher latitudes flight may not be possible during bigger fractions of the Martian year. Both Pathfinder and Centurion were found to be able to fly all year long during daylight hours with a 100 Kg - 100 W payload near the equator, and in spring and summer months in the ± 50° latitude range. Flight latitude and time of year play significant roles in determining payload mass for both aircraft. If the required duration of flight is restricted to summer months the payload mass is greatly improved and this effect is more pronounced the more northern the latitude. It is to be noted that no structural or aeroelastical analysis has been done in this paper, assuming that existing aircraft could behave as well in Martian environment as they do on Earth. Nevertheless it would be interesting to determine the effects of Martian low gravity on aircraft structures. If such an aircraft could be used by the first astronauts on Mars, it could greatly help to deploy planet-wide weather station and microrover networks, while providing a very detailed picture of soil chemistry and morphology of the terrain. The improvements in fuel cell and solar cell efficiency that will be made in the years that still part us from the first manned landing on Mars could really improve the performances of such an aircraft.

REFERENCES

1. Dryden ERAST Home Page. ("http://www.dfrf.nasa.gov/Projects/ERAST/ERAST.html").
2. *Aviation Week & Space Technology*, Jan. 13, 1997, pp. 92-97.
3. Dryden Pathfinder Home Page. ("http://www.dfrf.nasa.gov/Projects/Pathfinder/index.html").
4. *Aviation Week & Space Technology*, Nov. 4, 1996, pp. 29-30.
5. *Aviation Week & Space Technology*, Feb. 3, 1997, p.59.
6. *Aviation Week & Space Technology*, Mar. 17, 1997, p.36.
7. *Aviation Week & Space Technology*, May 4, 1998, p.54.
8. *Mars*, H. H. Kieffer, B. M. Jakosky, C. W. Snyder, M. S. Matthews editors, University of Arizona Press, Tucson, 1992.
9. *Introduction to Flight*, 3rd ed., John D. Anderson Jr., McGraw-Hill, 1989.
10. *The New Solar System*, 3rd ed., J. K. Beatty and A. Chaikin, Sky Publishing Co., 1990.
11. "The Mars Airplane Revived - Global Mars Surface Surveys", B. W. Augenstein, (AAS 87-270) in *The Case for Mars III*, AAS *Science and Technology Series*, Vol. 75, 1987.
12. *Race to Mars*, Frank Miles and Nicholas Booth editors, Macmillan, London, 1988.
13. "Effect of Power System Technology and Mission Requirements on High Altitude Long Endurance Aircraft", Anthony J. Colozza, NASA Lewis Research Center (Contract NAS3-25266), February 1994. (available online at "http://powerweb.lerc.nasa.gov/psi/DOC/erast.html").
14. "Design Considerations for a Mars Solar Energy System", David Atkinson and Owen Gwynne, in the *Journal of the British Interplanetary Society*, Vol. 45, pp. 183-194, 1992. Also published as paper AAS 97-380, in *From Imagination to Reality: Mars Exploration Studies of the Journal of the British Interplanetary Society* (Part II: Base Building, Colonization and Terraformation), R. M. Zubrin, ed., AAS *Science and Technology Series*, Vol. 92, 1997, pp. 117-138.
15. "Feasibility of Using Solar Power on Mars: Effects of Dust Storms on Incident Solar Radiation", Scott Geels, John B. Miller, Benton Clark, (AAS 87-266) in *The Case for Mars III*, AAS *Science and Technology Series*, Vol. 75, 1987.
16. "Solar-Powered Unmanned Aerial Vehicles", by K. C. Reinhardt, T. R. Lamp, J. W. Gels, A. J. Colozza, presented at the Intersociety Energy Conversion Engineering Conference, Washington D.C., August 1996. (available online at "http://powerweb.lerc.nasa.gov/psi/DOC/erast.html").

AUTONOMOUS ROVERS FOR HUMAN EXPLORATION OF MARS

John Bresina,* Gregory A. Dorais,† Keith Golden, David E. Smith and Richard Washington†

NASA Ames Research Center, Mail Stop 269-2, Moffett Field, California 94035-1000.
E-mail: {bresina, gadorais, kgolden, de2smith, richw}@ptolemy.arc.nasa.gov.

Autonomous rovers are a critical element for the success of human exploration of Mars. The robotic tasks required for human presence on Mars are beyond the ability of current rovers; these tasks include landing-site scouting, mining, as well as emplacement and maintenance of a habitat, fuel production facility, and power generator. These tasks are required before and also during human presence, the ability of rovers to offload work from the human explorers will enable the humans to accomplish their mission. Performing these tasks will require significant advances in rover autonomy and will require improvements in robustness, resource utilization, and failure recovery.

The Pathfinder mission demonstrated the potential for robotic Mars exploration, but at the same time indicated the need for more rover autonomy. The highly ground-intensive control with infrequent communication and high latency limited the effectiveness of the Sojourner rover. Advances in rover autonomy offer increased rover productivity without risk to rover safety.

Towards this end, we are developing an integrated on-board executive architecture that incorporates robust operation, resource utilization, and failure recovery. This work draws from our experience with the Deep Space One autonomy experiment, with enhancements to ensure robust operation in the face of the unpredictable, complex environment that the rover will encounter on Mars.

Our ultimate goal is to provide a complete agent architecture for rover autonomy. The complete architecture will include long-range mission and path planning, self-diagnosis and fault recovery, and continual monitoring and adjustment of execution resources. The architecture will enable robust operation over long ranges of time and distance, allowing rovers to perform complex tasks in a planned and opportunistic manner and serve as an intelligent, capable tools for human explorers.

1. INTRODUCTION

Human exploration of Mars in the relatively near future is becoming increasingly likely. NASA has been investigating possible manned missions, and there is widespread public interest in Mars due to factors such as the recent debate about evidence of life on Mars, the success of missions such as Mars Pathfinder, and advocacy by visionaries such as Robert Zubrin [Zubrin and Wagner, 1996].

However, colonizing Mars will not happen by itself; we need to make it happen. In order to make future missions safe, effective and affordable, we must anticipate and develop the technologies that will be needed. In this paper, we discuss the advances in Mars rover software that

---

* Recom Technologies.
† Caelum Research Corporation.

will be needed to support future missions, and we discuss the steps we are taking to realize those improvements. In particular, we argue that rovers need to be significantly more autonomous.

In the next section, we discuss anticipated missions leading to and following human presence on Mars, and explain why those missions require autonomous rovers. In Section 3, we discuss the current state of the art in rover autonomy and where we would like to see it extended. In Section 4, we discuss the work we are doing to improve rover autonomy.

2. FUTURE MARS MISSIONS

2.1 Precursor Missions

Prior to sending humans to Mars, we need to send machines: to continue learning about conditions on Mars, to test out technologies that will be needed for human missions, and to set up the facilities that will be needed by humans when they first arrive.

We need to learn more about the "Red Planet," both to keep the astronauts safe and to maximize the impact of the first human mission. For example, we still know little about the composition of the Martian soil. We need to learn more—both to guard against unforeseen chemical hazards and to determine whether activities such as growing plants, making bricks or extracting chemicals is feasible. Similarly, a more detailed study of the Martian atmosphere, winds, and dust storms will give us useful information to prepare us for such tasks as *in-situ* propellant production, deployment of solar arrays, and navigation.

Even more importantly, we need rovers to prepare the way for humans. Rovers may be needed to set up facilities for propellant production, to set up power plants, and to begin resource collection. After humans arrive, rovers are still useful to augment human capabilities and perform tasks in the hazardous Martian environment.

All of these tasks call for rover autonomy. These missions will run for a long time—some of the upcoming rover missions are expected to last a year—and communication with Earth will be infrequent and high latency. Controlling the rovers manually from Earth would incur a huge cost and would achieve much lower returns due to the time the rover is idle waiting for instructions. Given that rovers have a limited lifetime, such wasted opportunities translate to a much lower return on our investment.

2.2 First Human Exploration

When the first humans land on Mars, they will be in the company of rovers—some who were waiting for their arrival and some that arrived with them. These rovers will amplify the productivity of the crew by carrying out mundane or dangerous tasks that the limited crew will not be able to handle alone. Rovers can traverse long distances, gathering samples for the crew to analyze, and can act as "eyes and ears" of the crew, allowing them to "explore" distant terrain, safe from radiation and other perils.

In principle, these tasks could be accomplished by non-autonomous rovers, teleoperated by the crew on Mars or the operations team on Earth. In practice, the crew will be busy with more important tasks and will not have the time to continuously monitor the rovers and control their every move. Thus, autonomous rovers can greatly amplify the productivity of the small crew. It is also advantageous for the rovers to be responsible for their own well-being: to avoid getting damaged, and to diagnose and correct recoverable software and hardware failures. This way, the crew doesn't have to spend much time babysitting the rovers, and can concentrate on doing science and survival.

2.3 Colonization

Once humans settle on Mars, the tasks for rovers will become more complex. They will be needed for activities such as mining, building habitats and other structures, and maintaining the life-support system of the humans (which includes the rovers themselves). These tasks demand autonomy, since teleoperating rovers to perform such tasks would most likely be more difficult than performing the tasks themselves. The rovers will also need to respond quickly to hazards, such as falling objects.

3. CURRENT AND FUTURE ROVER AUTONOMY

3.1 Current Rovers

Current rover missions, such as Pathfinder, depend almost entirely on ground-based commanding and employ only enough on-board autonomy to safely follow uplinked commands. If anomalous situations arise, the rover waits for updated commands to be uplinked. In practice, this approach leads to high mission cost and missed science opportunities. The ground operations team for the Pathfinder mission had to adjust themselves to Mars time for the entire mission and had much more access to expertise in diagnosing and debugging problems than will be possible in future missions. With upcoming missions expected to last a year or more, expecting people to work such hours and maintain such vigilance is not feasible. Furthermore, since almost all of the intelligence behind the Sojourner rover was on Earth, the command sequences that Sojourner executed were quite fragile. There were many cases when something went wrong, forcing the humans on Earth to spend a day or more diagnosing and fixing the problem.

3.1.1 Robust Operation

The impressive success of the Pathfinder mission and in particular the Sojourner rover [Mishkin *et al.*, 1998] raises hopes of realizing the vision of rovers roaming the surface of Mars, performing tasks nearly or completely autonomously. At the same time, the mission points out the gaps between that vision and current reality.

Currently, a rover command sequence is specified to the lowest level of detail and leaves few, if any, choices to be made at execution time; hence it admits only one, or a very small number of, valid execution behaviors. This simplifies the execution process, but does not allow execution to be responsive to the dynamic status of the rover and the environment. This inflexibility can cause reduced productivity and execution failures.

For example, in Sol* 22 of the mission, Sojourner received the following challenging sequence instruction sequence.

1. Back up to the rock named Soufflé.
2. Place the arm with the spectrometer on the rock.
3. Do extensive measurements on the rock surface.
4. Perform a long traverse to another rock.

---

* Martian day.

At the next communication time, the news came in from Mars. The good news was that the sequence was executed to completion, including the longest traverse ever done in one day, a world's record for Mars. However, there was bad news with the good. The spectrometer data was useless, because the rover stopped short of the rock, and the spectrometer was left hanging out in mid-air rather than placed on the rock. The rock was never again visited.

3.1.2 Resource Utilization

The generation of uplink plans is based on estimated profiles, over time, of capacity and demand for each resource. However, there is great inherent uncertainty in the operating environment and its impact on performance of rover hardware components. For example, the power demand of a traversal is highly dependent on the incline, roughness, and traction of the terrain. Without an accurate estimate of power demand, battery capacity at a given time cannot be accurately predicted. Furthermore, battery capacity will also depend on how much time is spent in shade, which in turn depends on the surrounding landscape. In response to this uncertainty, worst case estimates of resource usage and availability are typically used; however, even tight worse cases bounds may be difficult to predict. Because demand estimates are too pessimistic, execution of such plans often results in reduced rover productivity through wasted time and lost opportunities. On the other hand, overly optimistic estimates of resource usage and availability result in higher risk to basic rover safety and broken plans.

3.1.3 Failure Recovery

Current rovers have very limited capabilities for recovering from faults or anomalous situations. The rover's response to most execution failures is to halt all activity and wait for the ground operation team to determine what went wrong and uplink a recovery plan. Depending on the nature of the anomalous situation and the quality of the downlinked information for the purposes of diagnosis, this ground-based recovery process can cost days of rover idleness and lost science opportunity. For example, a rover traverse may fail with a high wheel current combined with the wheel encoder showing no movement. In this case the wheel may be stuck on a rock or the encoder may be broken; the ground team will need to uplink diagnostic sequences to determine which is the case, and then recovery sequences to remove the rover from the rock if that is the problem. Valuable science opportunities are lost during this time. Even transient problems such as an overheated motor can cause plan failure, where perhaps a brief pause to cool down would be sufficient to remedy the problem.

3.2 The Next Generation

The next generation of rovers, the first of which is expected to launch in 2003, will be more flexible than Sojourner, but will be shy of full autonomy. Instead of simple command sequences, the rovers will execute complex contingency plans, which tell the rover explicitly what to do if something goes wrong. They will also execute plans more robustly, so minor problems such as motor overheating do not cause failure. Finally, they will be able to identify and diagnose internal faults, and recover from simple failures.

To imagine how these rovers will behave, consider again the problem of backing up to a rock, deploying a sensor, gathering data and moving on. We can imagine what a smart rover would need to do in this case. First of all, the operation of backing up on Sojourner was brittle because it used a simple "try three times" strategy to back up. We can imagine a more robust operation of backing up until contact, with some timeout to avoid an insurmountable problem. The backing up operation should include the ability to try different approach paths if obstacles block the planned route; the rover should be able to take a moment to let an overheating motor cool down rather than abandoning the operation; the rover should notice that a wheel seems to be malfunctioning and pulling the rover off-path, and shift control algorithms to compensate for that. Certainly, the rover should notice contact or not before doing hours of measurements, and have alternative plans in case it cannot make contact with the rock despite its best efforts. While it is performing measurements, the rover should monitor its energy level to make sure that it will have enough energy left to send its data to Earth at the beginning of the morning, and potentially cut short its monitoring if it ends up in the shade of a larger rock and cannot charge its battery enough to complete the task and communicate.

A smart rover that behaves as described above is beyond the current state of the art in deployed missions. However, the technology needed to construct such a rover is within reach using artificial intelligence technology in development today. We are working toward that end by using components of the Remote Agent, a system developed for autonomous spacecraft, and applying them to rovers.

3.3 Future Vision

In the future, we would like to see rovers that are capable of full autonomy. These rovers will take very high level instructions from human operators and will be able to achieve those goals with no further supervision, even in dynamic and uncertain environments. These rovers will be self-diagnosing and self-repairing, and will be capable of detecting gradual degradation, adjusting internal parameters accordingly, and subjecting themselves to preventive maintenance to avoid catastrophic failure. For example, motors are subject to wear out over time and solar panels accumulate dust and are gradually damaged by UV. Rovers will be able to automatically replan when unexpected problems or opportunities arise.

We are building toward this vision in our research on autonomy. While we are not there yet, it is reasonable to expect such capabilities in the rovers that accompany humans in exploration of Mars.

4. INCREASING ROVER AUTONOMY

To achieve our goal of greater rover autonomy, and to work toward the goal of full autonomy, we are focusing on the following capabilities:

- Robust operation via execution and monitoring of flexible, contingent mission plans

- Optimal resource utilization via continuous resource capacity assessment, demand prediction and dynamic allocation re-scheduling

- Advanced failure recovery via active sensing/testing and reasoning from first principles.

To realize these capabilities, we are applying components from the Remote Agent (RA) architecture to the rover domain.

In this section, we describe the RA and how it is being enhanced to meet the particular requirements of the rover domain. We describe how the RA is embedded in the overall system, provide an overview of the components of the RA, and describe its high level functioning.

4.1 Architecture Overview

The Remote Agent (RA) is an architecture designed for intelligent control of complex systems. The RA has been applied to the requirements of spacecraft control [Muscettola *et al.*,

1998, Bernard et al., 1998], but since it is composed of heterogeneous, state-of-the-art, general-purpose components, it is appropriate for many of the characteristics of rover control as well.

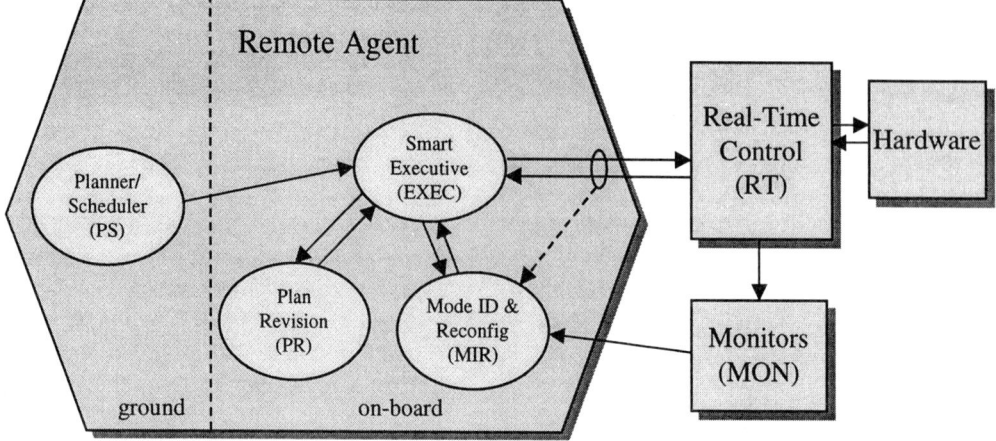

Figure 1 Remote Agent architecture embedded within rover real-time software.

The relationship between the RA and the system in which it is embedded is portrayed in Figure 1. When viewed as a black box, RA sends out commands to the *real-time control* software (RT). RT provides the primitive skills of the autonomous system, which take the form of discrete and continuous real-time estimation and control tasks. RT responds to high-level commands by changing the mode of a control loop or state of a device and sending a message back to RA when the command has completed.

In addition, the status of all RT control loops are passed back to RA through a set of *monitors* (MON). The monitors discretize the continuous data into a set of qualitative intervals based on trends and thresholds, and pass the results back to RA. The abstraction process is fast and simple, involving discretizing a continuous variable using thresholds on an absolute or relative scale.

The RA comprises three major reasoning components: a temporal planner/scheduler (PS), a smart executive (EXEC), and a model-based diagnosis and reconfiguration system (MIR). In the rover domain, PS is a ground-based planner, which is given high-level goals corresponding to the mission goals and generates a schedule. Some limited plan revision (PR) capabilities are on-board, and we envision that more of the planning work will migrate on-board as both computational power and the need for autonomy increase. The schedule is sent to EXEC on-board, which decomposes the general schedule into lower-level spacecraft commands that implement the schedule elements and respect the constraints between them. As described above, these commands are sent to RT, with results coming back via MON into MIR. MIR's *mode identification* layer, MI, infers the system state from the monitored information and updates the state for EXEC. If commands fail or schedule constraints are violated, EXEC tries to recover using retries or local recoveries, potentially calling MIR's *mode reconfiguration* layer, MR, to produce a recovery plan. PR adapts the plan as the situation changes.

4.1.1 Planner

Throughout a mission, detailed mission operations plans must be constructed, validated, and uplinked to a spacecraft or rover. Currently a mission operations plan takes the form of a rigid, time-stamped sequence of low-level commands. Unfortunately, there is uncertainty about many aspects of task execution: exactly how long operations will take, how much power will be consumed, and how much data storage will be needed. Furthermore, there is uncertainty about environmental factors that influence such things as rate of battery charging or which scientific tasks are possible. In order to allow for this uncertainty, current plans are based on worst-case estimates and contain fail-safe checks. If tasks take less time than expected, the spacecraft or rover just waits for the next time-stamped task. If tasks take longer than expected, they may be terminated before completion. In fact, all non-essential operations may be halted until a new command sequence is received. All of these situations result in unnecessary delays and lost science opportunities.

PS can actively plan for, and take advantage of, possible contingencies. Thus, if an operation takes longer than a certain amount of time, or the power remaining drops below a specified value, a different pre-planned sequence of operations can be performed. Building contingency plans is, in general, intractable, and so contingency planners tend to be slow [Draper et al., 1994, Pryor and Collins, 1996, Weld et al., 1998]. To overcome this problem, PS uses is *Just in Case* planning.

The Just in Case (JIC) technique was originally developed to generate contingent observation schedules for automated telescopes [Drummond et al., 1994]. The basic idea is to take an existing schedule and look for the places where it is most likely to fail. The JIC scheduler then generates alternative schedules for each of those situations. The JIC scheduler starts with a sequence of tasks, where each task must be performed in a certain temporal window. However, there is uncertainty in how long a particular task may take, and this can lead to potential failures of the schedule. For example, an execution failure could result if one task finished sufficiently late that the next task's start window has already passed.

The JIC approach is the most straightforward way to introduce contingency planning into mission operation planning. Using this idea, a plan is examined to determine the most likely places that it could fail. As with telescope scheduling, one source of potential failures is due to uncertainty in task duration –a task might take longer than expected, preventing its successor from starting in the required time window. Alternatively, a task might take less time than expected so the spacecraft or rover would sit idle until the beginning of the time window for the next task. If these failures were sufficiently likely a contingent plan could be built for these situations (just as in JIC scheduling for telescopes).

For spacecraft and rover operations, uncertainty about other resources can also lead to potential failures in a plan. For example, if a task utilizes more power than expected, or the battery has not charged as much as expected, there may be insufficient power available for the subsequent tasks. However, there may be enough power left to do other useful tasks. JIC could also be employed to generate contingent plans for such situations. Equally useful would be contingent plans for situations where more of a resource is available than expected. In this case, a JIC contingent plan could take advantage of the unexpected surplus to perform additional tasks, or perform tasks that are more power hungry.

4.1.2 Smart Executive

EXEC is responsible for interpreting the command sequence coming from the planner or an external source (e.g., ground control), decomposing high-level actions into low-level commands for the real-time system RT, checking run-time resource requirements and availability, and performing fault recovery in coordination with the fault recovery system MR.

The input to EXEC is a contingent command sequence. The contingencies are represented as a branching plan structure, plus a library of contingent plan fragments and suffixes that are invoked on plan failure, unexpected opportunities, or conditions such as resource shortfall or component degradation. At each point in time, EXEC may have a choice of multiple possible plan steps corresponding to the eligible plan branches or plan library elements. EXEC chooses the plan step with the highest estimated expected utility (computed over the remainder of the plan). This utility is initially computed by the ground planner PS, but may be updated by the plan revision component PR at run-time to reflect changes in resource availability, system state, or the environment.

Each of the plan steps may be a high-level action that needs to be decomposed into low-level commands for RT. This decomposition uses constructs for robustly executing commands, including catching failure conditions and performing local recoveries and retries. Thus low-level conditionality may be represented at a level below the granularity of the planner, reducing the computational complexity of the planning problem while guaranteeing the correct execution semantics. For example, the seemingly simple command "move to a point 5 meters straight ahead" may include the ability to pause if a motor overheats, try multiple routes if obstacles block the path, and switch to alternative control algorithms if a wheel becomes blocked.

EXEC extends its own robust activity mechanism using the fault recovery mechanisms of MIR. EXEC may request a recovery plan from MIR (see *Section 4.1.3* below), and it will integrate that plan in with its currently executing plan.

EXEC also monitors run-time usage and availability of resources. PS constructs its plans using expected resource availability profiles, but these may change at execution time. For example, a power profile for the day will depend on expected solar exposure for the solar arrays, but this may change based on long traverses on tilted terrain (more or less power available depending on the angle), dust storms, shadows, etc. A resource manager component takes the resource usage expectations for the plan's activities and the current best information for resource availability, and signals potential (future) and real (current) conflicts based on that information. EXEC responds to those signals as to other failures, using its fault response mechanism.

4.1.3 Mode Identification and Reconfiguration

MIR is a discrete, model-based controller that uses a single, declarative model of the rover for both mode identification (inferring the internal state) and reconfiguration [Williams and Nayak, 1996] (changing the system to a more desirable state). Like EXEC, MIR runs as a concurrent reactive process. MIR itself contains two components, one for *Mode Identification* (MI) and one for *Mode Reconfiguration* (MR).

MI is the sensing component of MIR's model-based reconfiguration capability. Based on sensor readings and the commands executed on the rover, MIR uses its model of the rover to infer the most likely current state. MI also provides a layer of abstraction to the executive: it allows the RA to reason about the state of the rover in terms of component modes, rather than in terms of low level sensor values.

MR serves as a *recovery expert* to EXEC, taking as input a *recovery request*, and returning a sequence of operations that, when executed starting in the current state, will move the executive into a state satisfying the properties required for successful execution of the failed activity.

MIR uses algorithms adapted from model-based diagnosis [de Kleer and Williams, 1987, de Kleer and Williams, 1989] to provide the above functions. MIR extends the basic ideas of model-based diagnosis by modeling each component as a finite state machine, and the whole rover as a set of concurrent, synchronous state machines. Modeling components as finite state machines allows MIR to effectively track state changes resulting from executive commands. Modeling the rover as a concurrent machine allows MI to effectively track concurrent state changes caused either by executive commands or component failures.

Interaction with the environment. In the past, MIR has been used to model systems, such as spacecraft, in which the environment outside the spacecraft can be effectively ignored. Spacecraft follow known trajectories, free of obstacles, in which the external environment can be reduced to a few simple variables, such as the relative position of the Earth. On rovers, interaction with the environment is central to many of the possible faults. Dust accumulates on the solar panels, the rover passes into the shadow of large rocks, or gets caught on small ones. In order to handle failures involving external factors, we need to add the capability to reason from first principles about the rover and its interaction with the environment. The proposed reasoning approach will use a combination of model-based deduction and hybrid simulation.

Active sensing and testing. In support of this reasoning, we will add the capability to perform active sensing and testing in order to narrow the candidate situation assessments (diagnoses) and in order to evaluate the utility of alternative recovery plans. The sensor information that MI can passively acquire is not always adequate to allow an unambiguous diagnosis. In general, it may be ambiguous whether a given component has failed, or the sensor responsible for measuring that component has failed. If an encoder indicates that a wheel drive motor isn't turning, or is turning too slowly, that could be a sensor error; the encoder may be skipping counts or entirely dead. The true state of the rover may be determined by performing experiments designed to eliminate certain diagnoses. For example, if the wheel does not appear to be turning, the rover could try backing up to see if the wheel is caught on a rock. It may also look at the motor currents to see if the motor appears to be stalled, which could indicate a stuck wheel or seized bearings. If it can't establish that the motor is stalled, it could try turning only that wheel, and use other sensors to detect movement. If so, that would indicate that the encoder has failed.

In support of this active testing, the rover can make use of its models, both to determine when there are multiple competing diagnosis, and to identify activities it can perform that will rule out or confirm certain hypotheses. Reasoning about the information to be gained by executing actions exceeds the ability of the original MIR system, but we are working to provide that capability.

5. CONCLUSIONS AND RELATED WORK

We believe that autonomous rovers will be an essential part of a program of human exploration of Mars. We have described our efforts underway to make rovers more autonomous, building on earlier work for autonomous spacecraft. Much work remains to be done, but we believe the ultimate goal of autonomous Mars rovers can be realized.

The motivation and examples that drive the rover-specific research come from the Pathfinder mission [Mishkin *et al.*, 1998], which will remain the reference against which all near-term rover work will be compared. The work described in this paper is directed towards future NASA missions to Mars, but it is first being tested on the NASA Ames Marsokhod rover [Christian *et al.*, 1997]. Other robotics work specifically designed for space applications includes the JPL Long Range Science Rover project [Volpe *et al.*, 1997], CMU Nomad project, and the LAAS IARES project [Chatila *et al.*, 1995].

The technology builds on the Remote Agent Experiment on the Deep Space One mission, where the Remote Agent is a technology experiment that will control the spacecraft during a week-long period [Muscettola et al., 1998, Bernard et al., 1998]. Other agent architectures for controlling real-time systems include CIRCA [Musliner et al., 1993] and 3T [Bonasso et al., 1995, Bonasso et al., 1997]. The attitude and articulation control subsystem (AACS) on the Cassini spacecraft [Brown et al., 1995] has explicit software modules for context-dependent command decomposition, resource management, configuration management, and fault protection, and provides an example of the state of the art in deployed spacecraft autonomy.

REFERENCES

[Bernard et al., 1998] D. E. Bernard, G. A. Dorais, C. Fry, E. B. Gamble Jr., R. Kanefsky, J. Kurien, W. Millar, N. Muscettola, P. P. Nayak, B. Pell, K. Rajan, N. Rouquette, B. Smith, and B. C. Williams. "Design of the remote agent experiment for spacecraft autonomy". In *Proceedings of the IEEE Aerospace Conference*, Snowmass, CO, 1998. IEEE.

[Bonasso et al., 1995] R. P. Bonasso, D. Kortenkamp, D. Miller, and M. Slack. "Experiences with an architecture for intelligent, reactive agents". In *Proceedings of IJCAI-95*, 1995.

[Bonasso et al., 1997] R. P. Bonasso, D. Kortenkamp, and T. Whitney. "Using a robot control architecture to automate space shuttle operations". In *Proceedings of IAAI-97*, pages 949-956, 1997.

[Brown et al., 1995] G. M. Brown, D. E. Bernard, and R. D. Rasmussen. "Attitude and articulation control for the Cassini spacecraft: A fault tolerance overview". In *14th AIAA/IEEE Digital Avionics Systems Conference*, Cambridge, MA, November 1995.

[Chatila et al., 1995] R. Chatila, S. Lacroix, T. Simeon, and M. Herrb. "Planetary exploration by a mobile robot: mission teleprogramming and autonomous navigation". *Autonomous Robots*, 2(4):333-344, 1995.

[Christian et al., 1997] D. Christian, D. Wettergreen, M. Bualat, K. Schwehr, D. Tucker, and E. Zbinden. "Field experiments with the Ames Marsokhod rover". In *Proceedings of the 1997 Field and Service Robotics Conference*, December 1997.

[de Kleer and Williams, 1987] J. de Kleer and B. C. Williams. "Diagnosing multiple faults". *Artificial Intelligence*, 32:100-117, 1987.

[de Kleer and Williams, 1989] J. de Kleer and B. C. Williams. "Diagnosis with behavioral modes". In *Proceedings of IJCAI-89*, pages 1324-1330, August 1989.

[Draper et al., 1994] D. Draper, S. Hanks, and D. Weld. "Probabilistic planning with information gathering and contingent execution". In *Proc. 2nd Intl. Conf, AI Planning Systems*, June 1994.

[Drummond et al., 1994] M. Drummond, J. Bresina, and K. Swanson. "Just-in-case scheduling". In *Proceedings of the 12th National Conference on Artificial Intelligence*, 1994.

[Mishkin et al., 1998] A. H. Mishkin, J. C. Morrison, T. T. Nguyen, H. W. Stone, B. K. Cooper, and B. H. Wilcox. "Experiences with operations and autonomy of the Mars Pathfinder microrover". In *Proceedings of the IEEE Aerospace Conference*, Snowmass, CO, 1998. IEEE.

[Muscettola et al., 1998] N. Muscettola, P. P. Nayak, B. Pell, and B. C. Williams. "Remote agent: To boldly go where no AI system has gone before". *Artificial Intelligence*, 103(1/2), August 1998. To Appear.

[Musliner et al., 1993] D. Musliner, E. Durfee, and K. Shin. "Circa: A cooperative, intelligent, real-time control architecture". *IEEE Transactions on Systems, Man, and Cybernetics*, 23(6), 1993.

[Pryor and Collins, 1996] L. Pryor and G. Collins. "Planning for contingencies: A decision-based approach". *J. Artificial Intelligence Research*, 1996.

[Volpe et al., 1997] R. Volpe, J. Balaram, T. Ohm, and R. Ivlev. "Rocky7: A next generation Mars rover prototype". *Journal of Advanced Robotics*, 11(4), December 1997.

[Weld et al., 1998] D. S. Weld, C. R. Anderson, and D. E. Smith. "Extending graphplan to handle uncertainty & sensing actions". In *Proceedings of AAAI-98*, pages 897-904, 1998.

[Williams and Nayak, 1996] B. C. Williams and P. P. Nayak. "A model-based approach to reactive self-configuring systems". In *Proceedings of AAAI-96*, pages 971-978, 1996.

[Zubrin and Wagner, 1996] R. Zubrin and R. Wagner, editors. *The Case for Mars: The plan to settle to Red Planet and why we must*. The Free Press, 1996.

OFFICIAL TELESCOPE SPONSOR

OF

THE MARS SOCIETY

BRING MARS INTO FOCUS

*National Geographic Books
is proud to sponsor the*

Founding Convention of the Mars Society and to introduce our spectacular new volume...

History was made during the summer of 1997 when Mars Pathfinder's arrival on the red planet was "greeted with the most attention since Apollo 11 touched down on the Moon" (*Time* magazine). National Geographic has captured that excitement in a book hailed by *Publishers Weekly* as "extraordinary... astonishing and unprecedented."

Combining dramatic digital images taken by Pathfinder and *Sojourner*, along with trailblazing images from Viking space missions of 21 years ago, this volume is a visual feast—including three-dimensional gatefold panoramic images and sweeping mosaics of the Mars surface.

Authoritative and compelling text by Paul Raeburn: former chief science correspondent for the Associated Press and currently senior science editor for *Business Week*.

Dr. Matthew Golombek: Mars Pathfinder project scientist and research scientist in the Earth and Space Sciences Division of the Jet Propulsion Laboratory at California Institute of Technology.

Available at this convention and in bookstores everywhere. Or call toll free **1-888-225-5647**, 24 hours a day.

- **224 oversize pages**
- **135 fascinating photographs**
- **3-D images and glasses included in the book**

Fisher Space Pen Co. • 711 Yucca Street • Boulder City, NV 89005 • 702.293.3011

MARS PEN

Guaranteed to write until man sets foot on MARS

Fully guaranteed by **fisher SPACE PEN**.